SILICICLASTIC DIAGENESIS AND FLUID FLOW: CONCEPTS AND APPLICATIONS

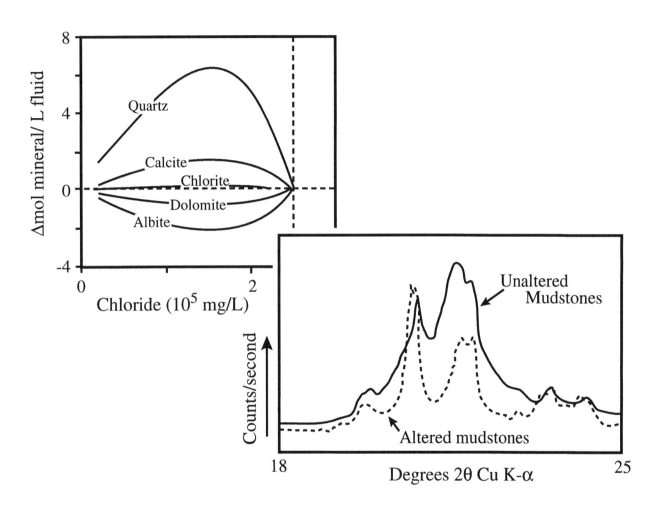

Edited by

Laura J. Crossey, University of New Mexico, Albuquerque, NM
Robert Loucks, ARCO Exploration & Production Technology, Plano, TX
and
Matthew W. Totten, University of New Orleans, New Orleans, LA

Copyright 1996 by
SEPM (Society for Sedimentary Geology)

Peter A. Scholle, Editor of Special Publications
Special Publication No. 55

Tulsa, Oklahoma, U.S.A.   *September, 1996*

A PUBLICATION OF

SEPM (SOCIETY FOR SEDIMENTARY GEOLOGY)

SEPM thanks the SEPM Foundation Inc. for its generous contributions toward this volume. Foundation funds were provided from the Bruce H. Harlton Publications Fund made possible by Allan P. Bennison.

ISBN 1-56576-032-8

Printed in the United States of America

INTRODUCTION

Research in the area of siliciclastic diagenesis has historically incorporated advances in related disciplines such as petrography and petrophysics, mineralogy, geochemistry, organic geochemistry, stratigraphy and basin analysis, and more recently, fluid flow. While the collection of papers herein covers a broad range of topics, an underlying theme is the importance of fluid flow in diagenetic processes. The mineralogy, texture and geochemistry of authigenic minerals provide constraints for fluid flow models, while formation waters provide modern snapshots of pore fluid evolution. Separated into two sections (Part I: Concepts and Part II: Applications), conceptual and practical applications are both represented. Part I begins with a study of chloride variations as a driving force in siliciclastic diagenesis by Jeff Hanor. The next two papers examine the influence of fluctuating hydrologic regimes on sandstone diagenesis: meteoric incursion in Tertiary units of the San Joaquin basin, California (Lee and Boles) and marine/freshwater interactions in cyclic Cretaceous deposits of the U. S. Western Interior (Loomis and Crossey). Taylor and Land utilize formation water chemistries to support their interpretation of diagenesis as influenced by deep basin-derived waters in offshore Texas Miocene reservoirs. Smith, Dunn and Surdam examine the utility of cation geothermometers in petroleum-rich environments. The final paper three papers in Part I focus on less permeable strata in siliciclastic systems: Surdam, Jiao and Yin emphasize the importance of the principles of multiphase fluid flow in explaining the sealing behavior of low-permeability units and concomitant diagenesis of sandstone bodies within hydrodynamically-isolated compartments; Hover, Peacor and Walter use high-resolution clay and organic matter examination of Devonian shales from the Illinois and Michigan basins to assess mechanisms of primary hydrocarbon migration; and Totten and Blatt examine high-grade shales and postulate that the illite-muscovite transition can be a potential silica source during latest stages of diagenesis.

Part II (Applications) consists of case studies, focusing on the implications of diagenetic processes for reservoir quality and hydrocarbon production. McLaughlin and co-workers utilize high-resolution stable isotope analyses of authigenic quartz to test a model for stratigraphic control of compactionally-de-

rived fluids in North Sea sandstones of the South Brae Oilfield. Dutton and others describe a case of preservation of sandstone porosity through early siderite precipitation (microbially-mediated) in the Val Verde basin of southwest Texas. Stone and Siever evaluate the growth of authigenic quartz and related effects on compaction, pressure solution and cementation of moderately- to deeply-buried sandstones in the Green River basin, Wyoming; noting that high fluid flow rates assist in silica transport. Malisce and Mazzullo examine pedogenic features in desert deposits of the Permian basin and develop criteria for their recognition. Hays, Walling and Tich utilize organic matter characterization and authigenic mineralogy to support a model for mass transport involving the formation of organo-metallic complexes during late-stage diagenesis. Pitman and Spötl continue on this theme with an assessment of the links between carbonate cementation in the St. Peter Sandstone and Mississippi Valley-type mineralization. Finally, Barrett and Mathias look at production-induced diagenesis during thermal heavy-oil recovery, relating changes in fabric and permeability to original grain size and sorting.

The review process is a critical element in producing a sound and informative volume. We would like to thank the many reviewers who provided thoughtful and critical input during the revision process. They include: Mary Barrett, James R. Boles, Knut Bjorlykke, Mark Copper, Tom Dunn, Per Egeberg, Steve Franks, Jeff Hanor, Jim Hickey, Andrew J. C. Hogg, David W. Houseknecht, Julie Kupecz, G. L. (Wendy) Macpherson, Ariel Malicse, Donald R. Peacor, Ed Pittman, Doug Shultz, William B. Simmons, Leta Smith, Ronald K. Stoessell, W. Naylor Stone, Ron Surdam, Tom Taylor, Thomas Tich and William C. Ward (among others). We also would like to thank the authors for their patience throughout the publication process. The SEPM Publication Committee made several helpful suggestions during development. Finally, we thank Dana Ulmer-Scholle for her continuous assistance and support.

Laura J. Crossey
Robert Loucks
Matthew W. Totten
February, 1996

CONTENTS

PART I
CONCEPTS

VARIATIONS IN CHLORIDE AS A DRIVING FORCE IN SILICICLASTIC DIAGENESIS

JEFFREY S. HANOR

Department of Geology and Geophysics, Louisiana State University, Baton Rouge, LA 70803

ABSTRACT: There is abundant evidence that the major solute composition of many saline waters in sedimentary basins is controlled by an approach toward metastable equilibrium with respect to multi-phase silicate and carbonate mineral assemblages. Where the composition of a fluid is rock buffered, the activity ratios of individual major cations to hydrogen ion, $a_i^{z+}/(a_{H^+})^{zi}$, are constrained within fairly narrow limits. Charge balance constraints in such systems require that changes in the concentration of aqueous anionic charge, which are normally equivalent to changes in chloride concentration, be accompanied by covariant changes in the aqueous concentration of hydrogen ion and, hence, of each of the major dissolved cations. These changes in cation concentrations require mass transfer of components between the fluid and ambient mineral phases and, equally as important, mass transfer between mineral phases. Theoretical mass balance calculations show, for example, that substantial quantities of quartz and calcite could be dissolved or precipitated in the absence of either temperature changes or flow-through of large volumes of fluid simply by changing the chloride concentration of fluids in rock-buffered systems. Chloride concentration can be varied in time and space by dissolution of evaporites, by dehydration reactions, and by hydrodynamic dispersion.

INTRODUCTION

In thermodynamically closed sedimentary systems in either equilibrium, metastable equilibrium, or kinetically-constrained disequilibrium, chemical reactions may be induced as a result of changes in temperature and/or pressure. In thermodynamically open systems, however, it is also possible to induce reaction by externally changing the chemical potential of one or more components in a phase within the system. One example is the precipitation of calcium carbonate from an aqueous fluid by loss of CO_2 through degassing:

$$Ca^{2+} + 2HCO_3^- \rightarrow CaCO_3 + H_2O + CO_2(g) \uparrow \quad (1)$$

Some of the many mechanisms which have been proposed for inducing diagenetic reactions in thermodynamically open siliciclastic sedimentary systems by externally changing the chemical potential of components include: loss or introduction of CO_2, introduction of organic acids, introduction of electron sources and sinks, and the introduction of dissolved sodium through dissolution of salt (e.g., see papers in McDonald and Surdam, 1984; McIlreath and Morrow, 1990; and Crossey et al., this volume). It is the purpose of this paper to explore the potential effects of another mechanism for driving diagenetic reactions in siliciclastic sequences: externally induced changes in the concentration of dissolved chloride.

It may intially seem counterintuitive that chloride should play a role in siliciclastic diagenesis because it is not explicitly present in a typical silicate hydrolysis reaction:

$$NaAlSi_3O_8 + 4H^+ + 4H_2O = Na^+ + Al^{3+} + 3H_4SiO_4^\circ \quad (2)$$

Implied in the above reaction, however, is the presence of one or more dissolved anions to balance the cationic charge of the aqueous solution. It will be shown that if aqueous anionic charge can be externally changed by adding or removing one or more thermodynamic components, such as H_2O or a dissolved salt, in an aqueous solution in equilibrium with a suite of mineral phases then the concentrations of all of the cations in solution will also change through mineral dissolution and precipitation.

Two aspects of this problem will be discussed in this paper: first, a review of the thermodynamic basis for concluding that changes in chloride or anionic charge should induce changes in

the absolute and relative concentrations of cations in rock-buffered systems; and second, an estimate of the potential magnitude of such effects on mineral precipitation and dissolution in sedimentary diagenetic systems where the concentration of chloride ion can be externally varied through the introduction of either NaCl or H_2O.

OBSERVED COMPOSITIONS OF SUBSURFACE WATERS

Chloride is by far the dominant anion and thus is the dominant contributor to anionic charge in subsurface sedimentary fluids having salinities in excess of approximately 20,000 mg/L (Fig. 1; Hanor, 1994a). At lower salinities, anionic composition is highly variable and can be dominated by chloride, sulfate, bicarbonate or acetate. The sources of dissolved chloride in more saline fluids include: connate dissolved chloride buried at the time of sediment deposition, chloride derived by refluxing of surface brines and chloride derived from subsurface mineral dissolution, principally of halite. The absolute concentration of

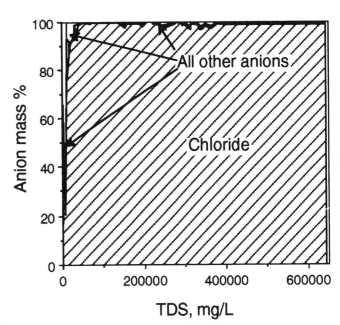

FIG. 1.—Plot of the variation in anionic composition of subsurface basinal waters worldwide as a function of salinity (modified from Hanor, 1994a).

Siliciclastic Diagenesis and Fluid Flow: Concepts and Applications, SEPM Special Publication No. 55

dissolved chloride in most pore waters is controlled primarily by physical mixing of fluids of different chloride concentration (Hanor, 1994b).

The cationic composition of saline sedimentary fluids on a global basis shows a strong correlation with dissolved chloride and total dissolved solids (TDS; Fig. 2). As has been noted previously (Hanor, 1988, 1994a,b) the observed 1:1 log-log variation of major monovalent cations such as Na^+ and K^+ with TDS and the 2:1 log-log variation of divalent cations such as Mg^{2+}, Ca^{2+}, and Sr^{2+} with TDS is consistent with these solutes having approached equilibrium with respect to ambient metastable assemblages of silicate-carbonate minerals. This conclusion is consistent with the work of Land et al. (1988), Michard and Bastide (1988), Smith and Ehrenberg (1989), Hutcheon et al. (1993) and others for various components in fluids in individual sedimentary basins.

CATION CONCENTRATIONS IN ROCK-BUFFERED FLUIDS OF VARIABLE m_{Cl^-}

In this section, we investigate why cation compositions in rock-buffered fluids are a predictable function of chloride concentration when chloride is the dominant anion. The concept that variations in salinity, dissolved chloride, or equivalents of anionic charge can change the absolute and relative concentrations of major dissolved cations was first described for high-temperature hydrothermal systems (Shikazano, 1976, 1978; Giggenbach, 1984, 1988) and has recently been invoked in the hydrothermal experiments of Savage et al. (1993). With the exception of work by this author and colleagues (Hanor, 1988, 1994a; Esch and Hanor, 1993) and Michard and Bastide (1988) the concept has not generally been applied to sedimentary systems. An intuitive derviation of these relations has been presented by Hanor (1988, 1994b), but it is useful to establish them on a more rigorous thermodynamic basis. In the discussion which follows, terms such as component, intensive variable, species, open system, chemical potential, and others will be used in their conventional thermodynamic sense (see Helgeson, 1970; Anderson and Crerar, 1993).

Thermodynamic Components in Siliciclastic-Fluid Systems

The compositions of most mineral phases in siliciclastic sediments are typically expressed in terms of thermodynamic oxide components, such as Na_2O. It is more common in sedimentary aqueous geochemistry, however, to express the aqueous equivalent of these chemical potentials in terms of ionic or aqueous components, e.g.,

$$\mu_{Na_2O} = (2\mu_{Na^+} - 2\mu_{H^+} + \mu_{H_2O}) \qquad (3)$$

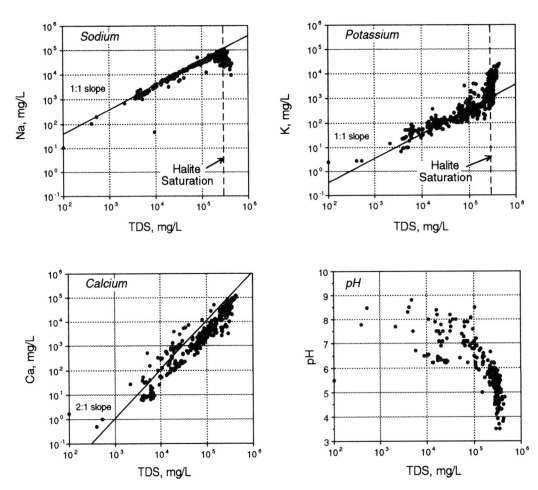

FIG. 2.—Scatter plots showing the coverience of dissolved Na, K, and Ca, and of pH as a function of total dissolved solids for typical waters in sedimentary basins worldwide (modified from Hanor, 1994a).

Here and in the discussion which follows throughout the rest of this paper, each of the chemical symbols for aqueous components, such as Na^+, is taken to represent a thermodynamic component, which may or may not correspond to an actual aqueous species or to the predominant aqueous species containing that element.

The chemical potentials of the seven components in the oxide-HCl system Na_2O-K_2O-Al_2O_3-SiO_2-H_2O-CO_2-HCl can thus be alternatively described by eight aqueous components, such as Na^+-K^+-Al^{3+}-SiO_2°-H_2O-HCO_3^--H^+-Cl^-. As alternative conventions, predominant aqueous species such as $H_4SiO_4^\circ$ and $Al(OH)_2^+$ can be used in place of SiO_2° and Al^{3+} as aqueous components. It may seem that an additional variable has been added in the list of aqueous components, but the constraint of electrical neutrality requires that the anionic charge contributed by bicarbonate and chloride be balanced by the sum of the charges contributed by the four cation components. Hence, there are still only seven independent intensive compositional variables.

Following the nomenclature of Reed (1982), there can exist in solution secondary aqueous species, such as $NaCl^\circ$, OH^-, and $AlCl^{2+}$. These are not new components. The chemical potentials and aqueous concentrations of each of these secondary species can be related to the chemical potentials and aqueous concentrations of the aqueous components through suitable mass action equations and equilibrium constants.

Minimum Number of Mineral Phases in a Rock-Buffered System

The Phase Rule, $f = c - p + 2$, where f = degrees of freedom in terms of intensive variables (P, T, and independently-controlled chemical potentials), c = number of components, and p = number of phases, can be used to predict the minimum number of phases present in a system in thermodynamic equilibrium (Anderson and Crerar, 1993):

$$p = c - f + 2 \qquad (4)$$

Consider, for example, the seven component system, Na_2O-CaO-Al_2O_3-SiO_2-H_2O-CO_2-HCl, containing an aqueous phase at an arbitrary P and T. If the composition of the aqueous phase is completely controlled by equilibrium with respect to coexisting solid phases of fixed composition, then there are two degrees of freedom (P and T), and the Phase Rule predicts that there should be at least six mineral phases present which are controlling the composition of the liquid. These phases could be four silicate minerals, a carbonate mineral and halite (NaCl). If there is no chloride-bearing mineral phase to buffer the chloride concentration of the aqueous phase, however, then there is an additional degree of freedom (μ_{HCl} or m_{Cl^-}). At an arbitrarily selected concentration of chloride, a minimum of only five mineral phases are required to fix the chemical potentials of all of the other components in the system.

Cation Concentrations

The following discussion presents a method for determining the functional dependence of cation concentrations on the concentration of chloride in a rock-buffered system where chloride is the dominant anion and its concentration is externally con-

trolled. We will again take for the purposes of discussion the seven component system Na_2O-CaO-Al_2O_3-SiO_2-H_2O-CO_2-HCl, recognizing that the procedure can be expanded to systems having any number of components. It is assumed that sufficient thermodynamic information is available to calculate activity coefficients for aqueous species and equilibrium constants for hydrolysis reactions. Actual implementation of the calculations requires solving an array of largely non-linear simultaneous equations. This can be done using an iterative approach.

We will assume for the purposes of discussion in the example which follows that the composition of the aqueous phase in this system is buffered by four silicate minerals and one carbonate mineral. We will further assume that the carbonate mineral is calcite, the most likely carbonate phase to be present in this system under typical diagenetic conditions. At a given P and T, thermodynamic constraints on the system Na_2O-CaO-Al_2O_3-SiO_2-H_2O-CO_2-HCl include the following:

1. There are four silicate mineral equilibrium or metastable equilibrium reactions of the general form:

$$Silicate\ Mineral_j + (\nu_{Na+,j} + 2\nu_{Ca^{2+},j} + 3\nu_{Al^{3+},j})H^+$$
$$+ (\nu_{H_2O,j}\ H_2O) = (\nu_{Na+,j}\ Na^+)$$
$$+ (\nu_{Ca^{2+},j}\ Ca^{2+}) + (\nu_{Al^{3+},j}\ Al^{3+}) + (\nu_{H_4SiO_4^\circ,j}\ H_4SiO_4^\circ)$$
$$(5)$$

where $\nu_{i,j}$ is the stoichiometric coefficient of aqueous component i in the hydrolysis reaction for mineral j. The composition of the minerals is such that each of the components Na_2O-CaO-Al_2O_3-SiO_2 is also a component in one or more of the mineral phases. For any given hydrolysis reaction, one or more of the stoichiometric coefficients, $\nu_{i,j}$, may be zero. Use of $H_4SiO_4^\circ$ as the silica component provides a more convenient means for accounting for changes in the activity of H_2O than does SiO_2°.

It is convenient in the development which follows to write the equilibrium reaction for calcite so the charge of Ca^{2+} is balanced entirely by hydrogen ion:

$$CaCO_3 + 2H^+ = Ca^{2+} + H_2O + CO_2^\circ \qquad (6)$$

where CO_2° is dissolved carbon dioxide. It is important to note that the presence of CO_2° does not require or imply the presence of a coexisting gas phase.

For each hydrolysis reaction involving a mineral of invariant composition, there is an equilibrium expression of the following form for the silicate minerals:

$$K_j = \left(\frac{a_{Na^+}}{a_{H^+}}\right)^{\nu_{Na^+,j}} \left(\frac{a_{Ca^{2+}}}{(a_{H^+})^2}\right)^{\nu_{Ca^{2+},j}}$$
$$\left(\frac{a_{Al^{3+}}}{(a_{H^+})^3}\right)^{\nu_{Al^{3+},j}} (a_{H_4SiO_4^\circ})^{\nu_{H_4SiO_4^\circ,j}}/(a_{H_2O})^{\nu_{H_2O,j}} \qquad (7)$$

where K_j is the equilibrium constant for the jth hydrolysis reaction, and of the following form for calcite:

$$K_{calcite} = \left(\frac{a_{Ca^{2+}}}{(a_{H^+})^2}\right) a_{H_2O}\ a_{CO_2^\circ} \qquad (8)$$

Defining the cation/hydrogen ion activity ratio, H_i, for each cation i as:

$$H_i = \left(\frac{a_i^{z_i}}{(a_{H^+})^{z_i}} \right) \qquad (9)$$

where z_i is the charge of cation i, we have from Equation 7 for silicate minerals:

$$K_j = (H_{Na})^{\nu_{Na^+,j}} (H_{Ca})^{\nu_{Ca^{2+},j}} (H_{Al})^{\nu_{Al^{3+},j}} (a_{H_4SiO_4})^{\nu_{H_4SiO_4^\circ,j}}/(a_{H_2O})^{\nu_{H_2O,j}} \qquad (10)$$

By including in the equation for the equilibrium constant for carbonate minerals the equilibrium constant for the hydrolysis of CO_2°,

$$CO_2^\circ + H_2O = H^+ + HCO_3^- \qquad (11)$$

and

$$K_{CO_2^\circ} = (a_{H^+} a_{HCO_3^-})/(a_{CO_2^\circ} a_{H_2O}) \qquad (12)$$

we thus have from equation 8

$$K_{calcite} = (H_{Ca})(a_{H^+} a_{HCO_3^-})/K_{CO_2^\circ} \qquad (13)$$

The activity of H_2O is not an independent variable in this system but is determined by the chemical potentials of the other components in aqueous solution, as shown by the Gibbs-Duhem relation (Helgeson, 1970):

$$d\mu_{H_2O} = -(1/55.5) \sum m_i d\mu_i \qquad (14)$$

During the initial iteration in solving for cation concentrations, the activity of H_2O can simply be assumed to be unity. The four silicate equilibrium expressions (Eq. 10) are then solved simultaneously to yield unique values for the three activity ratios, H_{Na}, H_{Ca}, H_{Al}, and for $a_{H_4SiO_4^\circ}$. These values are only weakly dependent on aqueous solution composition and as a first approximation could be taken to be nearly constant. Once the activities of all dissolved species are calculated in the steps which follow below, an improved value for the activity of H_2O can be determined, as for example from the calculated osmotic coefficient for water (Pitzer, 1987), and utilized in the next iteration.

2. Hydrogen ion concentration can now be determined as a function of dissolved chloride by substituting Equations 9 and 13 into the following aqueous charge balance equation for the system:

$$m_{H^+} + m_{Na^+} + 2m_{Ca^{2+}} + 3m_{Al^{3+}} = m_{Cl^-}$$
$$+ m_{OH^-} + m_{HCO_3^-} \qquad (15)$$

where m_i is the molality of ion i and by using activity coefficients, γ_i, to couple activities to molal concentrations. This yields the following fourth-order polynomial expression which can be solved by iteration for m_{H^+}:

$$(m_{H^+})(1 + H_{Na} \gamma_{H^+}/\gamma_{Na^+})$$
$$+ (m_{H^+})^2(2H_{Ca} (\gamma_{H^+})^2/\gamma_{Ca^{2+}})$$
$$+ (m_{H^+})^3(3H_{Al} (\gamma_{H^+})^3/\gamma_{Al^{3+}})$$
$$- (m_{H^+})^{-1}[(K_w a_{H_2O}/\gamma_{OH^-} \gamma_{H^+})$$
$$+ (K_{calcite}K_{CO_2^\circ}/(H_{Ca})(\gamma_{H^+} \gamma_{HCO_3^-}) = m_{Cl^-} \qquad (16)$$

where K_w is the equilibrium constant for the dissociation of H_2O. Values for the activity coefficients, γ_i, in this equation can intially be taken to be unity in the first iteration. Improved values for the activity coefficients for use in subsequent iterations can be calculated from suitable equations of state (e.g., Pitzer, 1987) once the molal concentrations of all species in solution have been calculated, as described below. The formation of aqueous complexes or secondary species, each of which could be included in Equation 16, decreases total absolute charge but does not change the charge balance. Equation 16 can now be solved for m_{H^+} at any given value of m_{Cl^-}. Once m_{H^+} is known, then the concentrations of all other cations can be calculated from their H_i ratios (Eq. 9). The concentrations of the species OH^-, HCO_3^-, and CO_2° can be calculated using suitable mass action relations (e.g., Equations 6 and 11).

Covariation in Cation Concentrations with Chloride

Most saline subsurface waters have Na^+ and Cl^- as the dominant components (Hanor, 1994a). It is a straightforward matter to show that as m_{Cl^-} increases in buffered waters dominated by sodium chloride, so must m_{H^+} in an approximately 1:1 manner. For example, by rearranging and simplifying equations 15 and 16 to eliminate the minor species we have:

$$(m_{H^+}) \approx m_{Cl^-}(\gamma_{Na^+}/H_{Na} \gamma_{H^+}) \qquad (17)$$

The numerical values of H_{Na} and the ratio $(\gamma_{Na^+}/\gamma_{H^+})$ are only weakly dependent on solution composition, and the principal variable on the right hand side of the equation is thus m_{Cl^-}. As m_{H^+} increases so must the concentrations of the other buffered cations in solution as shown from Equation 9

$$m_i^{z_i+} = H_i(\gamma_{H^+} m_{H^+})^{z_i}/\gamma_i \qquad (18)$$

The magnitude of these increases will depend on the magnitude of the cationic charge, z_i. Monovalent cations will thus covary in an approximately 1:1 manner with changes in m_{H^+} and m_{Cl^-} depending on activity coefficient ratios. Divalent cations, however, will covary in a ratio of approximately 10:1. These relations appear to hold for Na^+, K^+, Mg^{2+}, Ca^{2+}, and Sr^{2+} for many basinal waters worldwide up to approximately halite saturation (Fig. 2; Hanor, 1994a), when buffering by halide minerals apparently becomes a factor. Trivalent cations, such as Al^{3+}, should vary with chloride in a ratio of approximately 100:1. Little is known at the present time, however, of the actual variation in dissolved aluminum with salinity in basinal fluids.

Anionic Charge as a Master Variable

The master variable in the charge balance equation (Eq. 16) was chosen to be chloride simply for convenience, because chloride is by far the dominant anion in most saline subsurface waters in sedimentary basins, and chloride can thus be taken to be nearly equivalent to anionic charge. A more general relation for use in lower salinity waters or for saline waters of more exotic anionic composition is:

$$(m_{H^+}) + \Sigma(|z_i|(H_i/\gamma_i)(\gamma_{H^+} m_{H^+})^{z_i}) = \Sigma|z_A|m_A \qquad (19)$$

where $|z_A|$ and m_A are the absolute charge and the molality of anion A.

Example

The general techniques described above have been used by Hanor (1994b) to calculate the variation in the composition of aqueous fluid in the nine-component system Na_2O-K_2O-MgO-CaO-Al_2O_3-SiO_2-CO_2-H_2O-HCl in equilibrium with the hypothetical mineral assemblage quartz-muscovite-albite-K-feldspar-calcite-dolomite at an arbitrary temperature of 100°C, a pressure of 1 bar and a CO_2 fugacity of 10^{-3} bars (Fig. 3). Hydrolysis constants for mineral phases and for $CO_2(g)$ were taken from Bowers et al. (1984). Conversion between aqueous activities and concentrations was made using a Pitzer virial equation of state for aqueous solutions (Harvie et al., 1984; Pitzer, 1987). Note general similarities to observed fluid compositions (Fig. 2) up to halite saturation. See Hanor (1994a,b) for additional discussion.

MASS TRANSFER IN ROCK-BUFFERED FLUIDS
OF VARIABLE CHLORIDE CONCENTRATION

It is convenient to think of the concentration of chloride or anionic charge as a reaction path variable in diagenesis, changes in which must induce changes in the concentrations of rock-buffered cations and hence induce mass transfer between fluid and mineral phases. It must be remembered, however, that the concentration of chloride or anionic charge cannot be arbitrarily changed without changing the mass of at least one other aqueous component in the system. The concentration of chloride, for example, can be changed only by preferentially adding or removing a cation-chloride component, such as NaCl, KCl, $CaCl_2$, or HCl, by adding or removing H_2O, or by dispersive mixing of all aqueous components in fluids of differing chloride concentration. In the absence of any mineral sources or sinks for chloride, dissolved chloride will behave conservatively during dispersive mixing, that is, its concentration will be determined only by the relative proportions of various end-member waters in the mix.

The final fluid cation composition will be the same at a given value of m_{Cl^-} for a given mineral buffer assemblage no matter how the change in m_{Cl^-} was achieved. The details of the mass transfer reactions which produce this final fluid composition, however, may be significantly different. Here, we will explore the potential magnitude of the effects of the addition of a single cation-chloride component, NaCl, and the addition of H_2O on siliciclastic diagenesis. The effects of dispersive mixing, a process of first-order significance in sedimentary basins (Hanor, 1994b), will be dealt with in a subsequent paper.

Techniques

Reaction path modeling techniques have provided considerable insight into the degree of mass transfer which must accompany attainment or maintainance of equilibrium between fluid and mineral phases in diagenetic and hydrothermal systems (e.g., Helgeson et al., 1970; Wolery, 1979; Reed, 1982; Plummer et al., 1991; Lee and Bethke, 1994). The reaction path modeling techniques described by Reed (1982) involve numerically titrating an aqueous solution of given initial composition with a rock of given bulk chemical composition and tracking the mass transfer of components between aqueous and solid phases and the changes in aqueous solution composition and rock mineral composition which occur as reactions proceed under conditions of prescribed equilibrium. In the procedure which follows here, a volume of sediment consisting of fluid plus rock is instead numerically titrated with a single component of that fluid. It is assumed that the fluids must maintain compositions similar to the observed average compositions of subsurface saline fluids (e.g., Fig. 2), compositions which in turn are assumed to represent a global approach toward metastable equilibrium under variable PTX conditions. Two hypothetical examples will be presented: (1) introduction of the cation-chloride component, NaCl, simulating dissolution of halite and (2) the introduction of the component H_2O, simulating a dehydration reaction. The purpose of these calculations is not to simulate a specific field example but simply to investigate the potential magnitude of the mass transfer effects which could occur. The general computational procedures described here can be adapted to any buffered system or actual field example of interest.

The following conventions will be made in the mass transfer calculations described in this paper:

1. Calculations will be made for systems whose compositions can be described by the ten oxide-chloride components Na_2O-K_2O-MgO-CaO-Al_2O_3-SiO_2-CO_2-H_2O-HCl-X, or alternatively, the eleven aqueous components Cl^-, SiO_2^0, Al^{3+}, H^+, Na^+, K^+, Mg^{2+}, Ca^{2+}, HCO_3^-, H_2O, and X. For the purpose of the mass balance calculations, X represents all non-reactive or conservative components (such as Br^-) in the simulations and is the dependent aqueous component, as required by the Phase Rule.

2. Adding chloride or diluting chloride through the addition of H_2O causes the other components to change in concentration in a manner similar to the patterns of covariance observed in natural subsurface waters worldwide. Polynomial regressions have been made of the concentrations of major solutes, TDS, and pH as a function of dissolved chloride for the compositional

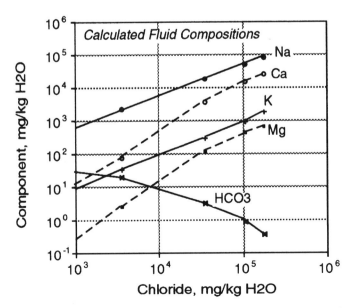

FIG. 3.—Calculated variation in major solute composition as a function of dissolved chloride for fluids buffered by a multiphase silicate-carbonate assemblage (see text; modified from Hanor, 1994a).

8

range 18,000 < Cl(mg/L) < 250,000 for approximately 350 published analyses of waters in sedimentary basins of widely varying geologic age and setting (Table 1). Sources of data are listed in Hanor (1994a). Most r^2 values for the regressions exceed 0.7, and the fits are thus surprisingly good in spite of the probable effects of variations in temperature, pressure, f_{CO_2}, and departure from equilibrium on fluid compositions.

In contrast to major cations, dissolved silica, here represented by the component SiO_2°, does not correlate with chloride in a significant way. The concentration of silica has been assumed to be constant for the purpose of the mass balance calculations described in this paper. This is a reasonable approximation, because silica is present in much lower concentrations than most of the other aqueous components, and variations in its aqueous concentration are not significant in terms of total mass transfer of the component SiO_2 between mineral phases in the present simulations. Little is known of the variation in concentration of Al^{3+} in formation waters. However, because it is normally present in very low concentrations, it too can be considered to be a constant.

The treatment of the aqueous component HCO_3^- is more problematical. Within individual published data sets, there is often a documented decrease in carbonate alkalinity with increasing chloride (e.g., see discussion in Hanor, 1994b). However, within the global data set referred to above, there is a wide scatter in alkalinity, particularly at high salinities, and regressions of alkalinity versus chloride yield fits with r^2 values no better than 0.3. This low degree of correlation is less severe for the purposes of this paper than it might seem because the overall rate of the change in alkalinity with chloride, which is the controlling factor in mass transfer, not absolute concentration, is far smaller than the rates of change of the other major solutes. Hence, even though dissolved bicarbonate can be a significant reservoir for CO_2, the size of this reservoir does not vary as significantly or systematically on a global basis as it does for the major cations. Mass transfer calculations run here on the assumption that HCO_3^- is constant yield results virtually identical to calculations using regressed values. It is the former set of calculations which will be discussed below. The whole problem of carbonate equilibria during mixing of saline waters of diverse composition and variable P_{CO_2} is dealt with in detail and in a more thermodynamically rigorous manner in another paper (Morse et al., in preparation).

3. Nine mineral species act as reversible sources or sinks for aqueous components other than chloride. These mineral species consist of selected combinations of the following common phases found in sedimentary rocks: quartz, kaolinite, muscovite, K-feldspar, albite, Na-beidellite, Ca-beidellite, anorthite, laumontite, chlorite, calcite, and dolomite. The stoichiometry of the mass transfer reactions involving these phases is given in Table 2.

4. No assumptions are made regarding the actual equilibrium state of the system, including the saturation state of individual mineral phases. It is assumed simply that the observed compositions of natural waters represent some averaged global approach toward metastable equilibrium involving variable mineral assemblages over a range of sedimentary PT conditions and that any of the above mentioned phases can reversibly exchange components with aqueous solutions having these compositions. Mass balance calculations similar to the ones presented below

could be done at specified PT conditions and under the contraints of true equilibrium involving specific mineral phases (e.g., Plummer et al., 1991; Lee and Bethke, 1994).

Mass Balance Constraints

For every independent aqueous component i in a system containing an aqueous phase and k mineral phases, whose composition can be described by c independent intensive components, let N_i be the total number of moles of i in the system and n_i be the number of moles of i in aqueous solution. For every mineral phase j, let n_j be the number of moles of j in the system.

For every mineral phase j there is a balanced mass transfer reaction of the form:

$$NaAlSi_3O_8 + 4H^+ \leftrightarrow Na^+ + Al^{3+} + 3SiO_2^\circ + 2H_2O \quad (20)$$

The use of the double arrow ($\leftrightarrow$) rather than an equals sign ($=$) in the above reaction is intended to stress the point that mass transfer does not necessarily imply existence of equilibrium between the mineral phase and aqueous species. The stoichiometric coefficients of the products on the right hand side are taken to be positive and those on the left hand side negative. Let ρ_{ij} be the ratio of stoichiometric coefficients of aqueous component i to the stoichiometric coefficient of phase j in the mass transfer reaction:

$$\rho_{ij} = v_i/v_j \quad (21)$$

The net change in number of moles of dissolved component i is related to the net change in the number of moles of all phases participating in mass transfer reactions and the change in the total number of moles of that component in the system by:

$$\Delta n_i = \sum_j (\rho_{ij} \Delta n_j) + \Delta N_i \quad (22)$$

If the system is closed with respect to i, then $\Delta N_i = 0$. If values for all Δn_j and ΔN_i are known, then it is obviously a straightforward matter to calculate values for Δn_i. If, however, it is the values for Δn_i which are known instead, then the unknown values of Δn_j can be calculated through matrix inversion of $j = 1$ through k simultaneous equations:

$$\rho_{ij} \cdot \Delta n_j = (\Delta n_i - \Delta N_i) \quad (23)$$

Results

Addition of chloride.—

In this section we will examine some representative results of simulations involving addition of chloride as the component NaCl. Figures 4, 5, and 6 show calculated changes in mineral abundance in three systems open with respect to NaCl. The starting fluid composition is water of approximately marine chlorinity (20,000 mg/L Cl). Results are presented as the net number of moles of mineral dissolved or precipitated at a given chloride concentration as the aqueous chloride concentration is increased through the addition of the component NaCl. There is a slight increase in fluid volume as the result the progressive addition of solute mass to the aqueous solution, and the mass transfer results are normalized with respect to a final unit volume of fluid. Positive slopes for the curves on these graphs (dy/

TABLE 1.—REGRESSION COEFFICIENTS

Parameter	Coefficients					
	a	b	c	d	e	r^2
Na, mg/L	2.1991E+03	4.9443E−01	1.8625E−07	−7.0381E−13	2.5175E−17	0.739
K, mg/L	2.2310E+02	2.2547E−02	7.9511E−07	−7.1724E−12	2.1381E−17	0.729
Mg, mg/L	−7.5877E+02	5.4647E−02	−4.6505E−07	2.0876E−12	−8.5095E−19	0.683
Ca, mg/L	−1.8415E+02	1.9851E−02	1.0896E−06	−5.8386E−12	2.9587E−17	0.874
TDS, mg/L	5.6644E+02	1.6346E+00	0	0	0	0.985
pH	8.0829E+00	−2.3236E−05	5.9337E−11	−9.5210E−17	0	0.738

The coefficients above are for equations of the form $y = a + bx + cx^2 + dx^3 + ex^4$, where y is the parameter shown and x is the concentration of dissolved chloride in mg/L.

TABLE 2.—STOICHIOMETRY OF MINERAL HYDROLYSIS REACTIONS

Albite	$NaAlSi_3O_8 + 4H^+ \leftrightarrow Na^+ + Al^{3+} + 3SiO_2^\circ + 2H_2O$
Anorthite	$CaAl_2Si_2O_8 + 8H^+ \leftrightarrow Ca^{2+} + 2Al^{3+} + 2SiO_2^\circ + 4H_2O$
Ca-Beidellite	$Ca_{.167}Al_{2.33}Si_{3.67}O_{10}(OH)_2 + 7.32H^+ \leftrightarrow .167Ca^{2+} + 2.33Al^{3+} + 3.67SiO_2^\circ + 4.66H_2O$
Calcite	$CaCO_3 + H^+ \leftrightarrow Ca^{2+} + HCO_3^-$
Chlorite	$Mg_5Al(AlSi_3O_{10})(OH)_8 + 16H^+ \leftrightarrow 5Mg^{2+} + 2Al^{3+} + 3SiO_2^\circ + 12H_2O$
Dolomite	$CaMg(CO_3)_2 + 2H^+ \leftrightarrow Ca^{2+} + Mg^{2+} + 2HCO_3^-$
K-feldspar	$KAlSi_3O_8 + 4H^+ \leftrightarrow K^+ + Al^{3+} + 3SiO_2^\circ + 2H_2O$
Kaolinite	$Al_2Si_2O_5(OH)_4 + 6H^+ \leftrightarrow 2Al^{3+} + 2SiO_2^\circ + 5H_2O$
Laumontite	$CaAl_2Si_4O_{12}4H_2O + 8H^+ \leftrightarrow Ca^{2+} + 2Al^{3+} + 4SiO_2^\circ + 8H_2O$
Muscovite	$KAl_2(AlSi_3O_{10})(OH)_2 + 10H^+ \leftrightarrow K^+ + 3Al^{3+} + 3SiO_2^\circ + 6H_2O$
Na-Beidellite	$Na_{.33}Al_{2.33}Si_{3.67}O_{10}(OH)_2 + 7.32H^+ \leftrightarrow .33Na^+ + 2.33Al^{3+} + 3.67SiO_2^\circ + 4.66H_2O$
Quartz	$SiO_2 \leftrightarrow SiO_2^\circ$

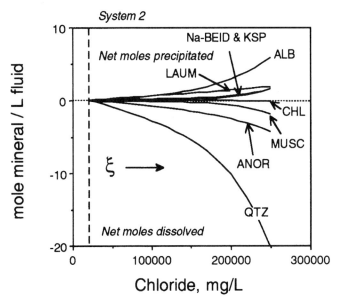

FIG. 5.—Calculated changes in mineral abundance in System 2, a system similar to System 1 (Fig. 4) in which CO_2 has been removed as a component. Laumontite has been added in place of calcite and dolomite.

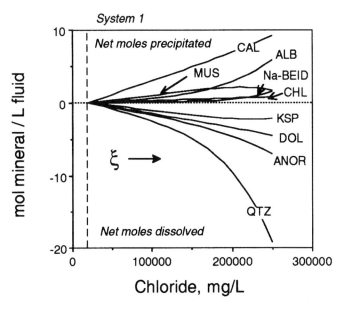

FIG. 4.—Calculated changes in mineral abundance in System 1, a ten phase system consisting of quartz—muscovite—K-feldspar—albite—Na-beidellite—anorthite—chlorite—calcite—dolomite—aqueous fluid in which the component NaCl is progressively added to a water of approximately marine salinity and the fluid evolves in composition according to the global average (Table 1). Results are normalized against final fluid volume. The reaction progress variable (x) is the aqueous concentration of dissolved chloride. The arrow shows the direction of compositional change. Positive values on the vertical axis represent net moles of mineral precipitated per liter of solution at a given chloride value, negative values represent net moles dissolved. Positive slopes ($dy/dx > 0$) indicate precipitation of the mineral phase in question with increasing chloride, negative slopes dissolution.

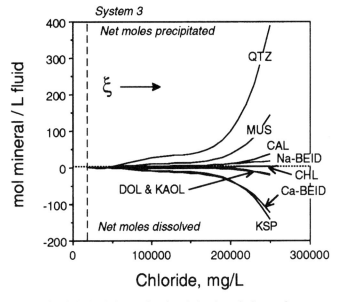

FIG. 6.—Calculated changes in mineral abundance in System 3, a system similar to System 1 (Fig. 4) in which kaolinite and Ca-beidellite have been substituted for albite and anorthite as mineral phases.

10 JEFFREY S. HANOR

dx > 0) reflect precipitation of the mineral phase in question with increasing chloride, negative slopes reflect dissolution.

System 1 contains the mineral phases quartz-muscovite-K-feldspar-albite-Na-beidellite-anorthite-chlorite-calcite-dolomite (Fig. 4). Much of the sodium which is progessively added to the system as the result of dissolution of NaCl is incorporated into newly-formed albite and Na-beidellite. These are hardly surprising results. What is of interest, however, is that production of albite and Na-beidellite and the constraints imposed by observed subsurface fluid compositions (Fig. 2) require changes in the relative proportions of all of the other mineral phases present as well. These include the formation of muscovite at the expense of K-feldspar and the formation of calcite and chlorite at the expense of anorthite and dolomite. Most profound in terms of change in mass is the destruction of quartz as the result of transfer of the SiO_2 component into newly formed silicates. The significant change in the abundance of quartz relative to the other silicates reflects the fact that (a) quartz has the stoichiometry of one of the components of the system, SiO_2, and (b) the other silicates contain from 2 to 3.67 moles of SiO_2 (Table 2). Because the aqueous solution is not a significant reservoir for the component SiO_2, the formation or destruction of a mole of albite, for example, requires the transfer of three moles of the component SiO_2 to or from one or more another mineral phases, such as quartz.

'The variations in the slopes of the curves in Figure 4 reflect changes in the amount of mineral precipitation or dissolution as a function of increasing chloride. For example, muscovite is precipitated up to a chloride concentration of approximately 200,000 mg/L. At higher salinities, however, muscovite begins to dissolve.

In *System 2*, CO_2 has been eliminated as a component, resulting in a carbonate-free system (Fig. 5). Calcite and dolomite thus no longer exist, and laumontite has been added to the mineral assemblage to satisfy the Phase Rule. The rest of the mineral assemblage is the same as that in System 1. As in System 1, the single most profound effect of the addition of NaCl in terms of changes in mineral mass is the destruction of quartz. In addition, albite and Na-beidellite are produced as a result of dissolution of NaCl. However, laumontite, not calcite, is now formed at the expense of anorthite. A small amount of chlorite is destroyed in System 2 rather than precipitated to account for the addition of Mg^{2+} into aqueous solution. In System 1, muscovite is formed at the expense of K-feldspar. In the carbonate-free system, the fate of these two phases is reversed.

In *System 3*, we return to a carbonate-bearing phase assemblage similar to System 1 in which kaolinite and Ca-beidellite have been substituted for albite and anorthite (Fig. 6). Some of the sodium added to the system thus goes into the production of Na-beidellite, which is now the only mineral reservoir for sodium. In results analogous to System 1, K-feldspar is destroyed at the expense of muscovite, and Ca-beidellite and dolomite are destroyed at the expense of calcite and chlorite. The production of muscovite and Na-beidellite, which are phases of high Al/Si ratios, are contributing factors to the destruction of kaolinite and, in marked contrast to System 1, to the massive precipitation of quartz. Although the net mass transfer between aqueous solution and solids is the same as in System 1, the absolute mass transfer between the solid phases is much greater.

Addition of H_2O.—

The chloride concentration of an aqueous solution can be decreased by the addition of the component H_2O. Such a process could occur in nature as a result of dehydration reactions. Simulations involving the addition of the component H_2O to the system require first calculating the change in concentration of each of the aqueous components that would be due solely to dilution and then calculating mass balances based on the difference between diluted values and calculated values (Table 1) at a given diluted chloride concentration. Results of one such simulation are shown in Figure 7. *System 4* contains the mineral phases quartz-muscovite-K-feldspar-albite-kaolinite-anorthite-chlorite-calcite-dolomite. In contrast to the previous simulations, the direction of the chemical evolution of the system is now from high chloride concentrations to low. The net changes in moles of mineral phases are normalized against final fluid volume. As presented in Figure 7, negative slopes to the curves of changes in mineral abundance represent mineral precipitation and positive slopes dissolution. In the initial stages of dilution, there is precipitation of quartz, calcite, and chlorite at the expense of dolomite and albite. There is a reversal in behavior at lower chloride concentrations, and albite and dolomite begin to be precipitated at the expense of the other three phases.

Changes in Volume

One of the potentially more important applications of mass transfer calculations is in predicting the change in rock porosity which may occur as a result of mineral dissolution or precipitation (Lee and Bethke, 1994). If the change in the number of moles of a mineral phase is known, then it is possible from the molal volume of the mineral to calculate the change in volume of that solid as a result of dissolution or precipitation. That information, in turn, can be used to calculate the net change in

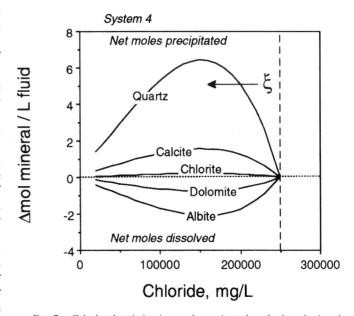

FIG. 7.—Calculated variation in net change in moles of selected mineral phases in a ten phase system when the component H_2O is progressively added to a saline brine and the fluid evolves in composition according to the global average (Table 1). The arrow shows the direction of compositional change.

porosity as a result of several simultaneous mass transfer reactions. Figure 8 shows the calculated relative change in mineral volumes in System 1 as a function of chloride. In this simulation, the porosity is created principally by the destruction of anorthite and is occluded principally by the precipitation of calcite. The relative contribution of each of the mineral phases to changes in total mineral volume varies with salinity. The net effect in this simulation is to increase porosity by increasing salinity. In an actual field setting, changes in porosity could also induce changes in permeability.

DISCUSSION AND CONCLUSIONS

The observed patterns of covariance of major cations and pH with chloride in saline waters in sedimentary basins worldwide is consistent with the conclusion that many of these fluids are attempting to achieve metastable equilibrium with multiphase suites of silicate and carbonate minerals under variable PT conditions. It is convenient to think of the concentration of chloride or anionic charge as a reaction path variable in diagenesis, changes in which must induce absolute and relative changes in the concentrations of rock-buffered cations and, hence, mass transfer between fluid and mineral phases. The concentration of chloride or anionic charge cannot be arbitrarily changed, of course, without changing the mass of at least one other aqueous component in the system.

It is possible using simple mass balance techniques to estimate how much of one mineral phase must be dissolved and another precipitate to acount for the observed evolution in fluid compositions accompanying changes in dissolved chloride. The most significant conclusion which can be reached from the preliminary mass balance simulations performed to date on hypothetical systems is that fluid-mineral mass transfer related to changes in chloride is not simply a matter of mineral dissolution and precipitation, but that it also requires significant transfer of mass between mineral phases. This is due in part to the fact that the mineral components SiO_2 and Al_2O_3 can be accomodated only sparingly in aqueous solution. If an aluminosilicate is dissolved in the process of transfering the components K_2O, MgO or CaO into aqueous solution, then one or more other aluminosilicates must precipitate to compensate for the release of SiO_2 and Al_2O_3. One phase whose relative abundance can significantly be affected by such processes is quartz. In all of the simulations described here, significant quantities of quartz are dissolved and precipitated, all in the absence of any temperature or pressure changes and without large volumes of fluid being involved.

The total problem of the potential effect of changes in dissolved chloride on mass transfer reactions is of course far broader than the examples discussed here. There is evidence, for example, that some subsurface brines have evolved from evaporated marine precursors (e.g., Moldovanyi and Walter, 1992), rather than having been generated by the dissolution of NaCl and subsequent reaction. In the case of evaporated marine waters, many solutes are already present in high concentration, and the subsequent mass transfer reactions between fluids and minerals will involve different pathways and masses of components. In addition, in systems where there is fluid flow and solute transport, each volume of sediment is open with respect to all mobile aqueous components, not just simply Na^+, Cl^-, or H_2O, as in the examples discussed here. These cases will be dealt with in a subsequent paper.

It is not the claim of this paper that changes in chloride concentration are a more important diagenetic agent in siliciclastic diagenesis than other mechanisms which have been previously proposed, but rather that spatial and temporal variations in chloride or salinity are additional factors which should be considered when interpreting observed diagenetic changes in siliciclastic sequences. The results of the theoretical calculations presented above suggest the effects may potentially be of large magnitude. Field and laboratory work is now required to determine the actual importance of this mechanism for inducing siliciclastic diagenesis.

ACKNOWLEDGMENTS

This work was supported in part by NSF Grants EAR 9019342 and EAR 9316627 and a grant from the Department of the Interior, U.S. Geological Survey, through the Louisiana Water Resources Research Institute. I would like to thank Laura Crossey, Lee Esch, Tom Taylor, and Glenn Wilson for their thoughtful reviews and helpful comments.

REFERENCES

ANDERSON, G. M. AND CRERAR, D. A., 1993, Thermodynamics in Geochemistry, the Equilibrium Model: New York, Oxford, 588 p.
BOWERS, T. S., JACKSON, K. J., AND HELGESON, H. C., 1984, Equilibrium Activity Diagrams: New York, Springer-Verlag, 397 p.
ESCH, L. AND HANOR, J. S., 1993, Reaction of NaCl brines with siliciclastic sediments: an experimental study: Geological Society of America, Abstracts with Programs, 1993 Annual Meeting, p. A322.
GIGGENBACH, W. F., 1984, Mass transfer in hydrothermal alteration systems—a conceptual approach: Geochimica et Cosmochimica Acta, v. 48, p. 2693–2712.
GIGGENBACH, W. F., 1988, Geothermal solute equilibria. Derivation of Na-K-Mg-Ca geoindicators: Geochimica et Cosmochimica Acta, v. 52, p. 2749–2765.

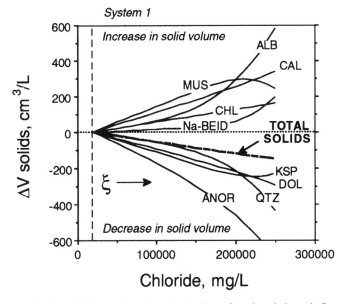

FIG. 8.—Calculated change in mineral volume for selected phases in System 1 (Fig. 4). Note the progressive decrease in the total volume of solids with increasing introduction of chloride.

HANOR, J. S., 1988, Origin and migration of subsurface sedimentary brines: Tulsa, Society of Economic Paleontologists and Mineralogists Short Course 21, 248 p.

HANOR, J. S., 1994a, Origin of saline fluids in sedimentary basins, *in* Parnell, J., ed., Geofluids: Origin, Migration, and Evolution of Fluids in Sedimentary Basins: Bath, Geological Society Special Publication 78, p. 151–174.

HANOR, J. S., 1994b, Physical and chemical controls on the composition of waters in sedimentary basins: Marine and Petroleum Geology, v. 11, p. 31–45.

HARVIE, C. E., MOLLER, N., AND WEARE, J. H., 1984, The prediction of mineral solubilties in natural waters: The Na-K-Mg-Ca-H-Cl-SO$_4$-OH-HCO$_3$-H$_2$O system to high ionic strengths at 25°C: Geochimica et Cosmochimica Acta, v. 48, p. 723–751.

HEKGESIB, H. C., 1970, Description and interpretation of phase relations in geochemical processes involving aqueous solutions: American Journal of Science, v. 268, p. 415–438.

HELGESON, H. C., BROWN, T. H., NIGRINI, A., AND JONES, T. S., 1970, Calculation of mass trsnfer in geochemical processes involving aqueous solutions: Geochimica et Cosmochimica Acta, v. 34, p. 569–592.

HUTCHEON, I., SHEVALIER, M., AND ABERCROMBIE, H. J., 1993, pH buffering by metastable mineral-fluid equilibria and evolution of carbon dioxide fugacity during burial diagenesis: Geochemica et Cosmochimica Acta, v. 57, p. 1017–1028.

LAND L. S., MACPHERSON, G. L., AND MACK, L. E., 1988, The geochemistry of saline formation waters, Miocene, offshore Louisiana: Gulf Coast Association of Geological Societies Transactions, v. 38, p. 503–511.

LEE, M-K AND BETHKE, C. M., 1994, Groundwater flow, late cementation, and petroleum accumulation in the Permian Lyons Sandstone, Denver Basin: American Association of Petroleum Geologist Bulletin, v. 78, p. 217–237.

MCDONALD, D. A., AND SURDAM, R. C., eds., 1984, Clastic Diagenesis: Tulsa, American Association of Petroleum Geologists Memoir 37, 434 p.

MCILREATH, I. A. AND MORROW, D. W., eds., 1990, Diagenesis: St. John's, Newfoundland, Geological Association of Canada, Reprint Series 4, 339 p.

MICHARD, G. AND BASTIDE, J.-P., 1988, Etude geochimique de la nappe du Dogger du Bassin Parisien: Journal of Volcanism and Geothermal Resources, v. 35, p. 151–163.

MOLDOVANYI, E. P., AND WALTER, L. M., 1992, Regional trends in water chemistry, Smackover Formation, southwest Arkansas: geochemical and physical controls: American Association of Petroleum Geologists Bulletin, v. 76, p. 864–894.

PITZER, K. S., 1987, Thermodynamic model for aqueous solutions of liquid-like density: Reviews in Mineralogy, v. 17, p. 97–142.

PLUMMER, L. N., PRESTEMON, E. C., AND PARKHURST, D. L., 1991, An interactive code (NETPATH) for modeling net geochemical reactions along a flow path: Washington D.C., U.S. Geological Survey Water-Resources Investigations Report 91–4078, 227 p.

REED, M. H., 1982, Calculation of multicomponent chemical equilibria and reaction processes in systems involving minerals, gases, and aqueous phase: Geochimica et Cosmochimica Acta, v. 46, p. 513–528.

SAVAGE, D., BATEMAN, K., MILODOWSKI, A. E., AND HUGHES, C. R., 1993, An experimental evaluation of the reaction of granite with streamwater, seawater, and NaCl solutions at 200°C: Journal of Volvanology and Geothermal Research, v. 57, p. 167–191.

SHIKAZANO, N., 1976, Thermodynamic interpretation of Na-K-Ca geothermometer in the natural water system: Geochemical Journal, v. 10, p. 47–50.

SHIKAZANO, N., 1978, Possible cation buffering in chloride-rich geothermal waters: Chemical Geology, v. 23, p. 239–254.

SMITH, J. T. AND EHRENBERG, S. N., 1989, Correlation of carbon dioxide abundance with temperature in clastic hydrocarbon reservoirs: relationship to inorganic chemical equilibrium: Marine and Petroleum Geology, v. 6, p. 129–135.

WOLERY, T. J., 1979, Calculation of chemical equilibrium between aqueous solution and minerals: the EQ3/6 software package: Livermore, Lawrence Livermore National Laboratory Report UCRL-52658, 41 p.

DEPOSITIONAL CONTROL ON CARBONATE CEMENT IN THE SAN JOAQUIN BASIN, CALIFORNIA

YONG IL LEE

Department of Geological Sciences, Seoul National University, Seoul 151–742, Korea

AND

JAMES R. BOLES

Department of Geological Sciences, University of California, Santa Barbara, California 93106

ABSTRACT: Carbonate cements in the early Miocene Temblor Formation at Kettleman North Dome oil field, on the western flank of the San Joaquin basin in California, formed in marine and mixed meteoric-marine pore waters. Arkosic sands were deposited in deltaic to shallow marine and deep marine environments. Carbonate cements preserve the degree of compaction at the time of cementation. Micritic calcite cements are interpreted to have formed at the sediment-sea water interface when intergranular pore space was about 40%. Later, dolomite cements formed during shallow burial, when intergranular porosities were about 30%. Coarse crystalline calcite cement and grain-replacement cement precipitated during deep burial, when intergranular porosity was less than 25%. These carbonate cements originated from three types of formation waters based on oxygen isotopic data. The micrite formed in nearly pure meteoric water at the sediment-water interface. The dolomite precipitated from mixed marine-meteoric water during shallow burial. Late calcite that formed during deep burial (70–120°C) precipitated from diagenetically modified marine waters.

Studies of six fields in the central and eastern San Joaquin basin indicate that carbonate cements originate from both marine pore waters and from meteoric incursion during deposition and uplift of the basin perimeter. Sediments deposited in non-marine to shallow marine environments in the basin flanks were subjected to meteoric water infiltration during shallow burial. Early carbonate cements with meteoric isotopic signatures are found at distances of up to 5 km (Round Mountain Field, eastern flank) to 15 km (North Kettleman Dome Field, western flank) from potential recharge areas. Meteoric recharge is also recorded late in the cement history, due to local uplift on the west flank. The uplift focused meteoric water into deep marine sands up to 15 km from the basin edge (North Belridge Field). In the central basin, sands deposited in deep marine environments were isolated from meteoric influence due to their distance from meteoric recharge areas and lack of hydraulic continuity with the basin flanks. Thus, these sands only contain carbonate cements with marine or evolved marine geochemical signatures (e.g., North Coles Levee Field).

INTRODUCTION

Three types of formation waters are recognized in the San Joaquin basin (Fisher and Boles, 1990), all of which are involved in calcite cementation. They are meteoric, mixed marine-meteoric, and diagenetically modified marine waters (Boles and Ramseyer, 1987; Fisher and Boles, 1990; Hayes and Boles, 1993; Taylor and Soule, 1993). These studies show that the occurrence and distribution of carbonate cements depends on the position of the sand body in the basin, initial sand porosity and permeability, sand continuity, and the basin hydrology. Understanding the origin of carbonate cement and its distribution pattern is crucial for prediction of hydrocarbon reservoir quality. In addition it allows us to interpret the evolution of basin water compositions and estimate fluid flux through time. Studies to date illustrate the complexity of cement origin and timing in what is generally regarded as a simple sedimentary basin.

Studies of six fields in the central and eastern San Joaquin basin indicate that carbonate cements originate from both marine pore waters and from meteoric incursion during deposition and uplift of the basin perimeter (Boles and Ramseyer, 1987; Hayes and Boles, 1993). Most of these sands were deposited in shallow marine environments. However, studies of marine deep sea fan deposits in the central basin found no evidence for meteoric mixing in the pore waters (Boles, 1987; Boles and Ramseyer, 1987; Schultz et al., 1989, 1991). The deep sea sands have undergone simple burial to their present levels. Early dolomite cementation followed by a long history of calcite cementation resulted from marine pore water, which was progressively modified from interaction with the rock during burial.

Less is known about the origin of carbonate cements in the western part of the San Joaquin basin compared to the eastern part. The only published study to date is the North Belridge Field on the southwest flank of the basin (Fig. 1), where Oli-

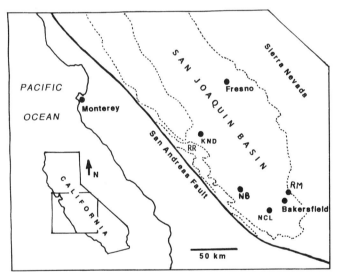

FIG. 1.—Location map of the Kettleman North Dome (KND) in the San Joaquin basin. The boundary is marked by dotted lines (simplified after Bent, 1988). RR: Reef Ridge outcrops in the western San Joaquin basin. Oil fields: NCL = North Coles Levee; NB = North Belridge; RM = Round Mountain and Mount Posa.

gocene marine sands have been cemented by four episodes of carbonate cementation (Taylor and Soule, 1993). The earliest cement is calcite derived from marine pore waters. A dolomite cement of meteoric origin formed next, followed by a calcite cement and a late-stage calcite fracture-fill. These late-stage calcite cements are interpreted to have formed during uplift of the basin flank in a mixed water derived from marine shale and meteoric sources.

The purpose of this study is two-fold: (i) to describe diagenetic carbonates at Kettleman North Dome on the western flank

Siliciclastic Diagenesis and Fluid Flow: Concepts and Applications, SEPM Special Publication No. 55

of the San Joaquin basin and (ii) to compare this area to other occurrences of diagenetic carbonates in the San Joaquin basin in order to understand the control of depositional facies and paleohydrology on carbonate cementation.

GEOLOGIC BACKGROUND

The San Joaquin basin of California (Fig. 1) contains more than 7,620 m (25,000 ft) of Tertiary sediments. The central basin produces hydrocarbons from Miocene and younger sediments, whereas the basin flanks has additional production from Oligocene and Eocene strata (Callaway, 1971). Kettleman North Dome oil field, discovered in 1928, is the northernmost of the Kettleman Hills, which consist of three anticlinal structures that lie along the west side of the central San Joaquin basin, California (Fig. 1).

The Temblor Formation is the main oil and gas producing zone in the field. The formation is 300–680 m thick (Merino, 1975) and is subdivided into Zones I through V, each separated by at least 15 m of shales (Fig. 2). Zone I was deposited in a non-marine to shallow marine setting; Zones II, IV and V were deposited in a shallow marine to deltaic setting and Zone III in a submarine fan setting (Kuespert, 1985).

Kettleman North Dome is a steep, doubly plunging, asymmetrical anticline. Based on the relative thickness of the units affected by the folding, Harding (1976) determined that the fold was formed within the last million years. This interpretation implies very recent hydrocarbon emplacement into the reservoir. At Kettleman North Dome, about 600–1,000 m of sediment have been removed by erosion since 1 Ma (Merino, 1975).

Sandstones of the Temblor Formation are similar to arkosic sands found elsewhere in the San Joaquin basin but have lower overall feldspar contents and, in some cases, higher proportions of rock fragments (Fig. 3). The main framework grains are quartz, plagioclase, potassium feldspar, and rock fragments, primarily of volcanic origin. Point-count analysis indicates the ratio of plagioclase to potassium feldspar ranges from 0.1 to 3.4 with an average of 1.2. Volcanic rock fragments are the main lithic component and vary from less than 20% of the rock in Temblor Zones I to IV and up to 45% in Zone V (Merino, 1975; Fig. 3). Mean point-count porosity of sandstones without carbonate cement is about 15%.

Our sample control is too sparse to determine the detailed distribution of cements in relation to Temblor stratigraphic zones. Overall carbonate cemented zones are relatively sparse and thin. Three types of carbonate cements are recognized: (i) micritic calcite, (ii) dolomite, and (iii) sparry calcite (Table 1). Available data indicate that micrite is found in Zone I, and that dolomite is found in Zone II. Sparry calcite is widespread in all zones except Zone III.

SUMMARY OF DIAGENETIC FEATURES

Temblor sandstones at Kettleman North Dome contain an assemblage of carbonate, clay, sulfide-sulfate, and zeolite minerals. Although most samples contain only a few dominant diagenetic minerals, it is possible to reconstruct the following paragenetic sequence: (i) precipitation of early calcite and dolomite cements, (ii) cementation by quartz and albite, (iii) formation of authigenic clay minerals (including kaolinite), (iv) precipitation of anhydrite cement, (v) precipitation of late calcite cement, and (vi) late-stage albitization and dissolution of plagioclase and associated precipitation of laumontite (Merino, 1975; Lee and Boles, in prep.). K-feldspar overgrowths and pyrite are minor diagenetic minerals. In addition to chemical

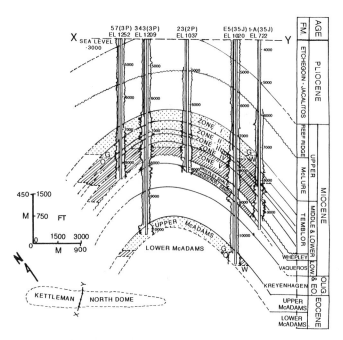

FIG. 2.—Cross section X-Y of Kettleman North Dome oil field (modified after Sullivan, 1966). This study focused on Zone I through V in the Miocene reservoir. Gas and oil-bearing strata shown by stipple and shaded pattern, respectively.

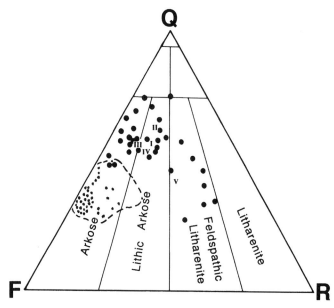

FIG. 3.—Modal composition of of Temblor sandstones at Kettleman North Dome. Classification scheme from Folk (1968). Letters I-V represent average sandstone compositions in each zone. Dashed line encloses composition field for sandstones in central and eastern San Joaquin basin (data from Hayes and Boles, 1993)

TABLE 1.—LOCATION OF TEMBLOR SANDSTONE CORE SAMPLES AT KETTLEMAN NORTH DOME FIELD

Well Number	Well Section	Township	Range	Core Interval Studied Meter (ft)	Number of Thin Sections	Temblor Zone	Number Micritic Calcite	Dolomite	Sparry Calcite
56	8	22	18	2533-3249 (8312-9747)	5	I to V		tr	tr
58	17	22	18	1951-2143 (6400-7030)	4	I to IV	1	2	
61	20	22	18	2155-2336 (7070-7663)	3	III to V		1	
55	20	22	18	2175-2478 (7136-8131)	4	I to III, V	1		1
V21	17	22	18	2005-2222 (6578-7291)	3	I, II, IV		1	tr
8	8	22	18	2406 (7893)	1	V			
F1	35	21	17	2099-2211 (6888-7253)	2	I, II		1	
H1	29	21	17	2264-2522 (7428-8275)	9	I to IV		4	2
V38	34	21	17	2280-2283 (7480-7490)	2	V			

diagenesis, the presence of sutured quartz-quartz contacts and stress-induced albitization (Boles, 1984) indicate that burial stress is a diagenetic reaction control.

CARBONATE CEMENTS

Timing of Cementation

Carbonate cements preserve the degree of compaction at the time of cementation in sandstones. Cement volume, when completely filling intergranular pore space, has been used as an indicator of cement timing in the San Joaquin basin (e.g., Boles and Ramseyer, 1987; Taylor and Soule, 1993). Zieglar and Spotts (1978) compiled porosity-depth trends from a number of reservoirs in the basin and our regional studies have shown these porosity changes largely due to compaction. High intergranular cement volumes (30 to greater than 40%) would indicate early cementation near the sediment-water interface when initial porosities in the San Joaquin basin would be near 40% (Zieglar and Spotts, 1978). Intergranular volumes less than about 30% would indicate precipitation during burial. Average porosity today in uncemented sandstone at Kettleman North Dome is near 15% indicating loss of about one third of original porosity due to compaction. This value is very similar to compaction loss in marine turbidite sandstones buried to about 2,700 m at North Coles Levee (Boles and Ramseyer, 1987). Intergranular cement volumes of about 15% would indicate cementation at least as late as maximum burial depths or perhaps during uplift.

Micritic Calcite Cement

Micritic calcite is characterized by high intergranular volume (39%) and its distinctive texture (Fig. 4). Localized homogeneous patches of microspar within the cement suggest recrystallization of the micrite (Folk, 1965). As discussed, high intergranular cement volume indicates formation early in the diagenetic history, perhaps at the sediment-water interface. In the single example of micrite found, quartz and plagioclase overgrowths are absent indicating the carbonate formed prior to these phases. The micrite grains show dark to brown dull cathodoluminescence with thin dull luminescent rims.

FIG. 4.—Photomicrograph of micritic calcite cement (grey areas) showing high cement volume. Sample 1–55, 2175.0 m. Crossed nicols. Cement is interpreted to have formed close to the sediment-water interface from nearly pure meteoric water (see text).

Dolomite Cement

Dolomite fills up to 30% of the intergranular volume (excluding mold-fill; see below). Small, euhedral, rhombohedral crystals (0.02 to 0.1 mm) form partial rims attached to detrital grains as well as euhedral to subhedral pore-filling crystals (Fig. 5). In some samples, dolomite cement has a bimodal size distribution, with larger crystals formed later than the smaller ones.

Calcite Spar Cement and Grain Replacement

Calcite spar occurs mainly as pore filling cement which forms a poikilotopic texture. Intergranular cement volumes are as high as 24% but the average is about 17%. These relatively low volumes suggest calcite spar precipitated at moderate to deep depth. Where albite overgrowths are present, the calcite spar post-dates the albite (Fig. 6). Fossil molds of former aragonitic(?) mollusks contain euhedral dolomite rhombs floating in a fill of sparry calcite, suggesting the spar post-dates the dolomite. Calcite replacement usually occurs in compacted rocks, in which grains show long to concavo-convex contacts.

FIG. 6.—Photomicrograph of sparry calcite cement. Calcite (Ca) pore-fill cement after albite overgrowths (arrows). Sample 1–58, 1950.7 m. Crossed nicols. Cement is interpreted to have formed during deep burial from marine water modified by water-rock interaction (see text).

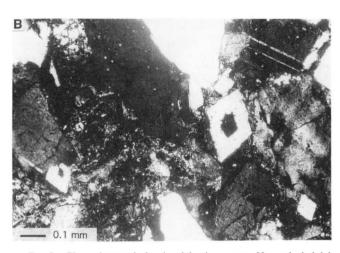

FIG. 5.—Photomicrograph showing dolomite cements. Note euhedral dolomite rhombs in (A) and (B). Some dolomites contain dissolved diamond-shaped hollow core as shown in (B). (A) Sample 2A-H1, 2342.1 m. Crossed nicols. (B) Sample 2D-H1, 2463.1 m. Crossed nicols. Dolomite cement is interpreted to have formed during shallow burial from mixed marine-meteoric water (see text).

Rarely, the calcite spar replaces plagioclase and occurs on the cleavage of deformed biotite flakes. Calcite spar cements precipitated prior to albitization of plagioclase as indicated by the preservation of fresh plagioclase in extensively cemented sandstones. The late calcite has a homogeneous, dull orange luminescence. Slightly darker, luminescent pore-filling and replacement calcite spar is also associated in some samples with dull orange luminescent calcite, but their paragenetic relationship is not clear.

ORIGIN OF CARBONATE CEMENT FROM ISOTOPIC AND TRACE ELEMENT DATA

Micritic Calcite

Isotopic and trace element analyses of the micritic calcite cement are listed in Tables 2 and 3, respectively. Microprobe analysis of 16 spots in the micrite-bearing thin section indicates relatively pure calcite with average compositions of 96.3 mole % Ca, 2.8 mole % Mg, and 0.8 mole % Fe + Mn (Fig. 7). An

intergranular volume of about 40% for the micrite cement suggests an origin near the sediment-water interface, based on the porosity-depth data of Zieglar and Spotts (1978). The following interpretation is hinged on this assumption.

If we assume crystallization in Miocene sea water with composition of $\delta^{18}O_{(SMOW)} = -1.0\%o$ (Woodruff et al., 1981), then using the fractionation equation of Friedman and O'Neil (1977), micrite with a $\delta^{18}O_{(PDB)}$ of $-4.5\%o$ is estimated to have precipitated at about 32°C. This is a relatively high temperature for cementation close to the sediment-sea water interface. The micrite has homogeneous patches of microspar, suggesting recrystallization of micrite (Folk, 1965). If the micrite was recrystallized, the oxygen isotope value of early calcite may represent the pore water condition at the time of recrystallization rather than that of initial precipitation. The $\delta^{13}C$ value is close to that of calcite precipitated from marine waters (e.g., Irwin et al., 1977).

The average 2.8 mole percent $MgCO_3$ in the early calcite cement is low compared to most calcite precipitated on the seafloor (3 to 17 mole percent; Videtich, 1985; Major and Wilber, 1991), even taking into account the dependence of Mg content in marine calcite cements on sea water temperature (Videtich, 1985; Mucci, 1987).

The micrite may have precipitated from a water of chiefly meteoric origin. The negative oxygen isotope value, high intergranular volume, and low Mg content favor this alternative interpretation. Calculated temperatures for micritic calcite cement are lowered if marine pore water is mixed with isotopically light non-marine water. It is likely that crystallization temperatures within a few meters of the sediment-water interface would be around 10°C. For the micrite analyzed here, this would correspond to water compositions of $\delta^{18}O_{(SMOW)} = -6$. This value is that of nearly pure meteoric water (present meteoric water in western San Joaquin area is -6 to -7, Kharaka et al., 1973). Temblor Zone I sediments, the only zone to contain micritic calcite cement, were deposited in non-marine to shallow marine environments (Kuespert, 1985), suggesting the original pore

TABLE 2.—STABLE ISOTOPE ANALYSIS FROM P. DOBSON AT UNOCAL RESEARCH LAB, BREA, CALIFORNIA

Well	Present Depth (m)	Temblor Zone	Cement type	IGV* (%)	Calcite PDB	Calcite PDB	Dolomite PDB	Dolomite PDB
55	2175.0	I	micrite	39.0	−4.5	−0.4		
	2478.3	V	spar	18.4	−8.9	−3.9		
58	1950.7	I	spar	3.5	−9.9	−10.8		
	2144.6	IV	spar	9.7	−11.5	−8.0		
61	2179.3	IV	spar	19.7	−8.9	−7.5		
F1	2210.7	II	spar	24.6	−8.8	−16.4		
H1	2264.0	I	spar	11.5	−10.8	−5.6		
	2342.1	II					−4.3	−6.4
	2346.9	II	spar	12.6	−13.5	−7.5	−3.0	−6.6

*IGV: intergranular volume (= cement + porosity) based on av. 600 point counts
See Boles and Ramseyer (1987) for analytical techniques and method of distinguishing calcite from

TABLE 3.—CARBONATE CEMENT COMPOSITIONS

Sample (depth:m)	Analytical Cement Type	Calcite N*	Ca	Mg	Fe	Mn	Sr	Dolomite N*	Ca	Mg	Fe	Mn	Sr
1-55 (2175)	micrite	16	96.32 (0.88)	2.78 (0.69)	0.48 (0.38)	0.41 (0.31)	0.02 (0.01)						
4-58 (2144.6)	spar	29	96.51 (0.87)	2.12 (0.51)	0.48 (0.23)	0.89 (0.22)	0.02 (0.01)						
4-61 (2179.3)	spar	8	94.14 (1.12)	3.39 (0.59)	1.12 (0.58)	1.03 (0.55)	0.32 (0.06)						
2-F1 (2210.7)	spar	15	95.99 (0.78)	2.64 (0.45)	0.53 (0.26)	0.82 (0.20)	0.02 (0.01)						
2A-H1 (2342.1)								30	51.56 (1.32)	45.56 (2.48)	2.48 (2.38)	0.36 (0.21)	0.05 (0.06)
2B-H1 (2346.9)	spar	6	96.19 (0.40)	2.66 (0.12)	0.36 (0.08)	0.49 (0.17)	0.30 (0.07)	22	52.93 (2.95)	41.63 (2.89)	4.75 (1.20)	0.50 (0.17)	0.18 (0.08)

*N = standard

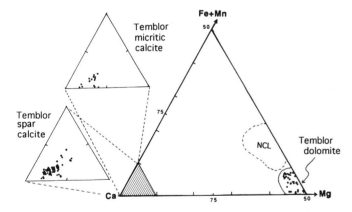

FIG. 7.—Carbonate cement composition (mole %) from Temblor sandstones, Kettleman North Dome. Data from microprobe analysis of 52 points from two samples (2A-H1 and 2B-H1). Refer to Boles and Ramseyer (1987) for microprobe operating conditions for analysis. Dotted area (NCL) represents early dolomite cements from North Coles Levee oil field for comparison (Boles and Ramseyer, 1987). Also shown are daughter triangles of early micritic calcite and late sparry calcite cements from Temblor sandstones (see Table 3).

fluid may have contained a significant meteoric water component.

Dolomite

Dolomite cements from Kettleman North Dome have $\delta^{18}O_{(PDB)}$ values of −3.0 and −4.3‰ and $\delta^{13}C_{(PDB)}$ values of

−6.4 and −6.6‰ (Table 2). The dolomite has up to 7 mole % Fe substitution and is slightly Ca rich with Ca/(Fe + Mg) ratios ranging from 1.0 to 1.2 (Table 3, Fig. 7). In general, these dolomites have less Fe substitution than dolomites from the central basin (Fig. 7). Some dolomite rhombs show diamond-shaped hollow cores due to dissolution of the center zone and this reflects varying trace element composition in the rhombs (Fig. 5B). The cores of dolomite rhombs are slightly enriched in Ca compared with the rim zones. Average Ca/(Mg + Fe) ratio in the cores is 1.13, whereas that in the rim zones is 1.05. The core contains more Mg and less Fe than the rim, with average Mg/Fe ratios of about 106 and 84, respectively.

Dolomite intergranular cement volumes of up to 30% suggest cementation before significant burial, but after micritic calcite precipitation. Compaction to intergranular volumes of 30% would suggest that the dolomite formed at a burial depth of about 600 m (see depth-porosity curve of Zieglar and Spotts, 1978). The isotopic composition of the pore fluid during dolomite crystallization can be estimated from calculating the crystallization temperature at 600 m. Bottomhole temperature measured in the Temblor Zone V reservoir near the northern crest of the structure is about 100°C (B. Bilodean, Chevron, pers. commun., 1992). Assuming a 10°C surface temperature, the measured bottomhole temperature indicates a geothermal gradient of 36°C/km, similar to gradients in the southern San Joaquin basin (Boles and Ramseyer, 1987). Modeling by Heasler and Surdam (1985) suggests that paleo-heatflow has nearly doubled since the time of Temblor deposition. Thus, the early geothermal gradient at the time of Temblor deposition was probably considerably lower than calculated present values and gradients near 20°C/km may be more appropriate.

Assuming the present day geothermal gradient of 36°C/km and a 10°C surface temperature, the temperature at 600 m burial depth is 32°C. Alternatively, a lower geothermal gradient of 20°C/km and a 10°C surface temperature correspond to a temperature of 22°C (for 600 m burial depths). If dolomite cement precipitated between 22°C and 32°C, the $\delta^{18}O$ of the water from which dolomite precipitated is calculated to be between $\delta^{18}O_{(SMOW)}$ of -5.3 and $-4.0‰$, respectively, using the fractionation equation of Land (1985). The present surface meteoric waters from the western San Joaquin hills contain $\delta^{18}O_{(SMOW)}$ values of -6 to $-7‰$ (Kharaka et al., 1973). The light oxygen isotope values of pore fluids during dolomite precipitation suggest mixing of meteoric and marine pore water. This interpretation is plausible considering that the sediments containing dolomite cements (Zone II) were deposited in a shallow marine to deltaic setting which may have been subject to meteoric influence during relative sea-level falls.

The $\delta^{13}C_{(PDB)}$ values of dolomite cements are -6.4 and $-6.6‰$. Isotopically light carbon found in carbonates from shallow sediment has been attributed to oxidation of organic matter by aerobic or sulfate-reducing bacteria (Curtis, 1978; Curtis and Coleman, 1986). Organic carbon for Temblor dolomite cements is interpreted as being derived from oxidation of organic matter. Oxygen isotope composition of formation waters indicates meteoric mixing, which may have allowed in situ oxidation of organic matter. Alternatively, bicarbonate depleted in ^{13}C could have been carried by meteoric waters which have passed through the soil and shallow vadose environments. Near the sediment-water interface, loss of isotopically light carbon to methane (methanogenesis) leaves a heavy carbon signature in the residual bicarbonate (Curtis, 1978; Curtis and Coleman, 1986), but this condition is not recorded in these samples.

Sparry Calcite

Sparry calcite cements have $\delta^{18}O_{(PDB)}$ values ranging from -8.8 to $-13.5‰$ and $\delta^{13}C_{(PDB)}$ values from -3.9 to -16.4 (Table 2). These values are significantly lower than for early micritic calcite. Microprobe analysis reveals that the average sparry calcite contains 95.8 mole % Ca, 2.7 mole % Mg, and 1.4 mole % Fe + Mn (Table 3; Fig. 7).

The intergranular volume (cement + porosity) in calcite-cemented sandstones ranges from 3.5% to 39% (Table 2), indicating a long history of calcite cementation during compaction. A trend of decreasing intergranular volumes is accompanied by decreasing calcite oxygen isotope composition (Fig. 8). The maximum intergranular volume of sparry calcite cement (24%) corresponds to a burial depth of about 1,700 m (Zieglar and Spotts, 1978). Using the present geothermal gradient, the burial temperature at this depth is around 70°C. If this is taken to be the burial temperature at the onset of late calcite cementation, the $\delta^{18}O_{(SMOW)}$ value of pore water is calculated to be $+0.8‰$ using the fractionation equation of Friedman and O'Neil (1977). This value is heavier than contemporaneous Miocene sea water ($\sim -1.0‰$; Woodruff et al., 1981), which rules out the possibility of meteoric water mixed into connate marine pore water.

The maximum temperature for calcite cementation is constrained by the interpretation that all late calcite formed prior

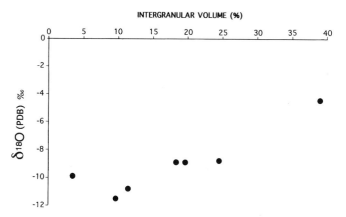

FIG. 8.—Correlation of $\delta^{18}O$ of calcite cement with intergranular volume. Data from Table 2. Data point at 39% intergranular volume is micrite, other data points are calcite spar. Plot shows decrease in $\delta^{18}O$ with inferred increase in burial depth (i.e., lower intergranular volume).

to albitization of plagioclase, a reaction known to occur between 120°C and 160°C in the San Joaquin basin (Boles and Ramseyer, 1988). If we use 120°C for the maximum temperature of crystallization for the calcite cementation, and assume the calcite cement with the lowest volume intergranular volume (3.5%) is the last sparry calcite to form, the pore water at that time would have been no heavier than $\delta^{18}O = +5.5‰$. This calculated oxygen isotope value is within present-day water values for Kettleman North Dome field (Kharaka et al., 1973). These calculated values are similar to the maximum $\delta^{18}O_{(SMOW)}$ values observed in modern pore waters from the central San Joaquin basin (Fisher and Boles, 1990). Positive oxygen isotopes within the San Joaquin basin have been attributed to extensive interaction between original marine pore water and sediment during burial. Evaporation can also create positive $\delta^{18}O_{(SMOW)}$ values, but the relatively dilute composition of all of these basin waters suggests they are not a result of extensive evaporation.

Late calcite cements have carbon isotopic composition of between -3.9 and $-16.4‰$, a wider range than found in dolomite cements. In intermediate to deep burial settings, the organic carbon derived from thermal maturation of kerogen becomes more prominent with depth, shifting the $\delta^{13}C$ composition of calcite to values as low as $-25‰$ (Curtis and Coleman, 1986). If HCO_3^- is derived from both inorganic and organic sources, the resulting calcite cements would have $\delta^{13}C$ values in the range of 0 to $-25‰$. The $\delta^{13}C$ values of late calcite cements from the Temblor at Kettleman North Dome fall into this range, suggesting mixing of marine and thermogenic carbon sources.

Peak generation of CO_2 from kerogen in sediments generally occurs at approximately 100°C (Lundegard and Land, 1986). The estimated crystallization temperatures of less than 120°C for sparry calcite cements in the Temblor is consistent with the expected temperature for thermogenic carbon from the surrounding shales (Wood and Boles, 1991). The more negative $\delta^{13}C$ values with depth suggest increasing proportions of thermogenic carbon in the sediments, relative to $HCO3^-$ derived from dissolution of shells.

Trace element compositions of sparry calcite cements are similar to other calcite in the basin, and is attributed to crystallization in evolved marine pore water during deep burial. At Kettleman North Dome, sparry calcite cements have Fe and Mg contents up to 7 mole % (average 3.3 mole %). Calcite from evolved marine fluids at North Coles Levee has up to 8 mole % impurity, mostly Mg and Fe (Boles and Ramseyer, 1987), similar to the sparry calcite in Temblor sands. In contrast, calcite cements of meteoric origin from the east flank are generally pure with >96 mole % Ca and <3 mole % Fe, Mg, and Mn (Hayes and Boles, 1993). The impure (up to 5 mole %) meteoric calcite is rich in Mn but never Mg or Fe.

METEORIC RECHARGE IN THE SAN JOAQUIN BASIN

Calcite cements of meteoric origin occur in shallow marine sandstones 5 km from the eastern side of the San Joaquin basin in the Mount Poso and Round Mountain Fields (Hayes and Boles, 1993; Fig. 1). Meteoric waters are presently found some 20 km from the eastern edge of the basin at depths of 1500 m in the Kern River Field (Fisher and Boles, 1990), although no cements have yet been reported which are attributable to these waters.

On the western side of the San Joaquin basin, the paleogeography during Miocene-Oligocene time is not known for certain due to uplift and erosion of the section. Meteoric recharge into the marine sandstones in the North Belridge Field (Taylor and Soule, 1993) may have originated near Cretaceous outcrops about 15 km southwest of the field. Similarly, at Kettleman North Dome, the Temblor Formation pinches out onto Cretaceous outcrop about 15 km to the southwest of the field; suggesting that meteoric recharge would have occurred somewhere between here and the field. In both cases, meteoric water is inferred to have flowed laterally a maximum of about 15 km into marine sediment.

Kharaka and Berry (1974) concluded that waters from the Temblor Formation are meteoric water, but have been concentrated and isotopically modified by shale membrane filtration. Total dissolved solids (TDS) levels observed for Temblor waters range from 19,200 mg/l to 40,800 mg/l with more than half above 30,000 mg/l (Kharaka and Berry, 1974; Table 3). Formation waters, having similar $\delta^{18}O$, δD, and TDS values in a marine turbidite setting without meteoric incursions, are interpreted as diagenetically modified connate waters by Fisher and Boles (1990). We believe present-day waters at Kettleman North Dome have an origin similar to those at North Coles Levee. Both are interpreted as marine pore water that has been modified by water-rock interaction.

Temblor sands were deposited mainly in a shallow marine to deltaic setting (Kuespert, 1985). Such a depositional setting is easily envisaged to have meteoric water invasion during deposition and/or shallow burial. Although we find evidence for localized meteoric water modification of pore water composition during shallow burial, late sparry calcite cementation during intermediate to deep burial is interpreted as a response to connate marine waters. An explanation for this phenomenon is that initial pore water, some of which were meteoric or brackish, was displaced by marine waters during continuous deposition of the overlying thick McLure (Monterey) Formation. Alternatively, the recharge elevation on the west side of the San

Joaquin Valley was not sufficiently high to maintain extensive meteoric water influx into the deeply subsiding Temblor sandstones. Initial reservoir pressures at Kettleman North Dome were above a hydrostatic pressure gradient (Kharaka and Berry, 1974), suggesting that the present reservoir is isolated from the influence of recent meteoric water influx.

COMPARISON OF DIAGENETIC CARBONATE CEMENTS IN THE SAN JOAQUIN BASIN

Diagenetic carbonate cements are common, though not volumetrically abundant, in sandstone reservoirs in the San Joaquin basin. Most of the basin has undergone relatively simple burial to present-day depths and one might expect that cementation would be restricted to a short and discrete interval of the burial history. However, this and previous studies have shown that carbonate cementation in the basin involves multiple events extending over much of the burial history. Calcite is the most important carbonate cement with some dolomite and minor siderite (Merino, 1975; Boles and Ramseyer, 1987; Hayes and Boles, 1993; Taylor and Soule, 1993; this study).

Shallow marine sandstones on the basin margins can be flushed with meteoric water during deposition due to sea-level changes. Flushing may also occur if the sands are hydraulically connected to strata exposed updip. Early carbonate cements on the western flank include calcite spar formed in marine waters (Taylor and Soule, 1993), micritic calcite formed in meteoric water (this paper) and dolomites formed during shallow burial in mixed marine-meteoric water (Taylor and Soule, 1993; this paper). On the eastern flank of the San Joaquin basin, early calcite cements precipitated from meteoric waters in fluvial sandstones and from mixed marine-meteoric waters in shallow marine sandstones (Hayes and Boles, 1993). The early calcite on both flanks formed at temperatures of 17–30°C. Due to nonmarine sedimentation and slight uplift of the eastern basin flank since 14–15 Ma (Callaway, 1971; Olson, 1988, Crowell, 1987), meteoric water invaded the underlying shallow marine sands.

In contrast to the flank facies, sands in the central basin, such as those at North Coles Levee (Fig. 1), are largely deep sea fan deposits encased in organic-rich marine silts and clays (Callaway, 1971). In the deeply buried rocks of the central basin, pore waters are diagenetically modified marine water with salinities similar to or slightly fresher than sea water (Fisher and Boles, 1990). Siderite and dolomite cements formed from these marine and diagenetically modified waters at temperatures below 40°C. However, neither the carbonate cements nor the present pore waters in the deep marine sandstones contain evidence of meteoric infiltration (e.g., Boles, 1987; Boles and Ramseyer, 1987; Fisher and Boles, 1990), despite the fact that the overlying Pleistocene section is non-marine fill. We believe this is due to lack of hydraulic continuity between these sediments.

Late-stage carbonate cements in the San Joaquin basin are mostly calcite and have formed over a broad range of burial depths. Precipitation temperatures are 70 to 170°C in the western flank; (Taylor and Soule, 1993; this study), 40 to 90°C in the central basin (Schultz et al., 1989), and 24 to 70°C in the eastern flank (Hayes and Boles, 1993). These temperatures indicate long periods of cementation after early diagenesis. Late-stage calcite cements in the western flank at North Belridge and Kettleman North Dome fields precipitated from evolved marine waters, as did calcite in the central basin at North Coles Levee. Pore waters were either initially trapped marine waters and/or marine waters derived from the overlying marine sediments. These late calcite cements from the western and central basin area are estimated to have formed in waters with oxygen isotopic compositions heavier than sea water. At North Belridge Field, calcite cements are calculated to have precipitated in water with $\delta^{18}O_{(SMOW)}$ between about $+2\%o$ and $+8\%o$ (Taylor and Soule, 1993). In the central basin, precipitation occurred in waters isotopically similar to present values of about $+4\%o$ (Schultz et al., 1989; Fisher and Boles, 1990). Taylor and Soule (1993) also recognized late calcite vein-fill that apparently formed during west flank uplift and invasion of meteoric water into the marine section.

East of Bakersfield (Fig.1), sediments were deposited in shallow marine and non-marine environments and have undergone burial to depths of more than 2 km (Callaway, 1971). The eastern flank, near the Round Mountain oil field, has been uplifted up to 300 m since 14 Ma (Olson, 1988). Late-stage calcite cements were derived from dilute marine waters in marine sandstones and from meteoric water in non-marine sandstones. The occurrence of meteoric calcite cement in marine sandstones at depths of up to 620 m suggests significant downdip infiltration of meteoric waters on the east flank (Hayes and Boles, 1993).

Trace element composition of calcite cements from the eastern flank (Round Mountain Field), western flank (North Belridge and Kettleman North Dome), and central basin (North Coles Levee) are shown in Figure 9. Eastern flank calcite shows a distinct compositional field rich in calcium and manganese and poor in magnesium compared to calcite cements originating in marine water (Fig. 9). Although calcite cements from the western flank and central basin have each been interpreted to have originated in marine water, they have different compositional fields. The western flank calcite contains more magnesium, whereas central basin calcite contains more iron. The

cause of this compositional difference is not clear, but each region has a somewhat distinctive composition field. Similarly, Hayes and Boles (1993) note that trace elements in calcite varied from field to field in the eastern part of the basin and suggested local controls, on a field-scale, on calcite composition. Calcite cements originating from marine water in the eastern basin overlap slightly with most calcium-rich calcite of the western flank and central basin. In terms of relative proportion of Fe-Mg-Mn (Fig. 9), calcite cements forming in meteoric water are Mn-rich relative to calcite from marine water (cf. Meyers, 1974; Mount and Cohen, 1984; Boles et al., 1985).

Oxidizing conditions preclude high iron content in carbonate cements due to the low Fe^{+2} concentration in the pore water. Relatively oxidizing conditions are inferred for early dolomite in Temblor sands at Kettleman North Dome, where $\delta^{13}C_{(PDB)}$ values are about -6 to $-7\%o$ and Fe contents are relatively low (Fig. 7). In contrast, dolomites at North Coles Levee precipitated at relatively reducing conditions as inferred from $\delta^{13}C_{(PDB)}$ values of $+5$ to $+8\%o$, which indicate reducing methanogenic conditions (Boles and Ramseyer, 1987). These dolomites contain significantly more iron than dolomites from Kettleman North Dome (Fig. 7). Compositional differences between dolomite in the two areas may be due to contrasting facies, which indirectly control oxygen content in the pore water. In the case of the Temblor sands, which were deposited in a sand-rich, shallow marine environment, conditions were apparently oxidizing as compared with North Coles Levee. In the latter situation, packets of deep water sands were isolated in marine silt-clay deposits and conditions were reducing due to the stagnation of the pore water.

CONCLUSIONS

Carbonate cements occur locally in the early Miocene Temblor Formation at Kettleman North Dome on the western flank

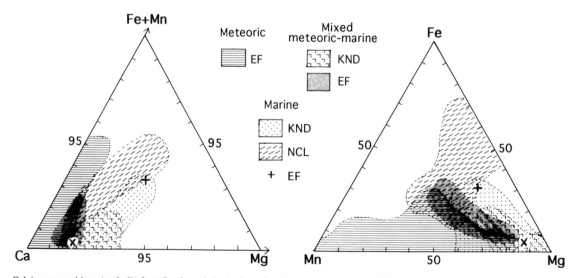

FIG. 9.—Calcite compositions (mole %) from San Joaquin basin. Data from Boles and Ramseyer (1987) and this paper. Compositional fields are distinguished by their inferred diagenetic origin. Calcites formed in marine or evolved marine pore fluids are marked by NCL (central basin), KND (western flank), and EF (eastern flank). Meteoric calcites are characterized by high Ca and Mn and low Mg contents relative to calcite from marine pore waters. Early micritic calcite composition in the western flank (this paper) is represented by the (x). Chemical compositions of calcites of mixed marine-meteoric origin overlap those of marine and meteoric origin calcites.

of the San Joaquin basin. Temblor sediments were deposited in deltaic-shallow marine to deep marine environments. Micritic calcite and dolomite cements formed early during shallow burial. Late calcite precipitated during deep burial, but prior to albitization of plagioclase.

At Kettleman North Dome, fluids with different compositions were involved in carbonate cement precipitation. Meteoric water in shallow marine to deltaic sandstones resulted in precipitation of early micritic calcite and dolomite cements. During late stages of burial, calcite cement precipitated at temperatures of 70–120°C from trapped marine water modified by water-rock interaction. Evidence for presence of modified marine fluids during late-stage calcite precipitation is inferred from isotopic and trace element composition. Organic carbon at late stages of burial was derived from thermal maturation of organic matter and incorporated into these calcite cements.

The type of and geochemical variation in diagenetic carbonates in the San Joaquin basin reflect depositional facies and local tectonics. Hydraulic continuity with other parts of the basin is controlled by depositional facies and the geometry of sand and silt beds. The basin flanks are made of sediments deposited in non-marine to shallow marine environments. These depositional environments are reflected in the presence of both meteoric and marine carbonate cements, which precipitated during shallow burial from trapped pore water and/or invasion of meteoric water. The formation of late carbonates on the sides of the basin depends on the overlying strata and on the subsidence and uplift history. In contrast, central basin sediments of the San Joaquin basin are comprised mainly of deep marine turbidites. These sands are effectively isolated from meteoric recharge from basin flanks, apparently due to lack of hydraulic continuity. As a result, carbonate cements formed from marine or evolved marine fluids ($+\delta^{18}O$) without the influence of meteoric water.

The long history of carbonate cementation in the San Joaquin basin preserves an important geochemical record of changing fluid composition and temperature during burial. This record is surprisingly complex considering the relatively simple burial history of the basin. The San Joaquin basin serves as an important note of caution to those that interpret cement histories in structurally complex, exhumed basins where basin geometry and geologic history are poorly known.

ACKNOWLEDGMENTS

The Temblor core samples from Kettleman North Dome oil field were kindly provided from Chevron Oil Field Research Company (COFRC) and California State University, Bakersfield. X-ray diffraction and part of microprobe analyses were done by E. Mountcastle. Y.I. Lee was supported by the fellowship from the Ministry of Energy and Resources, Korea while at UCSB. J. Boles acknowledges grants from Lawrence Livermore National Laboratory (8960–20), Petroleum Research Fund (18438-AC2) and National Science Foundation (EAR-17013). The manuscript benefitted from editorial comments of Society Economic Paleontologists Mineralogists reviewers and Dr. Stacey Zeck.

REFERENCES

BENT, J. V., 1988, Paleotectonics and provenance of Tertiary sandstones of the San Joaquin basin, California, *in* Graham, S. A., ed., Geology of the Temblor Formations, Western San Joaquin Basin, California: Society Economic Paleontologists Mineralogists Pacific Section, v. 44, p. 109–127.

BOLES, J. R. AND RAMSEYER, K., 1987, Diagenetic carbonate in Miocene sandstone reservoir, San Joaquin basin, California: American Association of Petroleum Geologists Bulletin, v. 71, p. 1475–1487.

BOLES, J. R. AND RAMSEYER, K., 1988, Albitization of plagioclase and vitrinite reflectance as paleothermal indicators, San Joaquin basin, California, *in* Graham, S. A., ed., Studies of the Geology of the San Joaquin Basin: Society Economic Paleontologists Mineralogists Pacific Section, v. 60, p. 129–139.

BOLES, J. R., LANDIS, C. A., AND DALE, P., 1985, The Moeraki boulders-anatomy of some septarian concretions: Journal of Sedimentary Petrology, v. 55, p. 398–406.

BOLES, J. R., 1984, Secondary porosity reactions in the Stevens sandstone, San Joaquin Valley, California, *in* McDonald, D. A. and Surdam, R. C., eds., Clastic Diagenesis: Tulsa, American Association of Petroleum Geologists Memoir 37, p. 217–224.

BOLES, J. R., 1987, Six million year diagenetic history, North Coles Levee, San Joaquin basin, California, *in* Marshall, J. D., ed., Diagenesis of Sedimentary Sequences: London, Geological Society Special Publication 36, p. 191–200.

CALLAWAY, D. C., 1971, Petroleum potential of San Joaquin basin, California, *in* Cram, I. H., ed., Future Petroleum Provinces of the United States-their Geology and Potential: Tulsa, American Association of Petroleum Geologists Memoir 15, p. 73–93.

CROWELL, J. C., 1987, Late Cenozoic basins of onshore southern California: complexity is the hallmark of their tectonic history, *in* Ingersoll, R. V. and Ernst, W. G., eds., Cenozoic Basin Development of Coastal California [Rubey Volume VI]: Englewood Cliffs, Prentice-Hall, p. 207–241.

CURTIS, C. D., 1978, Possible links between sandstone diagenesis and depth-related geochemical reactions occurring in enclosing mudstones: Journal of Geological Society of London, v. 135, p. 107–117.

CURTIS, C. D. AND COLEMAN, M. L., 1986, Controls on the precipitation of early diagenetic calcite, dolomite, and siderite concretions in complex depositional sequences, *in* Gautier, D. L., ed., Roles of Organic Matter in Sediment Diagenesis: Tulsa, Society Economic Paleontologists Mineralogists Special Publication 38, p. 23–34.

FISHER, J. B. AND BOLES, J. R., 1990, Water-rock interaction in Tertiary sandstones, San Joaquin basin, California, U.S.A.: Diagenetic controls on water composition: Chemical Geology, v. 82, p. 83–101.

FOLK, R. L., 1965, Some aspects of recrystallization in ancient limestone, *in* Pray, L. C. and Murray, R. C., eds., Dolomitization and Limestone Diagenesis: a Symposium: Tulsa, Society Economic Paleontologists Mineralogists Special Publication 13, p. 14–48.

FOLK, R. L., 1968, Petrology of Sedimentary Rocks: Austin, Hemphill's Bookstore, 170 p.

FRIEDMAN, I. AND O'NEIL, J. R., 1977, Compilation of stable isotope fractionation factors of geochemical interests: Washington, D.C., United States Geological Survey Professional Paper 440-KK, 12 p.

HARDING, T. P., 1976, Tectonic significance and hydrocarbon trapping consequences of sequential folding synchronous with San Andreas faulting, San Joaquin Valley, California: Americal Association of Petroleum Geologists Bulletin, v. 60, p. 356–378.

HAYES, M. J. AND BOLES, J. R., 1993, Evidence for meteoric recharge in the San Joaquin basin, California provided by isotope and trace element chemistry of calcite: Journal of Marine and Petroleum Geology, v. 10, p. 135–144.

HEASLER, H. P. AND SURDAM, R. C., 1985, Thermal evolution of coastal California with application to hydrocarbon maturation: American Association of Petroleum Geologists Bulletin, v. 69, p. 1386–1400.

IRWIN, H., COLEMAN, M. L., AND CURTIS, C. D., 1977, Isotopic evidence for source of diagenetic carbonates formed during burial of organic-rich sediments: Nature, v. 269, p. 209–213.

KHARAKA, Y. K. AND BERRY, F. A. F., 1974, The influence of geological membranes on the geochemistry of subsurface waters from Miocene sediments at Kettleman North Dome in Caifornia: Water Resource Research, v. 10, p. 313–327.

KHARAKA, Y. K., BERRY, F. A. F., AND FRIEDMAN, I., 1973, Isotopic composition of oil-field brines from Kettleman North Dome, California, and their geologic implications: Geochimica et Cosmochimica Acta, v. 37, p. 1899–1908.

KUESPERT, J., 1985, Depositional environments and sedimentary history of the Miocene Temblor Formation and associated Oligo-Miocene units in the vicinity of Kettleman North Dome, San Joaquin Valley, California, *in* Graham, S. A., ed., Geology of the Temblor Formations, Western San Joaquin Basin,

California: Society Economic Paleontologists Mineralogists Pacific Section, v. 44, p. 53–67.

LAND, L. S., 1985, The origin of massive dolomite: Journal of Geological Education, v. 33, p. 112–125.

LUNDEGARD, P. D. AND LAND, L. S., 1986, Carbon dioxide and organic acids: Their role in porosity enhancement and cementation, Paleogene of the Texas Gulf Coast, in Gautier, D. L., ed., Roles of Organic Matter in Sediment Diagenesis: Tulsa, Society Economic Paleontologists Mineralogists Special Publication 38, p. 129–146.

MAJOR, R. P. AND WILBER, R. J., 1991, Crystal habit, geochemistry, and cathodoluminescence of magnesian calcite marine cements from the lower slope of the Little Bahama Bank: Geological Society of America Bulletin, v. 103, p. 461–471.

MERINO, E., 1975, Diagenesis in Tertiary sandstones from Kettleman North Dome, California-I. Diagenetic mineralogy: Journal of Sedimentary Petrology, v. 45, p. 320–336.

MEYERS, W. J., 1974, Carbonate cement stratigraphy of the Lake Valley Formation (Mississippian), Sacramento Mountains, New Mexico: Journal of Sedimentary Petrology, v. 44, p. 837–861.

MOUNT, J. F. AND COHEN, A. S., 1984, Petrology and geochemistry of rhizoliths from Pliocene-Pleistocene fluvial and marginal lacustrine deposits, east Lake Turkana, Kenya: Journal of Sedimentary Petrology, v. 54, p. 263–275.

MUCCI, A., 1987, Influence of temperature on the composition of magnesian calcite overgrowths precipitated from seawater: Geochimica et Cosmochimica Acta, v. 51, p. 1977–1984.

OLSON, H. C., 1988, Middle Tertiary stratigraphy, depositional environments, paleoecology and tectonic history of the southeastern San Joaquin basin,

California: Unpublished Ph.D. Dissertation, Stanford University, Palo Alto, 305 p.

SCHULTZ, J. L., BOLES, J. R., AND TILTON, G. R., 1989, Tracking calcium in the San Joaquin basin, California: A strontium isotopic study of carbonate cements at North Coles Levee: Geochimica et Cosmochim. Acta, v. 53, p. 1991–1999.

SCHULTZ, J. L., BOLES, J. R., AND TILTON, G. R., 1991, Isotopic evolution of strontium in carbonate cements from the Stevens Sandstone: Implications for calcium mass transfer in the San Joaquin Basin, California, in Program and Abstracts to 66th Annual Meeting of AAPG-SEPM-SEG Pacific Section, Bakersfield, California, March 1991, p. 46.

SULLIVAN, J. C., 1966, Kettleman North Dome oil field: Summary of Operations, California Oil Fields, v. 52, p. 5–21.

TAYLOR, T. R. AND SOULE, C.H., 1993, Reservoir characterization and diagenesis of the Oligocene 64-Zone sandstone, North Belridge Field, Kern County, California: American Association of Petroleum Geologists Bulletin, v. 77, p. 1549–1566.

VIDETICH, P. E., 1985, Electron microprobe study of Mg distribution in recent Mg calcites and recrystallized equivalents from the Pleistocene and Tertiary: Journal of Sedimentary Petrology, v. 55, p. 421–429.

WOOD, J. R. AND BOLES, J. R., 1991, Evidence for episodic cementation and diagenetic recording of seismic pumping events, North Coles Levee, California, U. S. A.: Applied Geochemistry, v. 6, p. 509–521.

WOODRUFF, F., SAVIN, S. M., AND DOUGLAS, R. G., 1981, Miocene stable isotope record; a detailed deep Pacific Ocean study and its paleoclimatic implications: Science, v. 212, p. 665–668.

ZIEGLAR, D. L. AND SPOTTS, J. H., 1978, Reservoir and source bed history in the Great Valley, California: American Association of Petroleum Geologists Bulletin, v. 62, p. 813–826.

DIAGENESIS IN A CYCLIC, REGRESSIVE SILICICLASTIC SEQUENCE: THE POINT LOOKOUT SANDSTONE, SAN JUAN BASIN, COLORADO

JENNIFER L. LOOMIS AND LAURA J. CROSSEY

Department of Earth and Planetary Sciences, University of New Mexico, Northrop Hall, Albuquerque, New Mexico 87131

ABSTRACT: The Point Lookout Sandstone is a regressive, marine shoreline sequence deposited along the western margin of the Cretaceous epicontinental seaway. Other workers have identified 4th- and 5th-order parasequences within exposures of the Point Lookout Sandstone in the San Juan Basin, Colorado and New Mexico. Sandstone samples collected from outcrops and two cores drilled in the northwest corner of the San Juan Basin, Colorado have been characterized in terms of detrital mineralogy and diagenesis.

Variability in detrital mineralogy corresponds with depositional facies and position within the cyclical stratigraphy. Diagenetic reactions in sandstones include dissolution of feldspar, quartz overgrowth cementation, chlorite and kaolinite cementation, calcite cementation, occasional minor gypsum cementation, pyrite framboids, and authigenic Fe-rich dolomite overgrowths on intrabasinal dolomite grains. Early cements include chlorite, kaolinite, Fe-dolomite rims on dolomite grains, poikilotopic ferroan calcite and poikilotopic calcite. The extent of quartz overgrowth cementation and the abundance of Fe-dolomite rims are related to variability in initial sandstone composition. The spatial distribution of calcite cements reflect the stratigraphic geometry of the cyclical units within the Point Lookout Sandstone.

INTRODUCTION

The extensive surface exposure of the Upper Cretaceous (Santonian-Campanian) Point Lookout Sandstone along the northwestern perimeter of the San Juan Basin affords an excellent opportunity to examine three-dimensional diagenetic patterns at the surface, where diagenetic features can be spatially mapped in detail within the context of the stratigraphy. Stratigraphic studies of Point Lookout outcrop exposures along the western, northwestern (Devine, 1980, 1991; Katzman, 1991) and southeastern (Wright, 1984, 1986, 1988) edges of the San Juan Basin have delineated transgressive-regressive depositional cycles and identified their depositional settings.

Studies describing modern coastal hydrology can be used to understand what the paleohydrology might have been like during deposition of the Point Lookout Sandstone. Modeling of present-day coastal hydrology on Long Island (Collins and Gelhar, 1971) predicts that intercalated semi-permeable and permeable sediment layers form a layered aquifer system, where each aquifer layer contains a freshwater/seawater interface or mixing zone (Fig. 1). The exact position of this mixing zone relative to the shoreline depends not only on head conditions and the hydraulic conductivity of the semi-permeable layers

(Collins and Gelhar, 1971; Harrison and Summa, 1991), but also the rate at which the overall seawater-freshwater hydrologic regime responds to fluctuations in relative sea level (Meisler et al., 1984). Because the Point Lookout Sandstone consists of interlayered sandstones, siltstones and mudstones, it is reasonable to expect that such a layered seawater aquifer could have existed during deposition. Moreover, the mixing zone would have shifted in response to changes in relative sea level.

Knowledge of Point Lookout stratigraphy, Point Lookout sandstone compositions and the probable coastal paleohydrology permits evaluation of the degree to which early hydrologic conditions influenced early diagenesis within the cyclical sandstone sequence. Characterization and spatial mapping of diagenetic patterns across the extent of the three-dimensional outcrops will allow investigation as to whether or not early cementation patterns reflect the lateral migration of a seawater/freshwater interface. Results from samples collected from 4 different vertical sections are presented in this paper.

STRATIGRAPHIC SETTING

The Point Lookout Sandstone is located within the San Juan basin of northwestern New Mexico and southwestern Colorado (Fig. 2). The San Juan basin is one of a number of basins within the western interior of the United States which contains middle to Upper Cretaceous siliciclastic sediments. Specifically, in northwestern New Mexico and southwestern Colorado, basin stratigraphy consists of Upper Cretaceous coastal deposits (marine, deltaic and fluvial), all of which are overlain unconformably by Tertiary fluvial and lacustrine deposits (Fig. 3). The Point Lookout Sandstone is part of the Mesaverde Group, which in the San Juan Basin area is comprised of the Point Lookout, the Menefee Formation (which largely consists of fluvial shales with some sandstones and coals) and the Cliff House Sandstone (a transgressive marine sandstone). The major sandstones within the San Juan basin (Point Lookout Sandstone, Cliff House Sandstone and Pictured Cliffs Sandstone) are regressive and transgressive strandline sandstones which recorded fluctuations in sea level during the time of the inland Cretaceous seaway.

The Point Lookout Sandstone is a 100-m-thick regressive marine sandstone which is composed of meter-scale transgressive-regressive depositional cycles or 5th-order parasequences (the 5th-order parasequences described here correspond to the

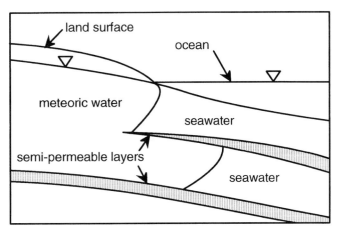

FIG. 1.—Schematic view of a simple layered aquifer along a marine coastline (modified from Collins and Gelhar, 1971). In this cross-section, the transition between meteoric water and seawater is depicted as a line or plane. In actuality, the transition is probably a mixing zone.

Siliciclastic Diagenesis and Fluid Flow: Concepts and Applications, SEPM Special Publication No. 55

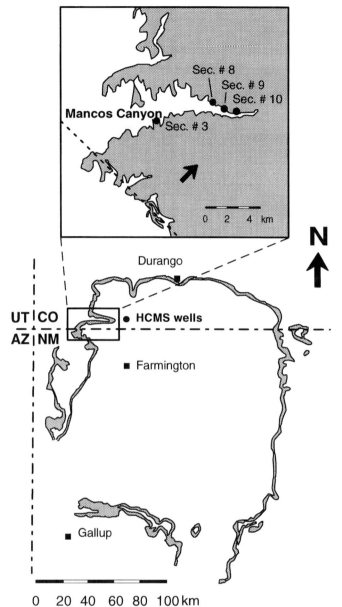

FIG. 2.—A map showing the San Juan basin, with the perimeter of the basin delineated by Point Lookout Sandstone outcrop. Mancos Canyon is shown in the northwest corner of the San Juan basin, with the two United States Geological Survey cores indicated immediately to the east of Mancos Canyon. The enlargement of Mancos Canyon shows the location of the 4 sections which have been sampled. The dashed northwest-southeast trending lines mark the positions of paleobeaches. The paleobeaches consists of foreshore and backshore (lagoonal and tidal) deposits. The southwesternmost line represents the exposed beach portion of the 4th-order parasequence which is the object of this study. The arrow points in the direction of the inland sea, indicating the direction of regressive progradation of the Point Lookout.

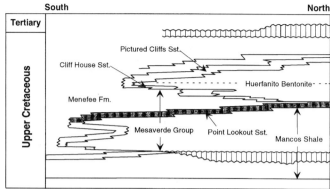

FIG. 3.—A stratigraphic cross section of the San Juan basin (modified from Law, 1991).

Cycle Stratigraphy and Hierarchy

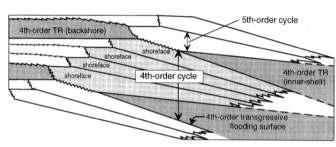

FIG. 4.—Generalized diagram which illustrates the 4th- and 5th-order parasequences as they occur in Mancos Canyon (modified from Katzman, 1991). Four 5th-order parasequences are grouped together to form one 4th-order parasequence. Also shown are the depositional facies which correspond to the different components of the 4th- and 5th-order cycles. TR = transgressive deposits.

marine parasequences defined by Van Wagoner et al., 1988). In the Point Lookout, four of these meter-scale 5th-order parasequences comprise a 4th-order parasequence (Fig. 4; Wright, 1986; Devine, 1991). The major regressive trend of the Point Lookout Sandstone corresponds to a 3rd-order sea-level fall (Devine, 1991) in the terminology of Vail et al. (1977). A typical 4th-order parasequence, as exposed along the northwestern

perimeter of the San Juan basin, is on the order of 1–10 m thick and over 15 km in length.

The regressive portions of both 4th- and 5th-order parasequences are shallowing-upward shoreface units. The transgressive units occur as either inner-shelf deposits or shallow backbarrier (lagoonal and tidal) deposits (Fig. 4). The boundaries of the transgressive sequences are abrupt; deposition was not continuous during the transition from regression transgression → regression. The transgressive event is recorded by the development of a transgressive surface which is laterally continuous over tens of kilometers. Landward, this transgressive surface is overlain by transgressive back-barrier deposits. Transgressive inner-shelf deposits were deposited on top of the transgressive surface toward the seaward limit. At intermediate depths the transgression is intermittently recorded by deposits and is largely distinguishable as a ravinement surface. In places immediately below this ravinement surface, the regressive shoreface sandstone contains reworked pebble clasts.

The four stratigraphic sections, from which the results presented in this paper were collected, are shown in Figure 5. The outcrop sections are located in Mancos Canyon (Fig. 2) and are described in detail by Katzman (1991).

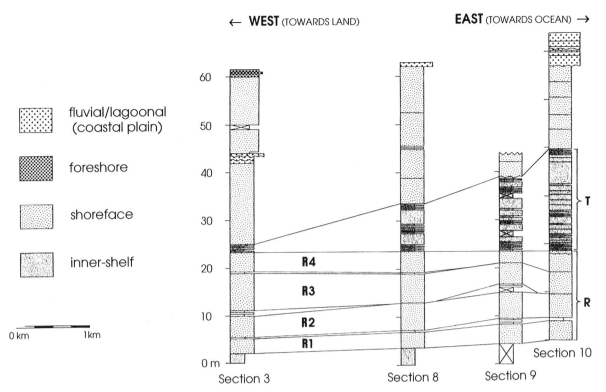

FIG. 5.—Simplified stratigraphic sections of the four vertical outcrop sections from Mancos Canyon. Both 4th- and 5th-order stratigraphic boundaries are correlated between the four outcrop sections, with R1, R2, R3 and R4 representing the 5th-order regressive units, R = 4th-order regressive sequence and T = 4th-order transgressive sequence. The 5th-order transgressive sequences between R1, R2 and R3 are laterally discontinuous throughout Mancos Canyon, sometimes showing up as a sandstone-sandstone contact.

METHODS

The principal study area is in Mancos Canyon, located along the northwestern perimeter of the San Juan basin in the Ute Mountain Indian Tribal Park (Fig. 2). Here, the Mesaverde Group has been exposed by erosion since the commencement of the Laramide orogeny during the Paleocene period (Fasset, 1985). This study area was chosen because it represents the one area in the San Juan basin where the Point Lookout is exposed over a large area, thereby allowing extensive sampling in three dimensions. Moreover, the Point Lookout, as it is exposed in the Four Corners area, has been well-characterized stratigraphically and sedimentologically by Katzman (1991) and Devine (1980, 1991). Sampling to date has been concentrated around the stratigraphic sections measured by Katzman (1991) and are from only one of the three 4th-order parasequences exposed in Mancos Canyon (the other two are the least exposed of the three). The four stratigraphic sections shown in Figure 5 cover approximately half of the lateral extent of the one 4th-order parasequence (the oceanward half). While sampling covers most of the entire length of the 4th-order parasequence, analysis of all the samples has not been completed. Fresh outcrop exposures have been sampled where possible in order to obtain the least-weathered samples.

Also included in this study are Point Lookout samples from two cores drilled by the U. S. Geological Survey. These cores

are located approximately 32 km east of the Mancos Canyon study area (see Fig. 2).

Petrographic analysis has been performed on 126 samples in thin section. A total of 38 thin sections have been analyzed from the two U. S. Geological Survey cores and 88 thin sections from outcrop samples. All thin sections were stained with Alizarin Red-S for calcite, potassium ferricyanide for Fe-carbonates, and sodium cobaltinitrite for potassium feldspars. A minimum of 500 point counts were made to quantify the abundances of diagenetic features.

X-ray diffraction aided in the identification of clays within the sandstones. Before preparing a clay mount, carbonate was removed (Moore and Reynolds, 1989) using a 1 N sodium acetate solution. The dissaggregated carbonate-free sandstone was then soaked in a sodium-pyrophosphate solution overnight to ensure deflocculation of the clays. The < 2-m-size fraction was separated via centrifugation, then rinsed with $MgCl_2$ and deionized water. Oriented mounts were prepared using the millipore filter method described by Drever (1973) and then saturated with ethylene glycol. Clay mineral identification procedures are described in Moore and Reynolds (1989).

Backscattered electron (BSE) images, collected from both a Hitachi S-450 scanning electron microscope (SEM) and a JEOL Super Probe 733 electron microprobe, were obtained from some of the iron-rich carbonates to investigate textures associated with authigenic ferroan dolomite rims in outcrop samples (dis-

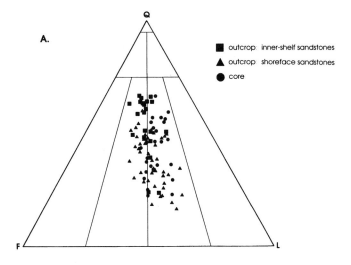

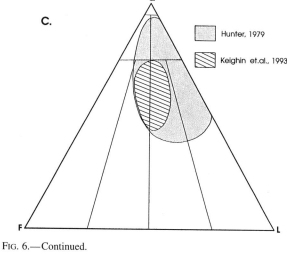

FIG. 6.—Continued.

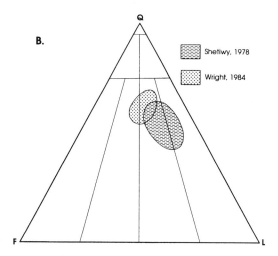

FIG. 6.—QFL diagrams depicting the composition of the Point Lookout sandstones as they occur around the San Juan basin and throughout the subsurface. The classification scheme shown is that of Folk (1980). (A) depicts the results of point counting from this study, and shows sandstones from both Mancos Canyon and the two United States Geological Survey cores. (B) shows the composition of Point Lookout sandstones from the southeastern portion of the San Juan basin (Wright, 1984; Shetiwy, 1978), while (C) shows Point Lookout from core (Hunter, 1979; Keighin et al., 1993). Hunters sandstone compositions are from Point Lookout core samples across the middle of the San Juan basin. Keighin et al. (1993) core samples are the same samples from the two United States Geological Survey cores used in this study (shown in (A)).

cussed later). Energy dispersive spectroscopy (EDS) was used on the SEM to obtain elemental spectra.

RESULTS

General Petrography

Point counting data from Mancos Canyon outcrop samples generally agree with outcrop data presented by Shetiwy (1978) and Wright (1984), as well as with core data presented by

Hunter (1979) and Keighin et al. (1993; Fig. 6). Most shoreface sandstone compositions are feldspathic litharenites and litharenites (Folk, 1980), with some samples falling under the lithic arkose category. Foreshore and fluvial/deltaic sandstones are either sublitharenites or subarkoses (Keighin et al., 1993).

The regressive sandstones within the Point Lookout typically have a coarsening-upward trend in grain size which corresponds to a depositional shallowing-upwards cycle (Sabins, 1964; Wright, 1984; Keighin et al., 1993). Variablility in detrital composition can be partially a function of the degree to which the detritus undergoes reworking in the depositional zone (Winn et al., 1984). In the Point Lookout Sandstone, the upper shoreface and foreshore sandstones are quartz-rich compared to middle and lower shoreface sandstones; the high energy associated with the surf-zone and shallow storm reworking not only abrades many of the lithics and feldspars, but also winnows out much of the fine-grained matrix material (Sabins, 1964). At greater water depths, in middle to lower shoreface, the lithic fragments are better preserved and the sandstones are generally more lithic and feldspar rich (Sabins, 1964).

Point Lookout sandstones are composed largely of monocrystalline quartz, polycrystalline quartz, some chert, plagioclase, K-feldspars, lithic fragments and intrabasinal dolomite. The lithic fragments are predominantly volcanic rock fragments and sedimentary rock fragments, with minor metamorphic rock fragments. The lithic fragments which are clearly volcanic in origin are mostly mafic. Some volcanic lithic fragments are severely altered to quartz, feldspar, zeolites in a few rare cases and clay minerals. Sedimentary lithics include siltstones and assorted shales/claystones, while schists and foliated polycrystalline quartz grains comprise the few metamorphic rock fragments.

The dolomite seen in the Point Lookout is the same "primary" dolomite identified in other Western Interior Cretaceous marine sandstones and described by Sabins (1964). In all cases, the dolomite occurs as a framework grain, having grain shapes which range between angular rhombahedra and rounded grains (Sabins, 1964). Such framework dolomite grains are referred to as intrabasinal dolomite in this study. Fluvial feeder channels which delivered detritus to the Point Lookout shoreline do not contain any detrital dolomite. The intrabasinal dolomite is therefore interpreted as having originated within the Western Interior seaway.

The dominant clay minerals in the sandstones are kaolinite, chlorite (both authigenic and detrital), and illite. Lesser amounts of smectite and interlayered illite/smectite are also present.

Diagenetic Features

Diagenetic features typical of marine sandstones, for example, the Brent Sandstone in the North Sea (Glasmann et al., 1989a, b) and the Cretaceous Shannon Sandstone further north in the United States Western Interior (Hansley and Whitney, 1990), are found in the Point Lookout Sandstone. The diagenetic features found in outcrop samples include the following: (1) calcite cement, (2) authigenic Fe-rich dolomite, (3) quartz overgrowths, (4) feldspar dissolution, (5) replacement of feldspar with clays, (6) kaolinite cement, (7) framboidal pyrite, (8) authigenic chlorite cement, 99) gypsum cement, (10) iron-oxide cement, (11) albitization, (12) feldspar overgrowths and (13) calcite dissolution (Keighin et al., 1993). Of these 13 diagenetic features, only the first 8 are noticeably abundant. The most volumetrically significant diagenetic features are chlorite cement, calcite cement, authigenic dolomite, feldspar dissolution and kaolinite cement.

There are three types of calcite, distinguishable by texture and composition: poikilotopic Fe-calcite, poikilotopic calcite, and sparry calcite. The poikilotopic Fe-calcite occurs both as large meter-scale concretions (up to 30% of the whole rock; Fig. 7A, see p. 93) and as millimeter-scale patches (Fig. 7C, see p. 93). Where the Fe-calcite is in the form of patches, it is usually associated with poikilotopic calcite. In a few locations, the poikilotopic Fe-calcite patches are present as discrete disseminated patches; the poikilotopic Fe-calcite is present in the center of the patches, with the non-ferrous poikilotopic calcite in optical continuity around the outer edges. The sparry calcite occurs as a later, pore-filling cement, post-dating the poikilotopic cements (Fig. 7D, see p. 93).

Authigenic dolomite occurs in two different forms: as Fe-rich rims on intrabasinal detrital dolomite cores (Fig. 7D) and as replacement rhombohedra within chert or polycrystalline quartz. The latter is not abundant and the former is limited by the abundance of primary intrabasinal dolomite.

Feldspar dissolution is pervasive and most abundant in the plagioclase feldspars. In some of the samples, kaolinite is present within the dissolved feldspars and is interpreted to be a direct product of plagioclase dissolution.

Chlorite is abundant in some of the samples, exceeding 10% of the whole rock. X-ray diffraction of the <2-m-size fraction confirms the petrographic identification; chlorite peaks are sharp and have high intensities compared to other clay components (illite, illite/smectite and kaolinite). In most instances, the chlorite found rimming pore spaces exhibits the radial form of growth and is clearly authigenic in origin (Figs. 7G, 7H, see p. 93). The presence of authigenic chlorite has been confirmed by SEM imaging. Chlorite also commonly replaces lithic grains. In many cases, the chlorite has replaced the matrix material within mafic volcanic rock fragments containing feldspar laths. The abundance of chlorite in the volcanic rock fragments, in conjunction with the low-abundance of any form of smectite

(as determined using X-ray diffraction) suggests that some of the smectite might have been altered to chlorite.

In addition to chlorite, illite is also an abundant clay mineral. The illite has an R3 ordering and is more abundant within the sandstones than interlayered illite/smectite or smectitic clay.

The remaining less-abundant diagenetic features have been consistently noted in sandstone samples, yet not studied in detail at this time. Poikilotopic gypsum cement occurs in both outcrop and core, with a patchy distribution throughout the sandstone. This same texture (spots of poikilotopic cement) is also seen with calcite. In both core and outcrop, framboidal pyrite occurs with the gypsum.

Weathering: Core Versus Outcrop

Because this study uses samples from outcrop, the possible effects of surface weathering on the diagenetic features must be considered. For this purpose, Point Lookout outcrop samples are compared to core samples. The only differences between outcrop and core samples which can be attributed to surface weathering are the following: (1) the apparent absence of authigenic Fe-rich dolomite overgrowths on primary dolomite grains in some outcrop samples, (2) the presence of oxidized, petrographically unidentifiable material in pore spaces and around some of the detrital grains throughout the outcrop samples and (3) the oxidized appearance of intergranular chlorite.

The authigenic Fe-dolomite rims take on a blue stain from the K-ferricyanide in the core samples, while in outcrop samples, no part of the dolomite takes on a stain in most samples studied. Backscattered electron images of outcrop dolomite grains reveal that the authigenic rims are present in varying degrees of preservation (Fig. 8). In most cases, the dolomite grains are surrounded by a rim of petrographically unidentifiable oxidized material (Figs. 7D, 8C). As determined by EDS, this material is comprised of mostly iron, some calcium and minor silica and is interpreted to be iron oxide. Where the outer edge of the dolomite has undergone dissolution, there is generally a rim of the oxidized iron material marking the original boundary of the dolomite grain.

The intergranular chlorite also shows signs of surface weathering. In some outcrop samples, replacive chlorite in lithic fragments (volcanic fragments) has maintained its green color, while intergranular chlorite cement ranges from green to orange-yellow.

In addition to the above late oxidation and dissolution features found exclusively in outcrop, there is also a collection of relatively late dissolution and alteration features which occur in both core and outcrop samples. If these features are significantly more abundant in outcrop, it is reasonable to attribute some of the dissolution and alteration in outcrop to surface weathering. These late features are the following: (a) a late episode of feldspar dissolution, (b) rock fragment dissolution, (c) calcite dissolution and (d) oxidation of pyrite.

The late feldspar dissolution is distinct from earlier feldspar dissolution in that the secondary pores inside the feldspar are clear of early authigenic cements. In the case of early feldspar dissolution, the secondary pores are filled with either early calcite cement or early chlorite cement.

Although present in core samples (Keighin, et al., 1993), calcite dissolution appears to be more abundant in outcrop sam-

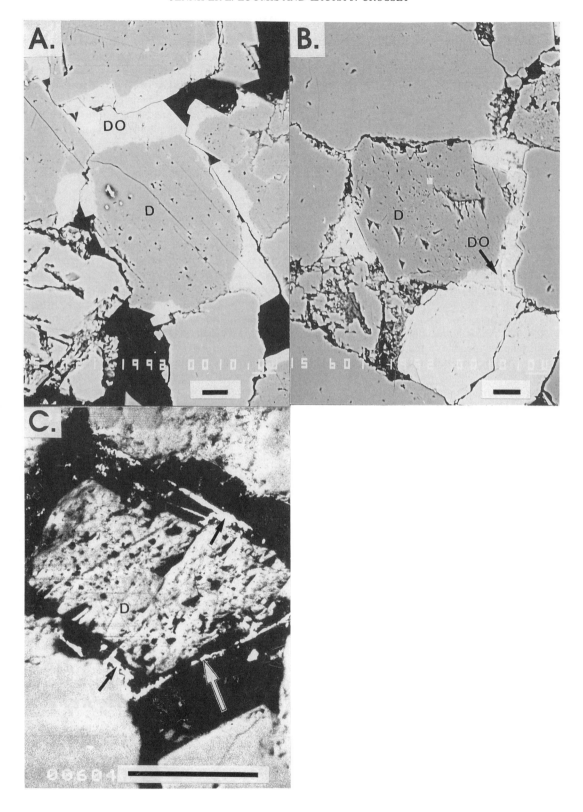

FIG. 8.—Backscattered electron images of dolomite grains, comparing authigenic dolomite between core and outcrop. (A) is a BSE image of dolomite from core (scale bar = 10 µm). The intrabasinal dolomite (D) has clean authigenic Fe-dolomite rims (DO). Intrabasinal dolomite from outcrop is shown in (B) (scale bar = 10 µm). This particular sample is an example of where surface weathering has not greatly modified the authigenic Fe-dolomite rims. The rims appear to be slightly dirty, but are otherwise no different from the rims on dolomite in core samples. The BSE image in (C) shows a dolomite grain which has been affected by surface weathering (scale bar = 50 µm). The intrabasinal dolomite grain has suffered partial dissolution around its edges, while the Fe-dolomite overgrowth has been completely dissolved away. Surrounding the intrabasinal dolomite grain, there is a rim of bright iron oxide material (arrows) marking the original shape of the authigenic Fe-dolomite rim.

ples. In outcrop samples suspected of having undergone calcite dissolution, the calcite (specifically sparry calcite) occurrence is constrained to small areas exhibiting poor sorting with high matrix content and in pore throats (i.e. in areas of relatively low permeability).

Pyrite oxidation is pervasive in both core and outcrop. Generally, the thin outer edges of pyrite framboids look like hematite: the light transmitted through the thin outer edges has a brick-red color. Oxidized rims on pyrite framboids are also visible in reflected light.

Quantification of feldspar and rock fragment dissolution allows comparisons to be made between core and outcrop. The purpose of such a comparison is to determine whether or not core and outcrop experienced the same degree of late-stage dissolution and alteration. Reconstructed and raw QFL compositions are used to quantitatively make the comparison. Raw QFL compositions use only unaltered feldspar and rock fragment data; all altered and partially dissolved detrital grains are omitted. Reconstructed QFL compositions, on the other hand, incorporate altered and dissolved lithic fragment and feldspar data in addition to all of the unaltered detrital quartz, feldspar and rock fragment data. For these comparisons, shoreface sandstone samples were specifically selected from both core and outcrop in order to minimize the effect of variable depth of deposition on sandstone composition. At least 300 point counts of the detrital grains were made on each sample.

In the QFL diagram shown in Figure 9, both the mean raw and mean reconstructed QFL compositions are shown for 19 core and 48 outcrop samples. The magnitude of the shift between the mean unreconstructed QFL composition and the mean reconstructed QFL composition is essentially the same for both the the outcrop samples and the core samples. This means that overall, a comparable proportion of lithic grains and feldspars has been altered in both the core and outcrop.

Because the two cores used in this study are relatively shallow (the Point Lookout Sandstone is between 210–400 m, deep), it is likely that shallow subsurface hydrologic processes are responsible for late-stage feldspar and rock fragment dissolution and alteration. The fact that Point Lookout Sandstone outcrop is a zone of recharge around the perimeter of the San Juan basin (Craigg et al., 1990), as well as a source of ground water, implies that near-surface pore fluids are moving through the sandstone and are probably responsible for late-stage alteration and dissolution in both core and outcrop.

In summary, late-stage fluids associated both with surface weathering and subsurface ground waters have modified detrital grains and some of the iron-rich authigenic mineral phases in both core and outcrop. Oxidation of the Fe-dolomite overgrowths and oxidation of chlorite cement are attributed to surface weathering because they are features which are not seen in core samples. In the case of the dolomite overgrowths, BSE images can be used to assess the size and abundance of the Fe-dolomite rims despite the effects of weathering. Calcite dissolution is another late-stage modification which will impact the assessment of calcite cementation throughout the Point Lookout Sandstone.

Paragenetic Sequence

The paragenetic sequence of diagenetic features in the Point Lookout Sandstone is fairly consistent throughout the core and

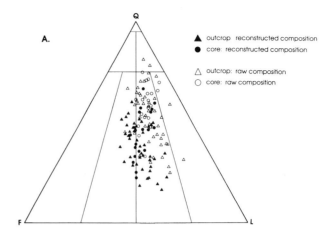

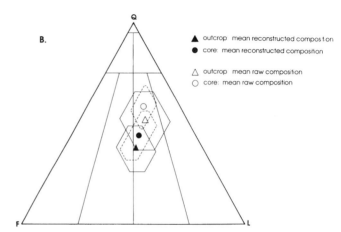

FIG. 9.—QFL diagrams which compare the degree of late-stage alteration between core and outcrop. (A) shows all of the raw QFL compositions and reconstructed QFL compositions from outcrop and core shoreface sandstones. The raw QFL compositions are based only on unaltered/unmodified detrital feldspars and lithic fragments; altered detrital QFL grains are omitted. In the case of the reconstructed QFL compositions, the altered feldspar grains and lithic fragment are included. In (B), the means of each QFL composition group (raw core, raw outcrop, reconstructed core, reconstruced outcrop) is shown. The solid polygons represent one standard deviation for the two outcrop means, while the dashed polygons show one standard deviation for the two core means.

outcrop samples studied to date. The general order in which the different diagenetic features formed (Fig. 10) is the following: feldspar dissolution → chlorite → kaolinite → Fe-dolomite overgrowths → poikilotopic Fe-calcite → poikilotopic calcite → sparry calcite → quartz overgrowths → feldspar dissolution

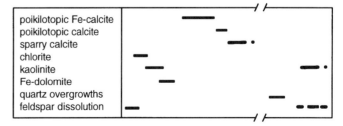

FIG. 10.—Paragenetic history of the Point Lookout Sandstone.

→ kaolinite. Pyrite is tentatively considered to predate sparry calcite cement. Observations of textural relationships were largely made from thin sections and some SEM imaging.

Early feldspar dissolution is distinguished from later episodes of feldspar dissolution by the presence of early cements in secondary pores of the feldspars. The early cements found in these partially dissolved feldspars are chlorite, poikilotopic Fe-calcite and poikilotopic calcite (Fig. 7D).

Where present, chlorite is the earliest cement in all samples studied to date. Authigenic chlorite rims are found lining kaolinite-filled pores (Fig. 7F, see p. 93) and calcite-filled pores (all three types of calcite mentioned previously; Fig. 7G). Chlorite is generally not found on Fe-dolomite overgrowths, nor on quartz overgrowths. In the case of quartz overgrowths, a rim of chlorite is sometimes distinguishable between the quartz core and the overgrowth (Fig. 7H). Where the chlorite occurs as large patches or thick rims, quartz overgrowths are either truncated against the chlorite or appear underdeveloped (i.e. the quartz overgrowth is anhedral and relatively thin).

More than one episode of kaolinite cementation is observed. Kaolinite precipitation is partially controlled by the dissolution of feldspar, as evidenced by the presence of kaolinite within feldspars and adjacent to dissolving feldspars (Fig. 7F). That some of the kaolinite is early is based upon the occurrence of kaolinite within poikilotopic calcite (both the ferrous and non-ferrous calcite cements). Where there is kaolinite in the poikilotopic calcites, the calcite does not occur as a large, single calcite crystal but rather as microcrystalline calcite. There are also places where calcite can be seen penetrating only the outer edges of a cluster of pore-filling kaolinite (Fig. 7E, see p. 93). These occurrences are largely restricted to intergranular pores. It is this kaolinite found within both types of poikilotopic calcite that is concluded to be early.

While Fe-dolomite overgrowths appear to be relatively early (i.e., before any calcite cementation), it is uncertain as to whether they preceeded or followed kaolinite precipitation. When seen adjacent to patches of kaolinite, most of the Fe-dolomite overgrowths have a euhedral form with no signs of having intersected with the kaolinite. On a few rare occasions, however, kaolinite has been seen within a Fe-dolomite overgrowth. In these instances, the dolomite has the same appearance as kaolinite-bearing poikilotopic calcite; the dolomite is not monocrystalline, but microcrystalline. For the most part, the Fe-dolomite overgrowths and kaolinite do not appear to have inhibited or affected the growth of each other. It is concluded that the order in which the Fe-dolomite overgrowths and kaolinite formed relative to each other varies throughout the Point Lookout sandstone.

Up to this point, the diagenetic features described have been described as "early" only because they are the first to have occurred in the Point Lookout Sandstone. The concretions are an important key to determining just how early the first diagenetic events are. Many of the concretions formed early, as evidenced by the preservation of an open grain spacing. Wilson and McBride (1988) have used the contact index (CI) and the tight packing index (TPI) to quantify the degree of compaction in Pliocene arkoses, lithic arkoses and feldspathic litharenites of the Ventura basin, California. The contact index is the average number of all types of intergranular contacts per grain, while the TPI is the average number of long contacts + con-

cavo-convex contacts + sutured contacts per grain. In the Point Lookout, the CI and TPI were determined for a sandstone in the middle of a concretion and a sandstone outside of a concretion. Inside the concretion, the CI = 2.8 and TPI = 0.7 (Fig. 7A). Outside of the concretion, CI = 4.2 and TPI = 2.2 (Fig. 7B, see p. 93). The relative differences between the two compaction indices inside and outside of a concretion demonstrate that some of the Point Lookout concretions formed prior to substantial compaction. In the sandstones studied by Wilson and McBride (1988), the Cl = 2.8 and TPI = 0.7 roughly correspond to depths of less than 1.5 km. There are, however, some concretions which appear to have experienced some compaction and grain re-orientation prior to calcite cementation.

DISCUSSION

Diagenesis Within the Context of the Cyclical Stratigraphy

When the occurrences and abundances of the various diagenetic textures are considered within the context of Point Lookout stratigraphy, several trends are noticeable in outcrop (Fig. 11). Specifically, some of the early diagenetic features have systematic distribution patterns across the three dimensional exposure of a single 4th-order parasequence. Based on the results from this study, there are two factors which can be shown to control the distribution of key diagenetic features: 1) depositional facies and 2) the position of the sandstone within the overall cyclical sequence of the Point Lookout Sandstone. In the case of the latter, the position of the sandstone influences the degree to which cycle hierarchy (i.e., the magnitude and duration of a shift in relative sea level) controls early diagenesis. In other words, the position of a sandstone within a 4th-order parasequene has a bearing on the extent to which the 4th-order sea-level fluctuation (as opposed to a 5th-order sea-level fluctuation) influenced early diagenesis.

The distribution of intrabasinal dolomite (and subsequently the authigenic Fe-dolomite overgrowths) and quartz overgrowths reflects a dependence upon depositional facies. As originally noted by Sabins (1964), the abundance of primary or intrabasinal dolomite grains increases with increasing water depth of the depositional environment. In Mancos Canyon, the same relationship between the intrabasinal dolomite and water depth occurs. The highest intrabasinal dolomite abundances are associated with the deeper inner shelf environments (Fig. 11A). The average abundance of intrabasinal dolomite in inner-shelf sandstones is 5.5% (21 samples with a standard deviation of 3.2%), and it is 2.9% (36 samples with a standard deviation of 2.0%) in shoreface sandstones. A Student t-test indicates that dolomite abundances in the shoreface sandstones are statistically different from those abundances found in the inner-shelf sandstones at a confidence level greater than 99%.

While the intrabasinal dolomite has been noted throughout the Wester Interior (e.g., Sabins, 1964; Hansley and Whitney, 1990), no definitive explanations have been provided regarding the source of this dolomite. The absence of the dolomite in San Juan basin fluvial feeder systems (the Menefee Formation), and the relationship between dolomite abundance and marine transgression suggests that the intrabasinal dolomite precipitated somewhere within the shelf environment and was then re-deposited. Transgressive re-working and re-deposition of sands

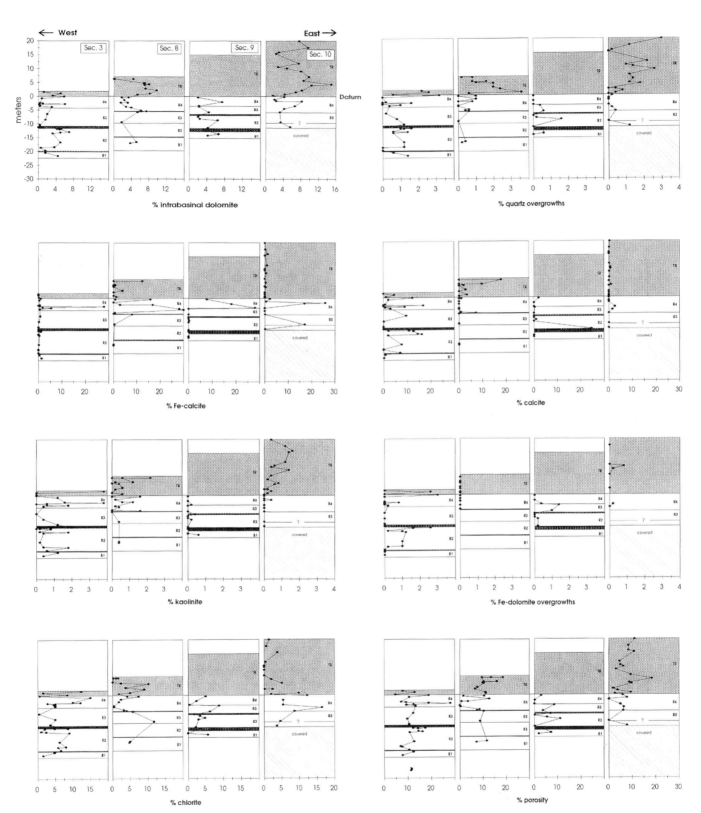

FIG. 11.:—Point counting results of the principal diagenetic features from outcrop. In each histogram, the four columns represent the four vertical sections from which samples were obtained. The base of the 4th-order transgressive sequence (the transgressive flooding surface) is the datum, with the shaded portions representing both 4th- and 5th-order transgressive sequences. The stratigraphic contacts in each section correspond to the stratigraphy shown in Figure 5. Note that data collection from the 4th-order transgressive sequence in Section 9 is incomplete at this time.

concentrated the intrabasinal dolomite. Stable isotopes would help to identify the zone of intrabasinal dolomite formation.

Quartz overgrowths also exhibit a relationship to depositional setting (Fig. 11B). While the range of abundance of quartz overgrowths is comparably low in both the inner-shelf and shoreface sandstones (0–3.6% in inner-shelf sandstones and 0–1.6% in shorefacesandstones), it is significant that there are more shoreface sandstones that have no observable quartz overgrowths. To see if there is a statistical difference between the shoreface sandstones and inner-shelf sandstones, a Student t-test was used to compare the two quartz overgrowth populations. The Student t-test indicates that the percent of quartz overgrowths is statistically different from the percent quartz overgrowths in the inner-shelf sandstones, with a confidence level exceeding 99.99%.

Detrital quartz abundances control to a large extent the occurrence of quartz overgrowths. Generally, the higher abundances of quartz overgrowths occur in the sandstones containing greater quantities of detrital monocrystalline quartz. The transgressive inner-shelf sandstones have a greater abundance of detrital monocrystalline quartz. According to Wright (1984) and Katzman (1991), subsidence of the marine deposits, along with decreased sediment input, is thought to have been responsible for an apparent rise in sea level. The source of the transgressive inner-shelf sands represent reworked shoreface sands (Katzman, 1991). The reworking of the shoreface sands preferentially broke down rock fragments and feldspars, thereby producing a relative increase in the abundance of quartz.

Unlike the quartz overgrowths and the dolomite, the distribution of the calcite cements is largely controlled by the position of the sandstone within the 4th-order parasequence and not by the depositional facies (Figs. 11C, 11D). Specifically, the poikilotopic Fe-calcite and poikilotopic calcite appear to be controlled by the 4th-order rise in relative sea level. The poikilotopic calcite is generally associated with the poikilotopic Fe-calcite and is not volumetrically abundant. The sparry calcite (which is not an early cement) does not have a systematic pattern of distribution throughout either the individual 5th-order parasequences or throughout the 4th-order parasequence.

Throughout Mancos Canyon area, the two shorefaces below the 4th-order transgressive sequence (R3 and R4 in Fig. 11) are well-cemented and marked by large Fe-calcite concretions. The complexity of concretion shapes, as well as the density of concretions throughout the vertical exposure, is related to the thickness of the overlying 4th-order transgressive sequence. At section 10, the 4th-order transgressive sequence is over 20 m thick, and the two underlying shoreface units contain concretions ranging between ~0.5 m and 3 m in diameter (Fig. 12). Moreover, the concretions do not form a single discrete horizon, but occur along numerous horizons throughout the vertical thickness of the two shoreface units. At section 3, the 4th-order transgressive sequence is 2 m thick. The concretions located in the two underlying shorefaces are between 1–4 m in diameter and positioned along a single horizontal zone within each of the two shorefaces. Where the 4th-order transgressive sequence pinches out at this section, the concretions are not as frequent, nor as large.

In addition to the concretions, the two uppermost shoreface units (R3 and R4) also contain disseminated patches of poikilotoic Fe-calcite and poikilotopic calcite in between the concre-

Fig. 12.—The outcrop exposure at Section 10. TR = 4th-order transgressive deposits (inner-shelf sandstones and shales); R4 and R3 are 5th-order parasequences (comprised only of amalgamated regressive shoreface sandstones in this photograph). (B) is a close-up photograph of (A). The stick in the bottom-center of (B) is 1.5 m in length. The calcite concretions are noticeable in outcrop because of their greater resistance to surface weathering relative to the surrounding poorly-cemented sandstone. The concretions shown here at Section 10 range between 0.5 m and 3 m in diameter.

tions. These disseminated patches are not constrained to just the two top shoreface units, but also occur elsewhere within the 4th-order parasequence in lesser quantities.

Other than an overall decrease in sparry calcite abundances towards the oceanward extent of the Point Lookout, the sparry calcite does not have an obvious pattern of distribution throughout the 4th-order parasequence. This may partly be a function of the late-stage calcite dissolution. It is the sparry calcite which tends to exhibit signs of what may have been extensive dissolution; more so than either of the poikilotopic calcite cements. Subsequently, the abundances of sparry calcite represent minimum values.

A Conceptual Model for Fluid Flow

The results presented in this paper not only describe and characterize the major diagenetic features present in Point Lookout Sandstone outcrop, but also demonstrate how the spatial distribution of some of the diagenetic features is related to cycle geometry and cycle hierarchy. These results are now considered in light of other studies in order to assess the role of paleohydrology. Cycle hierarchy and cycle geometry of sands and muds, along with the fluctutions of apparent sea level, controlled the lateral shifting of the meteoric and marine hydrologic regimes. This dynamic hydrology in turn controlled pore fluid chemistry throughout the parasequences. The following is a summary which relates the progressive deposition of the 4th-order parasequence to pore fluid chemistry and the early, prominent diagenetic features found in the regressive sandstones of the Point Lookout Sandstone (see Fig. 13).

Pre-transgression (i.e., before deposition of the 4th-order transgressive sequence):

(1) Early precipitation of chlorite throughout the shelf deposits under marine conditions.
(2) Kaolinite precipitation when a meteoric lens migrates seaward in response to stacked progradation of the shoreline during continued regression.

Early-transgression:

(3) Precipitation of ferroan dolomite on primary dolomite grains under sulfate-reducing conditions.

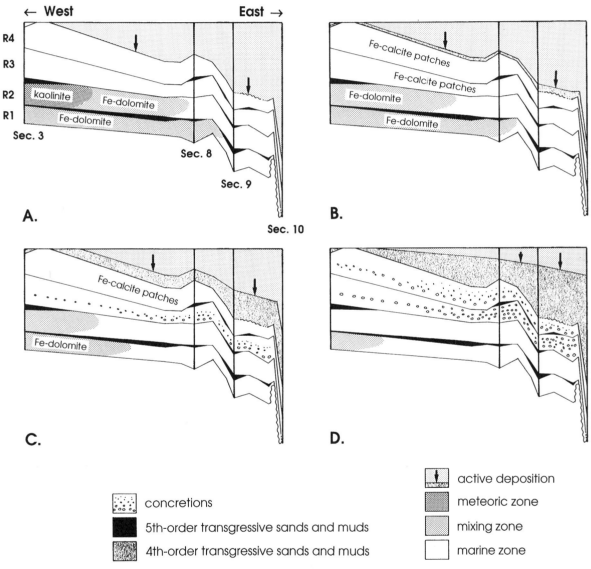

FIG. 13.—A conceptual model describing the precipitation of early cements throughout a 4th-order regressive sequence. Four different stages in time are shown. The stratigraphic cross section of the 4th-order parasequence is based on the 4 sections (Sections 3, 8, 9, and 10) which are part of this diagenetic study, as well as 7 other sections where the strata have been measured. Datum is the top of the 4th order transgressive sequence, or the maximum flooding surface. Sea level, relative to the surface of active deposition is not shown in this figure. (A) is at a time just before the 4th-order transgressive flooding. The active surface of deposition is the top of the uppermost shoreface unit, R4. In (B), there has been a relative decrease in sediment supply. With subsidence, relative sea level begins to rise. Deposition of the first transgressive mud/sand is shown in (B). Also, the meteoric/mixed/marine hydrologic regimes have shifted landward in response to the transgressive flooding. (C) shows a point in time where transgressive deposition of sands and muds has been ongoing and the hydrologic regimes have had more time to equilibrate with the higher relative sea-level position. Subsequently, the porewaters in the underlying 4th-order regressive sequence are becoming dominantly marine. The transgression is nearing its point of maximum flooding in (D).

Transgression and Post-transgression:

(4) Ferroan calcite cementation (poikilotopic Fe-calcite) under slightly reducing conditions over a long period of time within the increasingly marine-dominated hydrologic zone.
(5) Continued precipitation of poikilotopic calcite after Fe^{2+} concentrations decreased.

The above summary represents a simplification of processes occurring throughout the of the 4th-order parasequence over the time of its deposition. The following discussion elaborates on how the cyclical fluctuation in sea level and the subsequent cyclical deposition of marine facies may have controlled pore fluid chemistry over a two-dimensional cross section of the 4th-order parasequence.

Pre-transgression.—

At this stage in the depositional history, the uppermost shoreface in the 4th-order regressive sequence has prograded seaward and the relative sea-level is at its maximum lowstand level (Fig. 13A). Storm-dominated processes influenced sedimentation, as evidenced by the hummocky cross bedding observed throughout much of the shoreface facies (Katzman, 1991). Such storm activity maintained a flux of seawater through the pore spaces in the unconsolidated sands.

At some depth beneath the sediment/water interface, a lens of meteoric pore fluids moved oceanward in response to the oceanward progradation of the shoreline (Meisler et al., 1984; Bethke, 1988; Harrison and Summa, 1991; Bjorlykke, 1993, 1994). The extent to which the meteoric lens extended offshore depended on hydraulic head, the hydraulic conductivities of the various facies and the geometry of layered permeable and semi-permeable sediment bodies (Collins and Gelhar, 1971; Meisler et al., 1984; Harrison and Summa, 1991; Bjorlykke, 1993, 1994). In a layered aquifer, laterally-continuous permeable layers act as conduits for fluid flow when intercalated between semi-permeable layers (Collins and Gelhar, 1971). In Figure 13A, a possible ground water configuration is shown within the context of the 4th-order parasequence stratigraphy. The relatively impermeable 5th-order transgressive shale and sandstone units between the regressive sandstones would have created a layered aquifer, with lateral fluid flow occuring along the laterally continuous shoreface sandstones. The exact depth to the top of the meteoric lens and the extent to which the meteoric lens prograded oceanward within the 4th-order parasequence during regression is not certain.

In the meteoric hydrologic zone, there would have been early dissolution of lithic grains and feldspars, resulting in the precipitation of kaolinite (Bjorlykke, 1994). The relationship between feldspar dissolution and kaolinite precipitation is noted by the occurrence of both the early episode of kaolinite and the early episode of feldspar dissolution. The overall decrease in kaolinite abundances in the regressive shoreface sandstones oceanward (Fig. 11E) also supports a connection between kaolinite precipitation and early meteoric waters.

With the meteoric zone positioned toward the bottom of the 4th-order parasequence, there would have been a mixing zone transitional between the meteoric lens and early marine porewaters. It is in this mixing zone that Fe-dolomite precipitation is proposed to have occurred. Note that Figure 13 represents snapshots in time and that variations in pore water chemistry over time is part of a dynamic process linked to fluctuations in relative sea level and the subsequent deposition of cycles. In the case of Figure 13A, the meteoric zone is shown at its most oceanward extent; before the progradation of the meteoric lense to this point, the porewaters would have been those of a mixing zone and Fe-dolomite could have been the precipitating phase.

Dolomite precipitation in marine systems is largely controlled by the concentration of dissolved sulfate in pore fluids rather than the molar Mg^{2+}/Ca^{2+} ratio (Baker and Kastner, 1981; Kastner and Baker, 1983; Kastner, 1984). Two different processes that can decrease the sulfate concentration such that dolomite precipitation is no longer inhibited by the presence of sulfate include: (1) dilution of seawater by meteoric water and (2) microbial sulfate reduction in organic-rich sediments (Kastner, 1984). In the Point Lookout 4th-order parasequence, there is an overall increase in the frequency of occurrence and abundance of Fe-dolomite overgrowths toward land (Fig. 11F). This is not a simple reflection of intrabasinal dolomite content as intrabasinal dolomite is abundant in the oceanward sections (Fig. 11A). If microbial sulfate reduction in organic-rich sediments was the dominant process behind removing sulfate, then the abundance of Fe-dolomite would be expected to increase oceanward, where the 4th-order parasequence becomes progressively dominated by organic-rich muds (i.e., the Mancos Shale). Instead, the reverse is seen. The increase in Fe-dolomite overgrowths toward land suggests that a mixing zone may have played a role in the precipitation of the Fe-dolomite. The problem with calling upon a mixing zone to dilute sulfate concentrations, however, is that all of the ionic species in seawater which go to form the Fe-dolomite (Mg^{2+}, Ca^{2+}, Fe^{2+}, HCO_3^-) are diluted as well. One possible explanation is that neither process alone reduced the sulfate concentrations. In section 3 of Figure 11F, there appears to be a close association between Fe-dolomite overgrowths and intercalated sandstone/shale sequences (in particular, the two transgressive sequences between R1 and R2, and between R2 and R3). It is possible that dilution of seawater, in conjunction with microbial sulfate reduction and iron reduction in the transgressive muds, is what finally suppressed sulfate concentrations enough for dolomite to be able to precipitate in the adjacent sandstones.

Transgressive episode.—

As mentioned previously, subsidence of the marine deposits in conjunction with a significant decrease in sediment input was responsible for an apparent rise in relative sea level (Wright, 1984; Katzman, 1991). With both subsidence and marine flooding, the meteoric hydrologic regime and its transitional mixing zone retreated landward (Figs. 13B, 13C, 13D). The zone of Fe-dolomite precipitation correspondingly shifted landward along with the mixing zone. Further away from land, mixing zone pore fluids in shoreface sandstones became increasingly more saline. With the influx of seawater, sulfate concentrations increased and chemical conditions were no longer conducive for the precipitation of dolomite (Baker and Kastner, 1981; Kastner and Baker, 1983); this is where Fe-calcite started to precipitate instead of the Fe-dolomite. In zones of abundant Fe-calcite, Fe-dolomite overgrowths are sparse (Figs. 11C, 11F), implying that in the upper part of the 4th-order regressive sequence, conditions suitable for dolomite precipitation never existed prior to calcite cementation.

Precipitation of poikilotopic Fe-calcite is the dominant cementing event which appears to be a direct record of the changes brought about by a relative rise in sea level and the subsequent deposition of transgressive sandstones and shales. As mentioned previously in this paper, much of the poikilotopic Fe-calcite occurs in the form of meter-scale concretions and millimeter-scale patches, where there is a definite relationship between the size, shape and distribution of the concretions and the thickness of the overlying 4th-order transgressive sequence. A very similar form of Fe-calcite cementation has been studied by Wilkinson (1991) in the Bearreraig Sandstone Formation (a Jurassic marine shelf sandstone). According to Wilkinson (1991), the first generation of ferran calcite formed at shallow depths (few centimeters beneath the sediment-water interface) as millimeter-scale crystal clusters in marine porewaters during sulfate reduction. A second generation of ferran calcite precipated later in the form of concretions in meteoric or mixing zone pore waters; the Fe-calcite clusters served as the nuclei for the concretions (Wilkinson, 1991). The Bearreraig Sandstone concretions were found to be preferentially developed along discrete horizons. This spatial pattern of concretions was controlled by the location of the Fe-calcite clusters, which in turn formed during periods of negligible sedimentation (Wilkinson, 1991).

The timing of the concretions relative to the small patches of Fe-calcite is not known in the Point Lookout Sandstone; both have the appearance of having occurred before significant compaction. Because the patches are pervasive within the sandstones hosting the concretions, it is possible that they served as nucleation sites for the concretions. The patterns of concretionary growth across the shoreface sandstones immediately below the 4th-order transgressive sequence suggest a relationship between concretion growth and deposition of the overlying transgressive sandstones and shales. Where there is a thicker transgressive sequence of intercalated shales and sandstones, there are a greater number of concretion horizons in the two underlying regressive shorefaces. Toward land, where there are fewer shales and sandstones in a thinner transgressive sequence, only one concretion horizon in each of the two underlying regressive shorefaces has been observed. This pattern of concretion distribution is consistent with Raiswell's (1987) observation that low sedimentation rates play an important role in allowing continuous growth of concretions to occur. In the Point Lookout, the low sedimentation rates influencing the precipitation of concretions along a given horizon would have corresponded to the deposition of a mud in the overlying 4th-order transgressive sequence.

Studies conducted by others have aided in trying to explain some of the spatial diagenetic trends seen in a Point Lookout Sandstone 4th-order parasequence. The above scenario describing the manner in which the cyclical stratigraphy and paleohydrology may have influenced these diagenetic trends represents a preliminary explanation for the results presented in this paper. Additional research is needed in order to better understand the geochemical processes behind some of the diagenetic features. Questions which remain to be answered are the chemical conditions of Fe-dolomite precipitation, the exact nature in which the Fe-calcite concretions were formed, and the source of the ferran and non-ferran calcites. Nonetheless, by comparing the results here to other studies, it appears that the rel-

ative timing of these early diagenetic events, and their distribution patterns across the four stratigraphic sections, is consistent with the manner in which the meteoric, mixing zone and seawater hydrologic regimes would have been shifting in response to relative sea-level changes.

CONCLUSIONS

(1) Surface weathering has not obliterated the relative abundances of primary diagenetic features (with the possible exception of sparry calcite). Reconstructed compositions of both core and outcrop samples fall within the same field on a QFL ternary diagram suggesting that, with care, the effects of weathering can be accounted for and thereby factored out, allowing the study of outcrop samples to be useful for diagenetic history analysis. Backscattered electron imagery indicates that critical information regarding the extent of dolomite precipitation can be deduced despite later oxidation.

(2) The principal diagenetic features are poikilotopic Fe-calcite cement, authigenic Fe-rich dolomite overgrowths, poikilotopic calcite, quartz overgrowths, feldspar dissolution, chlorite cement, sparry calcite cement, kaolinite cement, framboidal pyrite, gypsum cement, and iron-oxide cement. Results indicate that the feldspar dissolution, kaolinite cement, Fe-rich dolomite cement and ferran calcite cements are the most volumetrically significant, and are also relatively early diagenetic features.

(3) Based upon petrographic relationships among different diagenetic features, the general order in which the principal diagenetic textures developed is: chlorite cement → feldspar dissolution → kaolinite cement → poikilotopic Fe-dolomite overgrowths → poikilotopic Fe-calcite precipitation → poikilotopic calcite → sparry calcite → additional feldspar dissolution.

(4) A few notable trends are apparent in terms of the distribution of diagenetic features within the cyclical stratigraphy of the Point Lookout. These trends include: the association of higher intrabasinal dolomite grain abundances in and around deeper inner-shelf sequences, the slightly higher concentration of quartz overgrowths in inner-shelf sandstones, and the spatial association of ferran-calcite concretions with the uppermost shoreface units beneath 4th-order transgressive deposits. Depositional facies controlled the distribution of quartz overgrowths and intrabasinal dolomite, while the distribution of the poikilotopic Fe-calcite and poikilotopic calcite was controlled by the position of the sandstone within the cyclical Point Lookout Sandstone stratigraphy.

(5) The relationship between poikilotopic Fe-calcite and 4th-order parasequence geometry is most clearly evidenced by the distribution patterns of large concretions. Toward the oceanward extent of the 4th-order parasequence, where the inner-shelf portion of the 4th-order transgressive sequence is at its thickest, there are numerous concretion horizons, and the concretions exhibit greater variability in terms of size and shape. Toward the landward extent of the 4th-order parasequence, there is only one horizontal concretion zone in each of the upper two shoreface units. Moreover, con-

cretion shape is less varied and the average concretion size is larger.

ACKNOWLEDGMENTS

This research project is funded by the American Chemical Society, Petroleum Research Fund. We would like to thank the U. S. Geological Survey core personnel at the core repository in Denver for use of core samples and thin sections; the Ute Mountain Tribal Park personnel and guides for access to Mancos Canyon; Robyn Wright-Dunbar, Rob Zech, Danny Katzman and Chris Timm for aid with field work and sampling; and Mike Spilde for his assistance on the UNM scanning electron microscope and electron microprobe. We would also like to acknowledge the use of the following analytical facilities at the Department of Earth and Planetary Sciences, University of New Mexico: x-ray diffraction lab, SEM lab and EPMA lab.

REFERENCES

BAKER, P. A. AND KASTNER, M. 1981, Constraints on the formation of sedimentary dolomite: Science, v. 213, p. 214–216.

BETHKE, C. M., HARRISON, W. J., UPSON, C., AND ALTANER, S. P., 1988, Supercomputer analysis of sedimentary basins: Science, v. 239, p. 261- 267.

BJORLYKKE, K., 1993, Fluid flow in sedimentary basins: Sedimentary Geology, v. 86, p. 137–158.

BJORLYKKE, K., 1994, Fluid-flow processes and diagenesis in sedimentary basins, *in* Parnell, J., ed., Geofluids: Origin, Migration and Evolution of Fluids in Sedimentary Basins: London, Geological Society Special Publication 78, p. 127–140.

CRAIGG, S. D., DAM, W. L., KERNODLE, J. M., THORN, C. R., AND LEVINGS, G. W., 1990, Hydrogeology of the Point Lookout Sandstone in the San Juan structural basin, New Mexico, Colorado, Arizona and Utah: Washington D. C., United States Geological Survey Hydrologic Investigations Atlas HA-720-G.

COLLINS, M. A. AND GELHAR, L. W., 1971, Seawater intrusion in layered aquifers: Water Resources Research, v. 7, p. 971–979.

DEVINE, P. E., 1980, Depositional patterns in the Point Lookout Sandstone, northwest San Juan Basin, New Mexico: Unpublished M.A. Thesis, University of Texas, Austin, 238 p.

DEVINE, P. E., 1991, Transgressive origin of channeled estuarine deposits in the Point Lookout Sandstone, northwestern New Mexico: A model for the Upper Cretaceous, cyclic regressive parasequences of the U.S. Western Interior: American Association of Petroleum Geologists Bulletin, v. 75, p. 1039–1063.

DREVER, J. I., 1973, The preparation of oriented clay mineral specimens for X-ray diffraction analysis by a filter-membrane peel technique: American Mineralogist, v. 58, p. 553–554.

FASSETT, J. E., 1985, Early Tertiary paleogeography and paleotectonics of the San Juan basin area, New Mexico and Colorado, *in* Flores, R. M. and Kaplan, S. S., eds., Cenozoic Paleogeography of West-Central United States: Denver, Rocky Mountain Section Society of Economic Paleontologists and Mineralogists, p. 317–334.

FOLK, R. L., 1980, Petrology of Sedimentary Rocks: Austin, Hemphill Publishing Company, 185 p.

GLASMANN, J. R., LUNDEGARD, P. D., CLARK, R. A., PENNY, B. K., AND COLLINS, I. D., 1989a, Geochemical evidence for the history of diagenesis and fluid migration: Brent Sandstone, Heather Field, North Sea: Clay Minerals, v. 24, p. 255–284.

GLASMANN, J. R., CLARK, R. A., LARTER, S., BRIEDIS, N. A., AND LUNDEGARD, P. D., 1989b, Diagenesis and hydrocarbon accumulation, Brent Sandstone (Jurassic), Bergen High Area, North Sea: American Association of Petroleum Geologists Bulletin, v. 73, p. 1341–1360.

HANSLEY, P. L. AND WHITNEY, C. G., 1990, Petrology, diagenesis, and sedimentology of oil reservoirs in Upper Cretaceous Shannon Sandstone Beds, Powder River Basin, Wyoming: Washingon D. C., United States Geological Survey Bulletin No. 1917-C, 31 p.

HARRISON, W. J. AND SUMMA, L. L., 1991, Paleohydrogeology of the Gulf of Mexico Basin: American Journal of Science, v. 291, p. 109–176.

HUNTER, B. E., 1979, Regional analysis of the Point Lookout Sandstone, Upper Cretaceous, San Juan Basin, New Mexico—Colorado: Unpublished Ph.D. Dissertation, Texas Tech University, Lubbock, 118 p.

KASTNER, M., 1984, Control of dolomite formation: Nature, v. 311, p. 410–411.

KASTNER, M. AND BAKER, P. A., 1983, Sedimentary Rocks, *in* Yearbook of Science and Technology, 1982/1983: New York, McGraw-Hill, p 406–408.

KATZMAN, D., 1991, Heirarchy of transgressive-regressive shoreline cycles and transgressive deposits in the Upper Cretaceous Point Lookout Sandstone, southwestern Colorado: Unpublished M.S. Thesis, University of New Mexico, Albuquerque, 71 p.

KEIGHIN, C. W., ZECH, R. S., AND WRIGHT-DUNBAR, R., 1993, The Point Lookout Sandstone: A tale of two cores, or petrology, diagenesis, and reservoir properties of Point Lookout Sandstone, Southern Ute Indian Reservation, San Juan Basin, Colorado: The Mountain Geologist, v. 30, p. 5–16.

LAW, B. E., 1992, Thermal maturity patterns of Cretaceous and Tertiary rocks, San Juan basin, Colorado and New Mexico: Geological Society of America Bulletin, v. 104, p. 192–207.

MEISLER, H., LEAHY, P. P., AND KNOBEL, L. L., 1984, Effect of eustatic sealevel changes on saltwater-freshwater relations in the northern Atlantic coastal plain: Washington D. C., United States Geological Survey Water Supply Paper No. 2255, 27 p.

MOORE, D. M. AND REYNOLDS, R. C., 1989, X-Ray Diffraction and the Identification and Analysis of Clay Minerals: New York, Oxford University Press, 332 p.

RAISWELL, R., 1987, Non-steady state microbiological diagenesis and the origin of concretions and nodular limestones, *in* Marshall, J. D., ed., Diagenesis of Sedimentary Sequences: Oxford, Geological Society Special Publication 36, p. 41–54.

SABINS, F. F., Jr., 1964, Symmetry, stratigraphy, and petrography of cyclic Cretaceous deposits in San Juan Basin: American Association of Petroleum Geologists Bulletin, v. 48, p.292–316.

SHETIWY, M. M., 1978, Sedimentologic and stratigraphic analysis of Point Lookout Sandstone, southeast San Juan Basin, New Mexico: Unpublished Ph.D. Dissertation, New Mexico Institute of Mining and Technology, Socorro, 262 p.

VAIL, P. R., MITCHUM, R. M., AND THOMPSON, S., 1977, Global cycles of relative changes in sea level, *in* Payton, C. E., ed., Seismic Stratigraphy—Applications to Hydrocarbon Exploration: Tulsa, American Association of Petroleum Geologists Memoir 26, p. 83–97.

VAN WAGONER, J. C., POSAMENTIER, H. W., MITCHUM R. M., JR., VAIL, P. R., SARG, J. F., LOUTIT, T. S., AND HARDENBOL, J., 1988, An overview of the fundamentals of sequence stratigraphy, *in* Wilgus, C. K., Hastings, B. S., St. C. Kendall, C. G., Posamentier, H. W., Ross, C. A., Van Wagoner, J. C., eds., Sea Level Changes: An Integrated Approach: Tulsa, Society of Economic Paleontologists and Mineralogists Special Publication 42, p. 39–46.

WILKINSON, M., 1991, The concretions of the Bearreraig Sandstone Formation: geometry and geochemistry: Sedimentology, v. 38, p. 899–912.

WILSON, J. C. AND MCBRIDE, E. F., 1988, Compaction and porosity evolution of Pliocene sandstones, Ventura basin, California: American Association of Petroleum Geologists Bulletin, v. 72, p. 664–681.

WINN, R. D., STONECIPHER, S. A., AND BISHOP, M. G., 1984, Sorting and wave abrasion: controls on composition and diagenesis in lower Frontier sandstones, southwestern Wyoming: American Association of Petroleum Geologists Bulletin, v. 68, p. 268–284.

WRIGHT, R., 1984, Paleoenvironmental interpretation of the Upper Cretaceous Point Lookout Sandstone: Implications for shoreline progradation and basin tectonic history, San Juan Basin, New Mexico: Unpublished Ph.D. Dissertation, Rice University, Houston, 404 p.

WRIGHT, R., 1986, Cycle stratigraphy as a paleogeographic tool: Point Lookout Sandstone, southeastern San Juan Basin, New Mexico: Geological Society of America Bulletin, v. 96, p. 661–673.

WRIGHT, R., 1988, Cycle stratigraphy: *in* Nummedal, D., Wright R., and Swift, D. J. P., eds., Sequence Stratigraphy of the Upper Cretaceous Strata of the San Juan Basin, New Mexico: Houston, 73rd Annual Meeting, American Association of Petroleum Geologists/Society of Economic Paleontologists and Mineralogists Field Guide, p. 134–143.

ASSOCIATION OF ALLOCHTHONOUS WATERS AND RESERVOIR ENHANCEMENT IN DEEPLY BURIED MIOCENE SANDSTONES: PICAROON FIELD, CORSAIR TREND, OFFSHORE TEXAS

THOMAS R. TAYLOR

Shell Development Company, Bellaire Research Center, Houston, Texas 77001

AND

LYNTON S. LAND

Department of Geological Sciences, University of Texas at Austin, Austin, Texas 78713

ABSTRACT: Anomalously high porosities (20–29%) in deeply buried (4.9–5.2 km) Miocene sandstones at Picaroon field, offshore Texas are largely a result of porosity enhancement by dissolution of calcite cement. Dissolution is not important in equivalent strata nearby, including Doubloon field, where reservoir porosities are generally less than 18%. Doubloon reservoir sands contain moderately saline (TDS = 63–74 g/l), "NaCl-type" water characterized by low concentrations of Ca and other cations, enrichment in $\delta^{18}O$ (+7.8‰ SMOW) and radiogenic Sr ($^{87}Sr/^{86}Sr$ = 0.71109). Formation waters of this type are common throughout the Gulf Coast Tertiary section. In contrast, Picaroon waters have salinities of 151 to 243 g/l TDS, Ca concentrations of 13 to 22 g/l, heavier $\delta^{18}O$ (+8.0 to +9.3‰ SMOW) and less radiogenic Sr ($^{87}Sr/^{86}Sr$ = 0.70992 to 0.71023). In addition, Picaroon water samples contain unusually high concentrations of other cations such as Sr, Ba, Fe, Pb and Zn. Picaroon sandstones contain late diagenetic fracture-filling ankerite, barite and sphalerite. Ankerite is in oxygen isotopic equilibrium with formation water at temperatures indicated by fluid inclusions (>147°C) and has a similar Sr isotopic composition ($^{87}Sr/^{86}Sr$ = 0.70970).

Picaroon brines are interpreted to be allochthonous to the Miocene section, because they have elemental and isotopic compositions similar to waters produced from Mesozoic reservoirs in south Texas and central Mississippi. The association of these waters with high quality reservoirs at Picaroon suggests a potential link between deep sources of fluids and carbonate dissolution. A model is proposed in which hot, acidic waters from the underlying Mesozoic section are injected along major faults into Picaroon sands, resulting in significant porosity enhancement. Fluid flow is likely episodic, driven by periodic buildup and release of geopressures.

INTRODUCTION

The extent to which diagenetic processes affect the reservoir properties of sandstones is important in hydrocarbon exploration and production. Due to the complexity of the physical and chemical aspects of diagenesis, reservoir quality remains one of the most difficult variables to predict. Although depth/temperature-related and regional diagenetic trends are of fundamental importance, it is often diagenetic anomalies that are of interest in exploration. Deviations from "normal" diagenetic trends can be associated with high quality reservoir sands in which compaction and/or cementation have been limited, or in which significant porosity enhancement has occurred.

Numerous studies have documented the dynamic nature of mass transfer which accompanies diagenetic processes in the Gulf of Mexico sedimentary basin (see Sharp et al., 1988). The recognition of important trends and variations in clay diagenesis (Burst, 1969; Hower et al., 1976; Freed and Peacor, 1989, 1992), feldspar dissolution and replacement (Boles and Franks, 1979; Milliken et al., 1981; Milliken et al., 1989), cementation (Boles and Franks, 1979; Fisher and Land, 1986; Taylor, 1990), and water chemistry (Morton and Land, 1987; Land and Macpherson, 1992) provide a foundation for interpreting controls on the reservoir properties of sands.

The depositional, structural, and diagenetic history of middle Miocene sandstones of the deep Corsair Trend (offshore Texas) have been described in previous work (Vogler and Robison, 1987; Taylor, 1990). In this study, we combine the results of previous petrographic and geochemical studies (Taylor, 1990) with detailed chemical analyses of produced formation waters in order to understand important diagenetic processes and aid in the delineation of the distances over which diagenetic processes operate.

GEOLOGICAL BACKGROUND

Picaroon and Doubloon fields, and prospect Plank are located in the offshore Texas Gulf Coast (Fig. 1) and are part of the

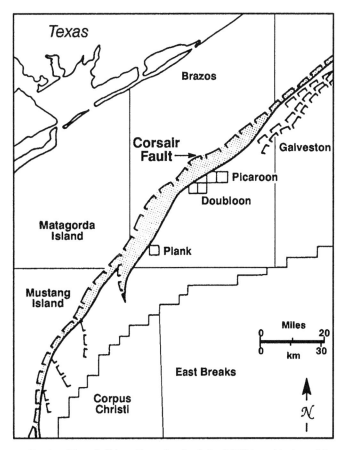

FIG. 1.—Map of offshore Texas showing federal OCS lease blocks and the approximate locations of the Corsair Fault, Picaroon field, Doubloon field, and prospect Plank.

Siliciclastic Diagenesis and Fluid Flow: Concepts and Applications, SEPM Special Publication No. 55

Corsair Trend. The Corsair Trend refers to a series of deep geopressured gas fields and is defined by a series of southeastward dipping, arcuate growth faults collectively known as the Corsair Fault (Vogler and Robison, 1987). The trend is located approximately 48 km from the Texas coast (Fig. 1) and extends approximately 145 km parallel to the shoreline.

Middle Miocene age sandstones at Picaroon, Doubloon, and Plank are similar in that they are presently at their respective maximum burial depths, which range from 4.0–5.2 km (13,200–17,000 ft) and are assumed to be at or near maximum temperature. The sands in all three locations are highly geopressured (max. 0.02 MPa/m or 0.94 psi/ft). Based on sedimentary structures and sequences observed in core, the reservoir sandstones are interpreted as deltaic deposits (Vogler and Robison, 1987), primarily from distributary mouth bar and delta front environments. They contain substantial amounts of detrital feldspar, carbonate, and lithic fragments (Taylor, 1990) and are classified as subarkoses, lithic arkoses and feldspathic litharenites (after Folk, 1974).

A rapid decline in porosity is observed in middle Miocene sandstones of the Corsair Trend between depths of approximately 1.8 and 2.7 km (6,000–9,000 ft; Fig. 2). Detailed petrographic work has shown that this is primarily due to the emplacement of large amounts of calcite cement on a regional scale (Taylor, 1990). The most porous reservoir sandstones from Picaroon field have anomalously high porosities (20–29%) for their depth; up to 15% higher than the regional average (Fig. 2). Statistical analysis of porosity data collected from Middle Miocene sandstones throughout the Brazos OCS area (Fig. 1) clearly indicates that the Picaroon sandstones are anomalous. Comparing porosities derived from over 500 thin-section point counts, the average thin-section porosity at Picaroon (17.6%) is higher than the regional average for all depth intervals greater than 1.8 km at the 99% confidence interval (Table 1). Furthermore, the average laboratory measured porosity for sandstone core material is 18.9% for Picaroon, 12.3% for Doubloon and 12.9% for Plank. The results of t-tests, using only laboratory data, indicate that Picaroon sandstones have significantly higher porosities (99% confidence interval) than age-equivalent sands from nearby Doubloon and Plank (Table 1).

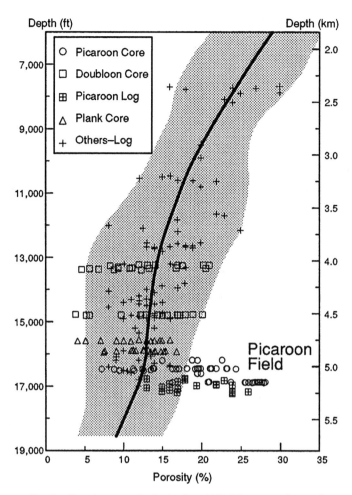

FIG. 2.—Porosity versus depth plot for middle Miocene sandstones from the Corsair Trend, Brazos OCS area, offshore Texas. Porosity data are taken from laboratory core analyses and density logs. The solid line represents the average trend (excluding Picaroon) and the shaded area represents the general range of porosity values. Reservoir sandstones from Picaroon field have up to 15% greater porosity than the average sandstone at depths of 4.9–5.2 km.

DIAGENESIS

Diagenetic reactions involving precipitation and dissolution of calcite ultimately determine reservoir quality of Corsair Trend sandstones. The paragenetic sequence established for sandstones from Doubloon and Plank represents a consistent regional pattern of diagenesis (Fig. 3). Textural evidence indicates that the formation of grain-coating chlorite, and the initiation of both quartz cementation and feldspar dissolution occurred prior to the precipitation of calcite cement (Fig. 4). The abundance of quartz cement is highly variable ranging from zero to approximately 13% (average ≅5%). In samples with little or no visible porosity, as much as 11% quartz overgrowth cement predates calcite cement. Throughout the Corsair Trend, calcite represents the most abundant pore-filling cement. Clay-poor sandstones with less than 10% porosity commonly contain 15 to 30% calcite cement. In all sands examined in this study, calcite cement is ferroan (detrital calcite is non-ferroan), unzoned under cathodoluminescence and exhibits poikilotopic

texture. Lesser amounts of ankerite cement (average ≅3%) formed after calcite cement.

The timing of both quartz and calcite cementation can be constrained using a combination of isotopic data and petrographic observations. Based on the observed regional porosity loss with depth attributable to calcite cementation, Taylor (1990) estimated that calcite cement was emplaced over a depth range of approximately 1.8 to 2.6 km. The non-radiogenic $^{87}Sr/^{86}Sr$ ratios (0.7083–0.7086) of calcite cement suggest a somewhat older Miocene marine source of carbonate (Taylor, 1990). The $^{87}Sr/^{86}Sr$ values also indicate precipitation prior to release of volumetrically significant radiogenic strontium associated with K-feldspar dissolution and illitization of mixed-layer clays during the burial diagenesis of mudrocks. Oxygen isotope modeling using data from Picaroon field shales predicts that at depths of 1.8 to 2.6 km, waters expelled from compacting shales would have $\delta^{18}O$ values of approximately +1 to +3‰ SMOW (Taylor, 1990). Calcite cement with $\delta^{18}O_{PDB}$ values of −4.5 to −10.7‰ (Table 2) would be in isotopic equilibrium with these waters at temperatures of approximately 55 to 100°C.

TABLE 1.—STATISTICAL SUMMARY OF POROSITY DATA FOR MIDDLE MIOCENE SANDSTONES—CORSAIR TREND, BRAZOS AREA, OFFSHORE TEXAS

Depth (km)	Depth (ft)	Porosity Mean	Standard Deviation	Max	Min	N	Data Source
		Brazos Area—excluding Picaroon					
1.2–1.8	4–6,000	21.3	9.7	33.4	1.3	25	Thin Section
1.8–2.4	6–8,000	13.8[1]	8.9	29.0	0.0	60	Thin Section
2.4–3.0	8–10,000	10.2[1]	5.7	27.4	1.0	59	Thin Secion
3.0–3.6	10–12,000	8.9[1]	5.6	24.6	1.3	64	Thin Section
3.6–4.2	12–14,000	8.8[1]	5.4	22.1	0.3	74	Thin Section
4.2–5.2	14–17,000	7.5[1]	5.1	19.6	0.0	137	Thin Section
		Picaroon					
4.9–5.2	16,202–16,898	17.6	7.1	29.6	0.0	92	Thin Section
		Doubloon					
4.0–4.5	13,219–14,819	12.3[2]	5.7	22.9	1.6	136	Core Analysis
		Plank					
4.7–4.8	15,578 -15,944	12.9[2]	3.1	16.9	2.9	113	Core Analysis
		Picaroon					
4.9–5.2	16,202–16,898	18.9	6.1	28.3	2.0	156	Core Analysis

[1]Results of T-test indicate that the mean porosity is less than Picaroon mean porosity (thin section) at $\alpha = 0.01$.
[2]Results of T-test indicate that the mean porosity is less than Picaroon mean porosity (core analyses) at $\alpha = 0.01$.

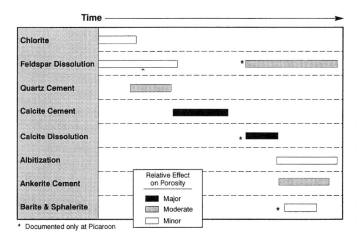

FIG. 3.—The relative timing of diagenetic events observed in middle Miocene sandstones of the Corsair Trend, Brazos OCS area, offshore Texas. Each event is coded according to its relative effect on sandstone porosity. Diagenetic events marked by an asterisk are observed only at Picaroon field.

Petrographic observations indicate that quartz precipitation preceded calcite cement (Fig. 4B). Minor amounts of quartz cement may have continued to form during and after calcite precipitation but conclusive textural evidence is lacking. The oxygen isotope composition of the bulk quartz fraction was measured on a suite of eight samples from Picaroon field for which the volume of detrital quartz and quartz cement were determined by thin section point counting. The estimated $\delta^{18}O$ of quartz cement, determined by linear extrapolation (Fig. 5), is 28.8‰ SMOW. Quartz cement would be in isotopic equilibrium with waters having $\delta^{18}O$ in the range of 0 to $+3‰$ at temperatures of 50 to 67°C. The isotopic data for both quartz and calcite cements are consistent with the observed paragenetic sequence and indicate quartz precipitation at intermediate burial depths.

The same general sequence of cementation events observed at Doubloon and Plank occurs in sandstones from Picaroon field (Fig. 3). However, the amount of calcite cement in Picaroon reservoirs averages 6.8% (n = 89) as opposed to 13.5% (n = 57) for Doubloon and Plank. Furthermore, abundant petro-

graphic evidence for a major episode of calcite dissolution is conspicuous in highly porous reservoir sandstones at Picaroon (see Taylor, 1990). Highly corroded remnants of both ferroan calcite cement and detrital calcite are common (Figs. 6A, B). Numerous examples of open pore space separating optically continuous remnants of calcite cement (Fig. 6B) can be found within a single thin section and are indicative of dissolution of poikilotopic cement. In contrast, calcite cement at Doubloon and Plank occurs as connected patches and commonly exhibits euhedral faces where abutting open pore space (Figs. 4C, D). In addition, detrital calcite at Doubloon and Plank shows no evidence of significant dissolution. At all locations, ankerite cement displays euhedral crystal faces where bordering open pores, exhibits no evidence of dissolution and at Picaroon encases corroded fragments of calcite cement (see Fig. 5E in Taylor, 1990). Fluid inclusion data indicate that ankerite formed at temperatures greater than 120°C, thereby constraining calcite dissolution to lower temperatures and to burial depths of approximately 2.6 to 3.0 km (Taylor, 1990). Extensive dissolution of both K-feldspar and plagioclase (along with a greater degree of albitization) is evident in the most porous sands from Picaroon, suggesting a second stage of feldspar alteration at Picaroon (Fig. 6C).

The importance of calcite cementation and dissolution in the Corsair Trend is evident in a plot of volume percent calcite cement versus intergranular porosity (Fig. 7). There is a roughly one-to-one, inverse linear relationship between the volume of calcite cement and intergranular porosity ($r^2 = 0.80$). In this sample set, the maximum intergranular porosities at Doubloon and Plank, where dissolution is not observed, are 18% and 15% respectively. The maximum intergranular porosity at Picaroon, where calcite dissolution is evident, is 28%. In addition, the average total porosity (laboratory core analyses) is 6% higher at Picaroon (Table 1). Based on these data it is estimated that calcite dissolution has increased porosity by 10–15% in the most porous Picaroon sandstones and by 6% in the average case.

FORMATION WATER CHEMISTRY

Previous Work

Extensive sampling and analyses of formation waters produced from the Oligocene Frio Formation in the onshore Texas

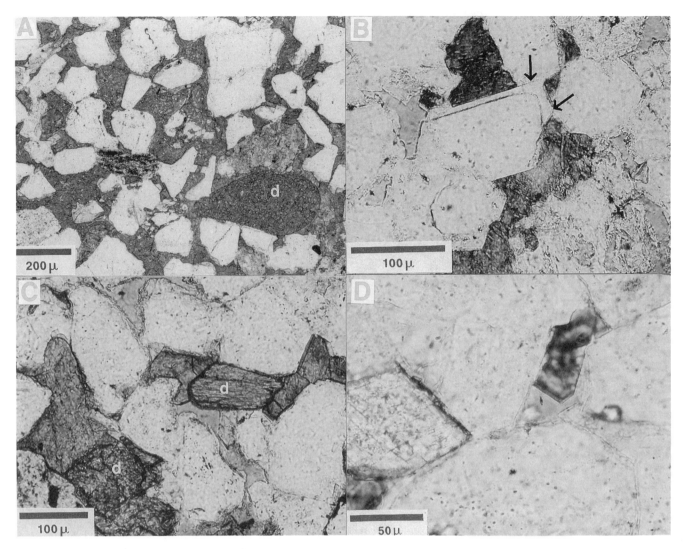

FIG. 4.—Thin section photomicrographs of calcite-cemented middle Miocene sandstones from Doubloon and Plank. (A) Calcite cement (stained) completely fills intergranular porosity in this sandstone. Detrital calcite (d) is composed of rounded grains of pre-existing limestone. (B) Quartz overgrowth cement joins three detrital quartz grains (arrows). Pore space adjacent to quartz cement is subsequently filled with calcite cement (stained). (C) Partially cemented reservoir sandstone (~15% porosity) from Doubloon field. Calcite cement (stained) and detrital calcite (d) show no evidence of significant dissolution. (D) High magnification view of calcite cement (stained) adjacent to open pore space (Doubloon field). Calcite cement displays euhedral crystal boundaries indicative of crystal growth into open pore space.

TABLE 2.—ISOTOPIC DATA FOR CARBONATES FROM PICAROON AND
DOUBLOON SANDSTONES.

Location	Mineral	$\delta^{18}O_{PDB}$	$\delta^{13}C_{PDB}$	$^{87}Sr/^{86}Sr$
Picaroon	Detrital Calcite	−6.4 to −6.9 n = 4	−1.6 to −2.0 n = 4	0.70760
Picaroon	Calcite Cement	−4.5 to −10.7* n = 6	−1.5 to −4.8* n = 6	0.7083−0.7086*
Doubloon	Calcite Cement	−5.6 to −8.9* n = 4	−1.6 to −2.1* n = 4	
Picaroon	Ankerite Cement	−7.4 to −8.8 n = 8	−2.1 to −2.9 n = 8	0.70963−0.70970 n = 2
Doubloon	Ankerite Cement	−8.1 to −9.7 n = 2	−2.4 to −2.6 n = 2	

*Calculated values corrected for inclusion of detrital calcite (Taylor, 1990).

Gulf Coast led to the recognition of three major water types (Morton and Land, 1987). Further sampling has confirmed that these same three water types are typical of the entire Gulf Coast Tertiary section (Fig. 8; Land and Macpherson, 1992).

(1) Na-acetate type waters are typical of shale-rich sections not associated with diapiric salt. They are characterized by low salinity (6,000–20,000 mg/l TDS) and the presence of variable concentrations of organic acid anions (<2,500 mg/l). Na-acetate waters presumably originated as interstitial seawater in marine shales and have subsequently undergone modification by both organic and inorganic reactions dur-

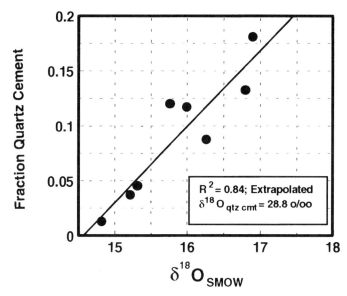

FIG. 5.—Isotopic composition of oxygen in the total quartz fraction versus fraction quartz cement (determined by thin-section point counting) for Picaroon sandstones. Extrapolation of linear regression analysis yields $\delta^{18}O_{SMOW}$ values of 14.7‰ for detrital quartz and 28.8‰ for quartz cement.

ing burial diagenesis. Low salinity is attributed to dilution with interlayer water released from mixed layer clays during the transformation of smectite to illite.

(2) NaCl type water is most commonly found in areas underlain by salt (Morton and Land, 1987) or in association with diapiric salt (Land et al., 1988). It is the most abundant water type in the Gulf, containing between approximately 30,000 and 70,000 mg/l Cl (Fig. 8). More than 90% of the total dissolved solids consist of Na and Cl. NaCl type waters are formed when fluids expelled from dewatering of mud-rich sediments dissolve salt.

(3) Ca-rich waters are the least abundant water type and are most commonly encountered in south Texas (Morton and Land, 1987). They are characterized by Cl concentrations generally greater than 40,000 mg/l and highly variable Ca concentrations, typically greater than 10,000 mg/l. Brines from Mesozoic reservoirs of the Gulf Coast are similar in composition, implying that Ca-rich brines in Tertiary-aged reservoirs can be allochthonous. Alternatively, some Ca-rich brines from the Frio Formation, characterized by very low $^{87}Sr/^{86}Sr$ (similar to volcanic plagioclase within the Frio sands) and anomalously low CO_2 in associated gases (due to calcite precipitation), have apparently evolved within the Tertiary section from an NaCl-type water by massive albitization of detrital plagioclase (Mack, 1990; Land and Macpherson, 1992).

The composition of formation water in Mesozoic units beneath the south Texas Tertiary and the offshore Corsair Trend is unknown. Highly saline brines from updip Mesozoic reservoirs in south Texas are characteristically Ca- and Sr-rich, have radiogenic $^{87}Sr^{86}Sr$ and have heavy $\delta^{18}O$ values. These waters are also enriched in B, Li, K, Br and in some cases Ba, Fe, Mn and Zn (Land and Prezbindowski, 1981). Waters from Mesozoic reservoirs of central Mississippi are also very saline, Ca-

and Sr-rich brines (Carpenter et al., 1974). Compared to south Texas Mesozoic waters they contain very high concentrations of Mg and Br, perhaps due to interaction with bittern salts (Land et al., 1994). In addition, many of the central Mississippi Mesozoic brines contain unusually high concentrations of Fe (max. = 490 mg/l), Mn (max. = 181 mg/l), Pb (max. = 110 mg/l) and Zn (max. = 575 mg/l)(Carpenter et al., 1974).

Picaroon and Doubloon Waters

Methods.—

Formation water analyses used in this study are from two sources. Water samples were collected periodically from platform separators during production and analyzed by a commercial laboratory for total alkalinity, Cl, Na, Ca, Mg, Ba and in some cases Sr (Table 3). In general, these commercial analyses accurately depict the bulk chemistry of the formation water. Multiple samples from individual wells taken over a period of three years show only minor variations in composition. The second set of data represent complete elemental and isotopic analyses of water samples collected at dedicated separators from three Picaroon wells and one Doubloon well (Table 4). Gas samples were also obtained from each well. Wells that were producing water at a rate of >100 bbl/day were selected for sampling, eliminating the possibility of significant dilution by low salinity water condensed out of the gas phase. All the water samples used in this study were obtained from wells completed over narrow intervals in middle Miocene sandstones. Analytical procedures are outlined in Land et al., (1988).

Comparison of Water Chemistry.—

Waters produced from Doubloon reservoir sands are moderately saline (TDS between 63,000 and 75,000 mg/l) and fall within the field defined by NaCl-type waters on a plot of 1000Ca/Cl versus 1000/Cl (Fig. 8). More than 93% of the salinity consists of Na and Cl. Like other NaCl-type waters, Doubloon waters are characterized by low concentrations Ca and other alkaline earth metals, very low concentrations of other cations such as Li, Mn, Zn and Pb, relatively high $\delta^{11}B$ (+30) and heavy $\delta^{18}O$ (Table 4). Overall, NaCl-type waters have radiogenic but highly variable $^{87}Sr/^{86}Sr$ ratios. Salt is extremely rare in the offshore Texas Miocene and no known occurrence of diapiric salt is present within 28 km of Doubloon field. However, middle Miocene sediments of the Corsair Trend were presumably deposited over a deep salt substrate which, due to loading created by rapid sedimentation, was mobilized and subsequently evacuated from the region (Bradshaw et al., 1993).

In contrast to the Doubloon waters, produced waters from Picaroon are extremely saline (TDS between 150,000 and 245,000 mg/l), Ca-rich brines, with Ca concentrations between 13,100 and 22,800 mg/l (Tables 3, 4). Like other Ca-rich brines with equal or greater salinities (see Land and Macpherson, 1992, Table 4), Picaroon waters are characterized by high concentrations of Sr (1,610–2,320 mg/l) and Ba (570–1,650 mg/l), heavy $\delta^{18}O$ (8.0 to 9.3‰), low $\delta^{11}B$ values (+15 to +22) and radiogenic $^{87}Sr/^{86}Sr$ ratios (0.7099 to 0.7102). Picaroon waters also have high concentrations of Li (85 to 225 mg/l), Fe (120 to 438 mg/l), Pb (1 to 6 mg/l) and Zn (25 to 194 mg/l). Natural gas from Picaroon field is CO_2-rich compared to gas associated with other Ca-rich waters from the Cenozoic section.

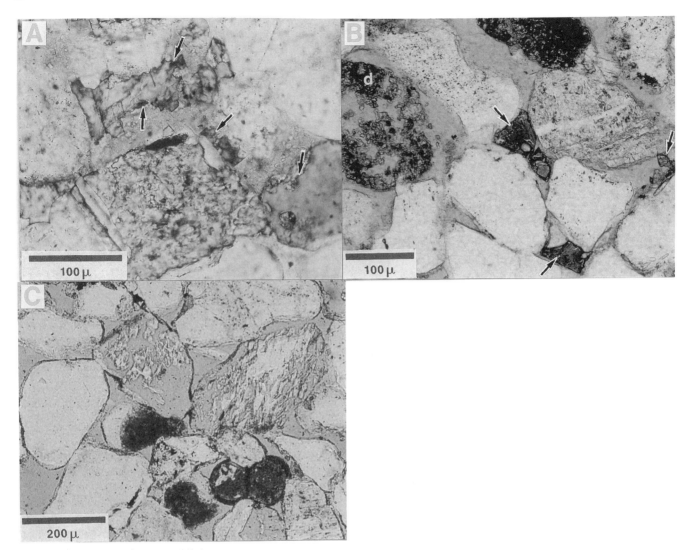

Fɪɢ. 6.—Thin-section photomicrographs of Picaroon reservoir sandstones showing evidence for calcite and feldspar dissolution. (A) Numerous small areas of calcite cement (stained) occur in the transition from cemented to highly porous sandstone at Picaroon field. Where in contact with open pore space, calcite cement displays irregularly shaped surfaces (arrows) indicative of partial dissolution. (B) Partially dissolved detrital calcite grains (d) and remnants of calcite cement (stained) are common in highly porous sandstones. Three small remnants of calcite cement (arrows), presently separated by pore space, are in optical continuity indicating that they were once part of a continuous poikilotopic network of cement. (C) Partial dissolution and albitization of detrital feldspar grains are commonly observed in the most porous sands.

Although formation waters from Picaroon are Ca-rich, they have higher Cl concentrations than most Ca-rich type waters from Gulf Coast Tertiary reservoirs and plot within the field defined by highly saline brines from Mesozoic reservoirs of the Gulf Coast (Fig. 8). Picaroon brines are also similar to many of these Mesozoic brines in terms of minor elements and isotopic signatures. Highly saline waters from Mesozoic reservoirs of south Texas contain similarly high concentrations of Ca, Sr, and Li, radiogenic Sr, low $\delta^{11}B$ and heavy $\delta^{18}O$ values (Table 5). In terms of Fe, Zn and Pb concentrations, Picaroon waters are similar to some brines produced from Mesozoic reservoirs of the Stuart City Trend (Land and Prezbindowski, 1981), central Mississippi (Carpenter et al., 1974; Kharaka et al., 1987) and Arkansas (Moldovanyi and Walter, 1992). The most significant dissimilarity between Picaroon and Mesozoic brines is in the concentration of Br (Table 5). Picaroon waters have low

Br concentrations (131–205 mg/l) compared with Mesozoic waters from south Texas (579–910 mg/l) and central Mississippi (800–2,300 mg/l). Relatively low Br, but Ca-rich brines are a common occurrence in south Texas and are attributed to water-rock interaction between water which has dissolved salt and silicate minerals at high temperature (Land, 1995).

Late Diagenetic Fracture-Filling Minerals.—

Strong evidence that the high concentrations of metals such as Fe, Zn and Pb in Picaroon brines are representative of *in situ* water chemistry and not a product of contamination (i.e., reaction with production hardware) is obtained from examination of late-diagenetic minerals. In core material from Picaroon field, several examples of mineralized fractures were encountered both within sandstone and shale. The veins contain coarse crystals of ankerite along with minor amounts of sphalerite

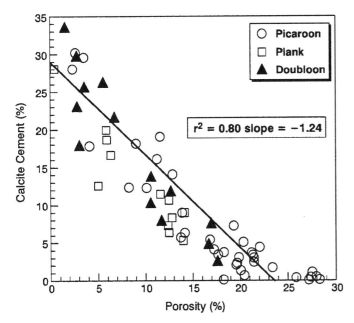

FIG. 7.—A negative, linear relationship exists between intergranular porosity and percent calcite cement in clay-poor sandstones from Picaroon, Doubloon, and Plank. Picaroon sandstones with more than about 15% porosity contain textural evidence of calcite dissolution. Evidence of dissolution is lacking in Doubloon and Plank sandstones.

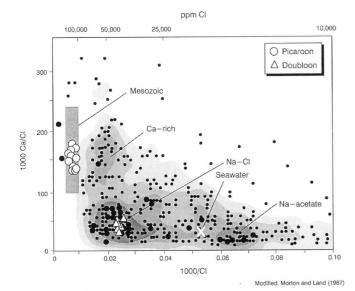

FIG. 8.—Weight ratio of Ca/Cl versus reciprocal Cl concentration (mg/L) for water samples from the Oligocene Frio Formation, onshore Texas (after Land and Macpherson, 1992). The data form three clusters corresponding to Na-acetate, Na-Cl, and Ca-rich water types. Very saline brines from Mesozoic reservoirs of south Texas and central Mississippi plot within the rectangle. Doubloon formation waters all plot within the field for Na-Cl type waters. Although also similar to some Ca-rich type waters, Picaroon brines all plot within the Mesozoic field.

(ZnS) and barite (Fig. 9). In thin section, sphalerite crystals contain clear and yellow colored zones implying a relatively pure (low-Fe) composition. Both sphalerite and barite form euhedral crystals (Fig. 9) that are intergrown with and surrounded by ankerite.

Fluid inclusion data yield a similar range of homogenization temperatures for barite (109–178°C) and ankerite (120–188°C). Melting temperatures indicate that the inclusions contain water with salinities of approximately 165,000–230,000 mg/l NaCl equivalent (Taylor, 1990), similar to the range for produced Picaroon formation waters. Fracture-filling ankerite with measured $\delta^{18}O$ values of -7.4 to $-8.8‰$ PDB would be in isotopic equilibrium with Picaroon formation waters at temperatures of approximately 147–185°C (Fig. 10), slightly less than the *in situ* reservoir temperature of approximately 190°C. Vein-filling ankerites have $^{87}Sr/^{86}Sr$ ratios of 0.70963 and 0.70970. Although the $^{87}Sr/^{86}Sr$ ratios for ankerites are less radiogenic than the values for formation waters, they are significantly more radiogenic than middle Miocene marine values (Fig. 11).

The occurrence of ankerite, barite and sphalerite as late diagenetic phases at Picaroon is consistent with the high concentrations of Fe, Ba and Zn in the sampled brines. The elemental, isotopic and fluid inclusion data suggest subsurface equilibrium between these late diagenetic minerals and water similar to the sampled formation waters.

Origin of Picaroon Waters

The elemental and isotopic compositions of produced formation water from Picaroon field strongly suggest that these brines are allochthonous to the Cenozoic section, having originated within underlying Mesozoic strata. Although not diagnostic of Mesozoic brines, the heavy $\delta^{18}O$ values and radiogenic $^{87}Sr/^{86}Sr$ ratios of Picaroon brines indicate extensive rock/water reactions involving crustal silicate minerals (Suchecki and Land, 1983). The boron isotope signatures of Picaroon brines also suggest extensive rock/water reactions involving ^{11}B-depleted crustal silicate minerals. Picaroon brines have highly depleted $\delta^{11}B$ and relatively high B concentrations, similar to brines from Mesozoic reservoirs of south Texas (Fig. 12). The isotopic composition of strontium and the relatively high CO_2 content of associated gas (>4%) argue against *in situ* albitization as a mechanism for the origin of Ca-rich Picaroon brines (Land and Macpherson, 1992).

The high concentrations of Cl, Na and Ca, as well as Sr, Ba, Zn, Pb and Li in Picaroon brines are the products of extensive mineral/water reactions and are characteristic of water produced from deeply buried Mesozoic strata of the Gulf Coast.

Unlike Mesozoic-sourced brines from the south Texas Frio Formation, which have undergone a loss of K, Li, B, Sr and Mg due to rock/water reactions within the Cenozoic section (Macpherson, 1992), allochthonous Picaroon brines have undergone only minor modification. The differences in extent of reaction following emplacement is attributed to the greater abundance of highly reactive volcanic detritus and the greater amount of time available for rock/water reactions to proceed within the much older Frio Formation (Diggs and Land, 1993) than is true of the offshore Miocene units. In comparison, discharge of allochthonous waters at Picaroon is apparently a rel-

TABLE 3.—PARTIAL ANALYSES OF FORMATION WATERS FROM PICAROON AND DOUBLOON FIELDS*

Well	Field	Cl	ALK	Na	Ca	Mg	Sr	Ba	TDS
A23 A-5	Doubloon	41170	240	24120	1930	NA	NA	31	67500
A23 A-5	Doubloon	41500	300	24140	2100	NA	NA	26	68060
A23 A-5	Doubloon	37660	330	24140	2300	NA	NA	19	64450
A23 A-5	Doubloon	38370	300	22460	1810	NA	NA	3	62950
A23 A-5	Doubloon	42240	450	24370	2130	NA	NA	34	69220
A23 A-5	Doubloon	45000	240	26820	1670	130	380	18	74260
A23 A-5	Doubloon	45000	330	27330	1350	140	300	16	74470
A19 B-2	Picaroon	109920	110	48750	18400	730	NA	1100	179000
A19 D-1	Picaroon	124260	90	58020	19200	250	NA	1380	203180
A19 D-2	Picaroon	135520	80	60670	22800	490	NA	1650	221210
A19 D-5	Picaroon	92910	80	38840	17200	730	NA	1250	151010
A20 JA-1	Picaroon	101530	160	48280	13220	1200	NA	1230	165610
A20 JA-3	Picaroon	103490	80	49670	13210	1180	NA	1160	168780
A20 JA-3	Picaroon	99620	160	47130	13100	1270	NA	1400	162670
A20 JA-3	Picaroon	107040	50	50890	14350	1050	NA	1430	174800

*All concentrations in mg/L. ALK is total alkalinity as bicarbonate. TDS = total dissolved solids.

TABLE 4.—DETAILED ANALYSES OF PRODUCED FORMATION WATERS; PICAROON AND DOUBLOON FIELDS

Well Field	A19 B-1 Picaroon	A19 D-2 Picaroon	A19 B-2 Picaroon	A23 A-5 Doubloon
Cl	147420	145360	112290	38200
I	11	6	6	16
H_4SiO_4	158	163	229	NA
Br	183	205	131	73
Total Alk.	210	230	250	330
Na	66900	65200	52300	21500
K	1980	1320	963	1010
Ca	21200	22800	16600	1700
Mg	720	1040	896	175
Li	225	143	85	16
Sr	2320	2190	1610	395
Ba	1690	1150	573	24
Fe	438	312	127	90
Mn	65	36	15	0.4
Pb	6	3	1.1	0.1
Zn	194	110	25	0.01
B	183	132	132	106
TDS	243903	240400	186233	63635
$\delta^{18}O$ (‰)	8.0	8.8	9.3	7.8
$^{87/86}Sr$	0.7099	0.7099	0.7102	0.7111
$\delta^{11}B$ (‰)	15	20	22	30
$\delta C \Sigma C$ (‰)	#1.7	−3.7	−2.9	−3.1
$\delta C\ C_1$ (‰)	−31.5	−31.7	−31.8	−37.6
$\delta C\ C_2$ (‰)	−23.7	−24.7	−24.1	−26.7
CO_2 (%)	4.7	4.5	4.7	2.7
C_2 (%)	1.4	1.3	1.2	2.3

*Concentrations are in mg/L except Br (mg/Kg). $\delta^{18}O$ is reported relative to SMOW, $\delta^{13}C$ relative to PDB, and $\delta^{11}B$ relative to NBS 951. Total Alk. is total titration alkalinity as bicarbonate. $\delta C \Sigma C$ is the carbon isotopic composition of CO_2 gas except for sample marked by # which is total dissolved carbon. $\delta C\ C_1$ is the carbon isotopic composition of methane, and $\delta C\ C_2$ is the carbon isotopic composition of ethane. %C_2 and %CO_2 are the volume percent ethane and CO_2, respectively in coproduced natural gas.

atively recent event. Based on burial history and the timing of fracture-filling ankerite, barite and sphalerite, emplacement of Mesozoic-sourced fluids at Picaroon began no earlier than 6 Ma ago.

Long distance migration is required to introduce waters from the Mesozoic into the Miocene section at Picaroon. Although not penetrated by drilling, the Mesozoic section is believed to be approximately 3.0 to 4.5 km (10,000–15,000 ft) below the Corsair fault at Picaroon (Fig. 13). The main Corsair growth fault may eventually intersect Mesozoic units far downdip to the southeast (Bradshaw et al., 1993). The sedimentary section between the Corsair fault and Mesozoic strata has not been

penetrated by drilling and is poorly imaged on seismic profiles (see Fig. 3 in Vogler and Robison, 1987), although some indications of faulting are seen. These faults, along with the Corsair fault itself, are apparently the conduits for migration of brines from the Mesozoic section into the Picaroon reservoir (Fig. 13).

Based on subsurface heat flow data from the Eugene Island area (offshore Louisiana), Anderson et al. (1991) demonstrate that despite the presence of high thermal conductivity diapiric salt, their observed heat flow anomalies require substantial advective heat transfer by active fluid flow. Fluid flow velocities of several meters/year are required, far in excess of rates estimated for sustained compactional fluid flow (Harrison and Summa, 1991). Theoretically, release of geopressures can produce episodic fluid flow along preexisting faults, effectively transferring large volumes of water and heat (Hunt, 1990; Anderson et al., 1991). The geothermal gradient at Picaroon (32–34°C/km) is similar to the anomalous areas of the Eugene Island area. Furthermore, because no diapiric salt is present at Picaroon, the elevated geothermal gradient is best explained by injection of hot fluids from deeper in the basin with faults acting as conduits. As indicated by the occurrence of anomalous brines only at Picaroon, conduits connecting deeper parts of the basin to the Corsair fault system are not uniformly distributed.

Model for Calcite Dissolution

The recognition of carbonate dissolution and allochthonous water at Picaroon field, and the absence of both features in a nearly identical setting less than ten kilometers away at Doubloon, suggest that extensive calcite dissolution in the Miocene sandstones is the result of a localized hydrodynamic and chemical process. Any viable model for porosity enhancement by calcite dissolution must include both a source of fluids capable of dissolving carbonate and a mechanism of delivering such fluids.

As previously recognized (Hutcheon and Abercrombie, 1990; Land and Macpherson, 1992), silicate hydrolysis or "reverse weathering" reactions are potentially important sources of acid during burial metamorphism of sediments. In these types of reactions, two examples of which are shown below, cations

TABLE 5.—COMPARISON OF FORMATION WATERS FROM MESOZOIC RESERVOIRS OF SOUTH TEXAS AND CENTRAL MISSISSIPPI WITH PRODUCED WATERS FROM PICAROON FIELD

Well	1*	2*	3*	4#	5#	6#	Picaroon
Cl	189000	153300	99461	103200	141700	181900	147420
SO4	125	191	52	NA	NA	NA	NA
H4SiO4	NA	212	159	NA	NA	NA	158
Br	858	910	579	950	2090	1210	183
Total Alk.	350	330	247	NA	NA	NA	210
Na	77500	69670	45564	47300	46500	70500	66900
K	5600	4640	2480	595	6800	960	1980
Ca	28600	21000	13425	13200	30600	34300	21200
Mg	800	1810	1193	1150	2970	2200	720
Li	322	255	168	9	NA	30	225
Sr	4980	2030	2462	422	1300	2450	2320
Ba	1120	26	50	20	66	143	1690
Fe	669	NA	1.6	278	98	468	438
Mn	209	11	NA	6	8	NA	65
Pb	NA	NA	NA	10	2	57	6
Zn	679	NA	0.5	575	5	251	194
B	296	285	250	NA	NA	NA	183
TDS	311108	254713	166107	167715	232139	294469	243903
$\delta^{18}O_{SMOW}$	3.4	10.2	10.6	NA	NA	NA	8.0
$\delta^{11}B$	13.4	19.4	18.3	NA	NA	NA	15
$^{87/86}Sr$	0.7090	0.7100	0.7089	NA	NA	NA	0.7099

Well Names: [1] General Crude 1 C Strawn, Karnes Co. Texas; [2] Shell 1 C Wishert Person, Karnes Co. Texas; [3] Gulf 9 E Tartt et al. Fashing, Ataskosa Co. Texas; [4] Glen W. Brogden 1-L, Leflore Co. Mississippi; [5] Lee Pierce, Jasper Co. Mississippi; [6] Central Oil 5–7, Smith Co. Mississippi.
*Data from Land and Macpherson, 1992;
#Data from Carpenter et al., 1974.

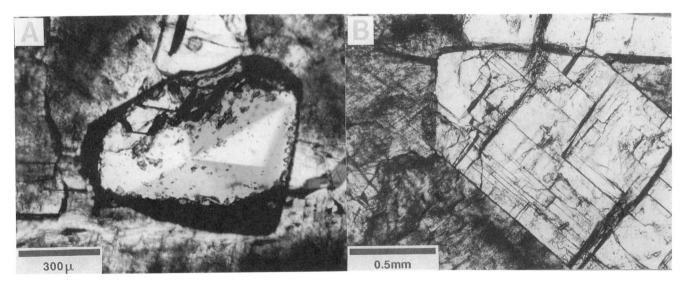

FIG. 9.—Thin-section photomicrographs of fracture-filling minerals from Picaroon sandstones and mudrocks. (A) Sphalerite (ZnS) occurs within veins and is surrounded by fracture-filling ankerite. (B) Coarse euhedral crystals of barite (BaSO$_4$) surrounded by ankerite.

are fixed in mineral phases, and protons are released into solution.

$$3Al_2Si_2O_5(OH)_4 + 7Mg^{2+} + 9H_2O$$

$$\rightarrow 2Mg_{3.5}Al_3Si_3O_{10}(OH)_8 + 14H^+ \quad (1)$$

$$Al_2Si_2O_5(OH)_4 + 4H_4SiO_4 + 2Na^+$$

$$\rightarrow 2NaAlSi_3O_8 + 9H_2O + 2H^+ \quad (2)$$

Although hypothetical, reactions 1 and 2 represent the formation of stable phases observed in Gulf of Mexico sediments during deep burial diagenesis and the accompanying decrease

in the cation/H$^+$ ratio which must occur. In addition to the well documented transformation of illite/smectite to illite, mudrocks experience a decrease in kaolinite and an increase in chlorite (reaction 1) with progressive burial diagenesis (Hower et al., 1976; Boles and Franks, 1979; Awwiller, 1993). Over a roughly similar depth/temperature range, significant amounts of authigenic albite form in mudrocks (Milliken, 1992) and in sandstones (Boles, 1982; Milliken et al., 1989) at the expense of detrital feldspars.

Reverse weathering reactions apparently are predominant in deeply buried mudrocks within the evaporite-rich Mesozoic section of the Gulf Coast. Where salt is present and Cl is essentially the only anion in solution, the acid produced from

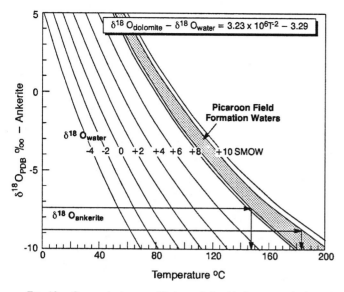

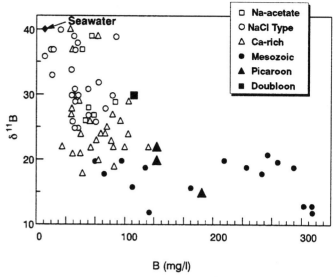

FIG. 10.—Oxygen isotope equilibrium relationship between ankerite and water as a function of temperature (dolomite equation of Sheppard and Schwarcz, 1970). Measured $\delta^{18}O$ for ankerite cement and Picaroon formation waters would be in equilibrium between 147° and 184°C, within the 120° to 188°C range indicated by fluid inclusions in ankerite.

FIG. 12.—$\delta^{11}B$ (NBS 951) versus B concentration for three major water types found in Cenozoic reservoirs of the Gulf Coast and for brines from Mesozoic (Cretaceous) reservoirs of south Texas (after Land and Macpherson, 1992). Picaroon brines have relatively high B concentrations and depleted $\delta^{11}B$ similar to Mesozoic sourced brines.

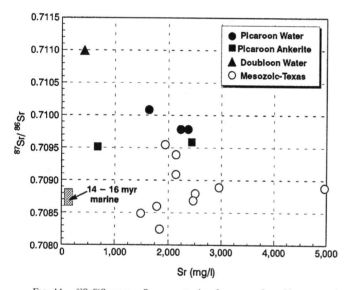

FIG. 11.—$^{87}Sr/^{86}Sr$ versus Sr concentration for waters from Picaroon and Doubloon fields and for brines from Mesozoic (Cretaceous) reservoirs of south Texas containing more than 100,000 mg Cl/l (Land and Macpherson, 1992). Also plotted are two analyses of ankerite cement from Picaroon core.

reverse weathering reactions is HCl. Ca-bearing minerals (calcite or anorthite) react in the presence of HCl, leading to the formation of Na-Ca-Cl CO_2-rich brines as a natural consequence of rock/water interactions during burial metamorphism (Land and Macpherson, 1992).

The a_{H+} in brines residing in the Mesozoic section is presumably controlled by water/rock reactions involving silicates (reverse weathering reactions). With increasing temperature, metastable equilibrium between fluids and silicate minerals buf-

fers the pH to progressively lower values (Hutcheon and Abercrombie, 1990; Hutcheon et al., 1993). As a model for porosity enhancement at Picaroon, we propose that hot acidic fluids from the underlying Mesozoic section are episodically injected along faults into Miocene sands (Fig. 13). These fluids react with the lower temperature reservoir sands resulting in dissolution of calcite and, to a lesser degree, feldspars. The amount of calcite dissolved depends on the volume and duration of fluid flow. Volumetrically significant calcite dissolution would be favored by repeated episodes of fluid injection triggered by periodic release of geopressures (Sibson, 1987; Hunt, 1990; Anderson et al., 1991; Wood and Boles, 1991).

The mass transfer required to remove as much as 6–15% calcite from sandstone reservoirs at Picaroon may to a first approximation, appear unreasonably large. However, in the 300 m of Miocene sediments that include the Picaroon reservoirs, a maximum of approximately 25% is sand; the remainder composed of mudrock. Even within the sands, dissolution is not homogeneously distributed; the effects being most apparent in thick (~10–35 m), clay-poor sands (aquifers) that comprise approximately one-half of the total sand present.

At Doubloon and Plank where calcite dissolution did not occur, the distribution of porous and permeable sand is nonuniform due to both depositional factors and to cementation. A similar nonuniform permeability distribution certainly existed at Picaroon prior to dissolution. If we assume, by analogy to the Doubloon and Plank sands, that the thickest (>10 m) sands at Picaroon had substantially higher porosity (15–18%) and permeability (10–100 md) than the thinner, highly cemented and/or clay-rich sand units (porosity ~2–15%; permeability ~0.02 to 10 md), then the bulk of the fluid flowing into the system must pass through the thickest, most permeable sands. The amount of calcite dissolution a given sand unit would experience is proportional to fluid flux, setting up a positive feed-

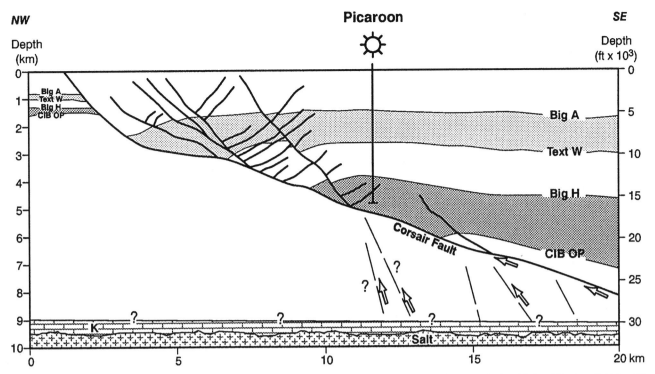

FIG. 13.—Schematic NW-SE cross section depicting large-scale structural features and regional foraminiferal zones in the vicinity of Picaroon field (modified after Vogler and Robison, 1987). Although not penetrated by drilling, Mesozoic carbonates (K) and the Louann Salt are thought to occur approximately 3–4 km beneath the Corsair Fault at Picaroon field. Emplacement of brines from Mesozoic strata into Miocene sands occurred presumably with the main Corsair Fault and other faults serving as conduits.

back mechanism whereby dissolution is concentrated in those specific units (Chen and Ortoleva, 1990). With continued flow, an increasing proportion of fluid would be attracted to the relatively porous and permeable sands, bypassing the low porosity/permeability units. Eventually, essentially all fluid flow would occur in the more porous/permeable sands (Chen and Ortoleva, 1990). Using the lithologic proportions described above for a 300-m section at Picaroon, simple mass balance calculations indicate that removal of 6–15% calcite from the thick sand units (12.5% of the total section) at Picaroon requires mobilization of only 1.5–2% total bulk volume calcite.

CONCLUSIONS

The observed differences in water chemistry and sandstone diagenesis between two fields with nearly identical depositional and burial histories demonstrates the importance of focused fluid flow in clastic diagenesis. The empirical association of allochthonous brines from underlying, presumably Mesozoic units, with calcite dissolution at Picaroon implies both a chemical and hydrodynamic mechanism for generation of volumetrically significant secondary porosity. Within deeply buried Mesozoic strata, rock/water reactions produce highly saline Na-Ca-Cl brines. Among these reactions, reverse weathering of silicates provides an important source of protons, significantly lowering the equilibrium pH of the waters. As a model for Picaroon field, we propose that major growth faults provide conduits for emplacement of these hot, acidic waters into Miocene sands, leading to reservoir enhancement by calcite dissolution.

ACKNOWLEDGMENTS

The authors gratefully acknowledge Shell Development Company for permission to publish this paper. We thank colleges A. D. Allie and G. L. Macpherson for help in obtaining water samples and N. R. Braunsdorf and T. N. Diggs for discussions and comments from which this paper benefited.

REFERENCES

ANDERSON, R. N., HE, W., HOBART, M. A., WILKINSON, C. R., AND NELSON, H. R., 1991, Active fluid flow in the Eugene Island area, offshore Louisiana: Geophysics: The Leading Edge of Exploration, v.10, p. 12–17.

AWWILLER, D. N., 1993, Illite/smectite formation and potassium mass transfer during burial diagenesis of mudrocks: a study from the Texas Gulf Coast Paleocene-Eocene: Journal of Sedimentary Petrology, v. 63, p. 501–512.

BOLES, J. R., 1982, Active albitization of plagioclase Gulf Coast Tertiary: American Journal of Science, v. 282, p. 165–180.

BOLES, J. R. AND FRANKS, S. G., 1979, Clay diagenesis in Wilcox sandstones of southwest Texas: implications of smectite diagenesis on sandstone cementation: Journal of Sedimentary Petrology, v. 49, p. 55–70.

BRADSHAW, B. E., IBRAHIM, A. B., AND WATKINS, J. S., 1993, Structural and stratigraphic evolution of the Corsair-Wanda fault system: Brazos OCS area, Gulf of Mexico: American Association of Petroleum Geologists Program with Abstracts, p. 79.

BURST, J. F., 1969, Diagenesis of Gulf Coast clayey sediments and its possible relation to petroleum migration: American Association of Petroleum Geologists Bulletin, v. 53, p. 73–93.

CARPENTER, A. B., TROUT, M. L., AND PICKETT, E. E., 1974, Preliminary report on the origin and chemical evolution of lead- and zinc-rich oil field brines in central Mississippi: Economic Geology, v. 69, p. 1191–1206.

CHEN, W. AND ORTOLEVA, P., 1990, Reaction front fingering in carbonate cemented sandstones, in Ortoleva, P., Hallet, B., McBirney, A., Meshri, I., Reeder, R., and Williams, P., eds., Self-Organization in Geological Systems:

Proceedings of a Workshop held 26–30 June 1988, University of California Santa Barbara: Earth Science Reviews, v. 29, p. 183–198.

DIGGS, T. N. AND LAND, L. S., 1993, Geochemical constraints on the origins and timing of emplacement of allochthonous formation waters, Frio Formation (Oligocene), South Texas: American Association of Petroleum Geologists Program with Abstracts, p. 92.

FISHER, R. S. AND LAND, L. S., 1986, Diagenetic history of Eocene Wilcox sandstones, south-central Texas: Geochimica et Cosmochimica Acta, v. 50, p. 551–561.

FOLK, R. F., 1974, Petrology of Sedimentary Rocks: Austin, Hemphill Publishing, 182 p.

FREED, R. L. AND PEACOR, D. R., 1989, Variability in temperature of the smectite/illite reaction in Gulf Coast sediments: Clay Minerals, v. 24, p. 171–180.

FREED, R. L. AND PEACOR, D. R., 1992, Diagenesis and the formation of authigenic illite-rich I/S crystals in Gulf Coast shales: TEM study of clay separates: Journal of Sedimentary Petrology, v. 62, p. 220–234.

HARRISON, W. J. AND SUMMA, L. L., 1991, Paleohydrology of the Gulf of Mexico basin: American Journal of Science, v. 291, p. 109–176.

HOWER, J., ESLINGER, E., HOWER, M. E., AND PERRY, E. A., 1976, Mechanism of burial metamorphism of argillaceous sediments: 1. Mineralogical and chemical evidence: Geological Society of America Bulletin, v. 87, p. 725–737.

HUNT, J. M., 1990, Generation and migration of petroleum from abnormally pressured fluid compartments: American Association of Petroleum Geologists Bulletin, v. 74, p. 1–12.

HUTCHEON, I., SHEVALIER, M., AND ABERCROMBIE, H. J., 1993, pH buffering by metastable mineral-fluid equilibria and evolution of carbon dioxide fugacity during burial diagenesis: Geochimica et Cosmochimica Acta, v. 57, p. 1017–1027.

HUTCHEON, I. AND ABERCROMBIE, H., 1990, Carbon dioxide in clastic rocks and silicate hydrolysis: Geology, v. 18, p. 541–544.

KHARAKA, Y. K., MAEST, A. S., CAROTHERS, W. W., LAW, L. M., LAMOTHE, P. J., AND FRIES, T. L., 1987, Geochemistry of metal-rich brines from central Mississippi Salt Dome basin, U.S.A.: Applied Geochemistry, v. 2, p. 543–562.

LAND, L. S., 1995, Na-Ca-Cl saline formation waters, Frio Formation (Oligocene), south Texas, USA: Products of diagenesis: Geochimica et Cosmochimica Acta, v. 59, p. 2163–2174.

LAND, L. S., EUSTICE, R. A., AND MACK, L. E., 1994, Reactivity of evaporites during burial: an example from the Jurassic of Alabama: Geological Society of America Abstracts with Programs, p. A220.

LAND, L. S. AND MACPHERSON, G. L., 1992, Origin of saline formation waters, Cenozoic section, Gulf of Mexico sedimentary basin: American Association of Petroleum Geologists Bulletin, v. 76, p. 1344–1362.

LAND, L. S., MACPHERSON, G. L., AND MACK, L. E., 1988, The geochemistry of saline formation waters, Miocene, offshore Louisiana: Transactions—Gulf Coast Association of Geological Societies, v. 38, p. 503–511.

LAND, L. S. AND PREZBINDOWSKI, D. R., 1981, The origin and evolution of saline formation water, Lower Cretaceous carbonates, south-central Texas, U.S.A: Journal of Hydrology, v. 54, p. 51–74.

MACK, L. E., 1990, Sr as a tracer of diagenesis in Cenozoic sediments of the Northern Gulf of Mexico sedimentary basin: Unpublished Ph.D. Dissertation, University of Texas at Austin, Austin, 198 p.

MACPHERSON, G. L., 1992, Regional variations in formation water chemistry: major and minor elements, Frio Formation fluids, Texas: American Association of Petroleum Geologists Bulletin, v. 76, p. 740–757.

MILLIKEN, K. L., 1992, Chemical behavior of detrital feldspars in mudrocks versus sandstones, Frio Formation (Oligocene), South Texas: Journal of Sedimentary Petrology, v. 62, p. 790–801.

MILLIKEN, K. L., MCBRIDE, E. F., AND LAND, L. S., 1989, Numerical assessment of dissolution versus replacement in the subsurface destruction of detrital feldspars, Oligocene Frio Formation, South Texas: Journal of Sedimentary Petrology, v. 59, p. 740–757.

MILLIKEN, K. L., LAND, L. S., AND LOUCKS, R. G., 1981, History of burial diagenesis determined from isotopic geochemistry, Frio Formation, Brazoria County, Texas: American Association of Petroleum Geologists Bulletin, v. 65, p. 1397–1413.

MOLDOVANYI, E. P. AND WALTER, L. M., 1992, Regional trends in water chemistry, Smackover Formation, Southwest Arkansas: geochemical and physical controls: American Association of Petroleum Geologists Bulletin, v. 76, p. 864–894.

MORTON, R. A. AND LAND, L. S., 1987, Regional variations in formation water chemistry, Frio Formation (Oligocene), Texas Gulf Coast: American Association of Petroleum Geologists Bulletin, v. 71, p. 191–206.

SHARP, J. M. J., GALLOWAY, W. E., LAND, L. S., MCBRIDE, E. F., BLANCHARD, P. E., BODNAR, D. P., DUTTON, S. P., FARR, M. R., GOLD, P. B., JACKSON, T. J., LUNDEGARD, P. D., MACPHERSON, G. L., AND MILLIKEN, K. L., 1988, Diagenetic processes in northwestern Gulf of Mexico sediments, in Chillingarian, G. V. and Wolf, K. H., eds., Diagenesis II: New York, Elsevier, p. 43–113.

SHEPPARD, S. M. F., AND SCHWARCZ, H. P., 1970, Fractionation of carbon and oxygen isotopes and magnesium between co-existing metamorphic calcite and dolomite: Contributions to Mineralogy and Petrology, v. 26, p. 161–198.

SIBSON, R. H., 1987, Earthquake rupturing as a mineralizing agent in hydrothermal systems: Geology, v. 15, p. 701–704.

SUCHECKI, R. K. AND LAND, L. S., 1983, Isotopic geochemistry of burial-metamorphosed volcanogenic sediments, Great Valley sequence, northern California: Geochimica et Cosmochimica Acta, v. 47, p. 1487–1499.

TAYLOR, T. R., 1990, The influence of calcite dissolution on reservoir porosity in Miocene sandstones, Picaroon Field, offshore Texas Gulf Coast: Journal of Sedimentary Petrology, v. 60, p. 322–334.

VOGLER, H. A. AND ROBISON, B. A., 1987, Exploration for deep geopressured gas: Corsair trend, offshore Texas: American Association of Petroleum Geologists Bulletin, v. 71, p. 777–787.

WOOD, J. R., AND BOLES, J. R., 1991, Evidence for episodic cementation and diagenetic recording of seismic pumping events, North Coles Levee, California, U.S.A: Applied Geochemistry, v. 6, p. 509–521.

CATION GEOTHERMOMETRY AND THE EFFECT OF ORGANIC-INORGANIC DIAGENETIC REACTIONS

LETA K. SMITH, THOMAS L. DUNN, AND RONALD C. SURDAM

Institute for Energy Research, P.O. Box 4068, University of Wyoming, Laramie, Wyoming 82071

ABSTRACT: Predicted cation ratio geothermometry temperatures, using equations of Na-K, Na-K-Ca, Mg-Na-K-Ca and Mg-Li, were compared between oilfield and geothermal settings. Geothermometers in oilfield waters yielded less consistent temperature predictions compared to geothermal waters in the same temperature range. Scatter of predicted temperature in oilfield waters is greatest in the temperature interval where carboxylic acid anions (CAAs) are in greatest concentration. CAAs are not present in geothermal systems. Temperature prediction improves in those oilfield waters where CAAs are present and account for less than 80% of total alkalinity.

The assumptions of cation ratio geothermometry are violated to varying degrees in oilfield waters where CAAs are abundant. These assumptions are: (1) cation ratios are controlled by exchange between solid aluminosilicates. However, CAAs affect mineral solubility by forming complexes with the cations. Therefore, the ratios of cations in solution differ from those values expected when cation exchange between aluminosilicate minerals is the only control on the cation ratios. Furthermore, concentrations of Ca and Mg are strongly controlled by carbonate equilibria, which in turn is strongly affected by the presence of CAAs; (2) aluminum is conserved in solid phases. However, CAAs form stable complexes with Al, increasing Al-silicate solubility and mobilizing Al; thus Al may not be conserved in mineral phases; (3) neither H^+ nor CO_2 enter into the net reactions (i.e., pH is buffered by aluminosilicate hydrolysis). However, acetate (the dominant CAA found in oilfield waters) is an effective buffer of pH in feldspathic rocks. Also, at higher temperatures, decarboxylation of CAAs increases the P_{CO2} of oilfield waters.

The consistently worse temperature prediction of cation ratio geothermometers in oilfield waters in the 80–120°C temperature range is another indication that organic-inorganic diagenesis is an important control on oilfield water chemistry.

INTRODUCTION

Purpose and Scope

This study is part of a continuing effort to evaluate the relative importance of carboxylic acid anions (CAAs) in diagenesis, an importance that has been debated since Willey et al. (1975), Curtis (1978) and Carothers and Kharaka (1978) reported on the presence and implications of CAAs in oilfield water. This analysis was undertaken following our observation that cation ratio geothermometers do not work particularly well in oilfield waters—even those empirical equations that were developed using oilfield waters (Fournier and Truesdell, 1973; Fournier and Potter, 1979; Kharaka and Mariner, 1989). Interestingly, the evolution of the field of cation ratio geothermometry and the development of new equations has largely been to accommodate oilfield waters that give anomalous results relative to the geothermal systems.

The importance of CAAs is evaluated here by comparing geothermometry applied to geothermal and to oilfield systems. This study demonstrates that cation ratio geothermometers do not work in most oilfield waters, and we conclude that it is because the assumptions upon which the geothermometers are based are invalid in oilfield waters, in part because of the presence of organic acid anions.

Cation Ratio Geothermometer Equations

The rationale behind cation ratio geothermometry and the methodology of equation development are explained well by Fournier and Truesdell (1973), Fournier et al. (1974) and Kharaka and Mariner (1989). Fournier (1981) gives an excellent review of the application of geothermometry to geothermal resources. More recently, Land and Macpherson (1992a) and Kharaka and Mariner (1989) discuss the application of various types of cation ratio geothermometers to oilfields.

In theory, any temperature-dependent exchange reaction can be used to develop a cation ratio geothermometer equation. However, exchange reactions among sodium-, potassium-, calcium-, lithium- or magnesium-bearing aluminosilicates are most commonly used (Fournier and Truesdell, 1973; Truesdell,

1976; Fournier, 1981; Arnorrson et al., 1983; Kharaka and Mariner, 1989).

For a simple cation exchange reaction such as

$$K^+ + \text{Na-feldspar} = \text{K-feldspar} + Na^+, \qquad (1)$$

the expression for the equilibrium constant is

$$K_{eq} = \frac{[\text{K-feldspar}][Na^+]}{[\text{Na-feldspar}][K^+]}, \qquad (2)$$

with the brackets indicating activities of the species. This equilibrium constant is temperature-dependent as given by the van't Hoff equation,

$$\frac{d \log K_{eq}}{d(1/T)} = -\frac{\Delta H^\circ_{(T)}}{4.5758}, \qquad (3)$$

where T is absolute temperature and $\Delta H^\circ_{(T)}$ is the standard heat of reaction at the given temperature. Assuming that the standard heat of reaction is independent of temperature, the right-hand side of equation 3 becomes a constant. A simplified equation can be written which relates the cation ratio to the temperature by integrating equation 3, substituting in the expression for the equilibrium constant from equation 2, and assuming unit activity for the solid species in equation 2:

$$\log\left(\frac{[Na^+]}{[K^+]}\right) a + b\left(\frac{1}{T}\right). \qquad (4)$$

The constants a and b can be determined empirically by plotting $1/T$ vs. $\log([Na^+]/[K^+])$.

If the exchange reaction, the equilibrium constant, and the standard heat of reaction are all known, theoretical curves in the form of equation 4 can be derived without the necessity of empirically determining a and b. These theoretical curves are used primarily to assess the reactions controlling the equilibrium and not necessarily for determining temperature. In practice, the equilibrium constant and the standard heat of reaction for a particular exchange reaction may not be known. Even the controlling exchange reactions may not be well understood. For example, Kharaka and Mariner (1989) developed the magnesium-lithium geothermometer equation, which works as well or better than previously published cation ratio geothermometers, without specifying the solid reaction phases.

Therefore, cation ratio geothermometer equations are typically empirically-derived, as described above, by plotting a cation ratio from a water analysis against the temperature of the water in the subsurface (which was determined by some other means).

Four types of cation ratio geothermometers, all of which were empirically-derived, were included in this study: Na-K, Na-K-Ca, Mg-Li and Mg-corrected Na-K-Ca. Nine equations were examined which are listed in Table 1 along with their reference.

Assumptions Of Cation Ratio Geothermometers

The assumptions upon which the geothermometers are based must be examined to understand where they are applicable. Fournier and Truesdell (1973) discussed these assumptions: (1) heat of reaction is independent of temperature; (2) activity coefficients for the aqueous species cancel from the net equilibrium constant expression; (3) cation ratios are controlled by cation exchange between pure solid phases; (4) excess silica is present; (5) aluminum is conserved in solid phases; and (6) hydrogen ions involved in hydrolysis cancel out of the net chemical reaction equations. A discussion of why some of these assumptions, in particular 3, 5 and 6, do not apply to oilfield waters is given after a presentation of the comparative results.

Data Set Description

The data set consists of 334 published water analyses. Table 2 lists the sources of these data. Two hundred fifty-seven of the analyses are of oilfield water samples, of which 127 include analyses for organic acid anions. Seventy-seven analyses are from geothermal water wells. All of the samples had downhole temperature measured by another means; in other words, samples whose temperature had been determined by another type of geothermometer were not included. Because published data were used, the accuracy of the reported temperatures could not be evaluated. However, most important is that the error in the temperature determination method not be different between the geothermal and oilfield datasets. Temperature in both geothermal wells and oilfield wells is commonly determined from extrapolation of bottomhole temperatures that are measured between successive mudlogging runs during drilling. Fertl and Overton (1982) note in their discussion of geothermal well logging that the oil industry well logging procedures are a "mature industry" compared to the more recently developed geothermal well logging industry. Therefore, the accuracy of measured temperature in oilfield waters is likely as good as or better than in geothermal waters. Also, Land and Macpherson (1992a) conclude, in their geothermometry study of Gulf Coast brines, that faulty temperature data are not a problem.

Within the limits of published data, the data were chosen to represent a variety of geologic and geographic settings. The oilfield water analyses represent fourteen oil-producing sections, Jurassic through Pleistocene, from eight sedimentary basins (Table 2). The geothermal water analyses represent eleven geothermal regions (Table 2). Also, water analyses from geothermal wells were chosen to include as many low-temperature

(<250°C) samples as could be found in the literature. It is important to note that none of the geothermal water analyses used in this study included analyses for CAAs, nor could any be found. An important assumption of this study is that CAAs do not dominate the alkalinity of geothermal waters. Carboxylic acid anions are known to originate from the maturation of petroleum source rocks (Vandenbroucke, 1980; Rouxhet et al., 1980; Vandergrift et al., 1980; Surdam et al., 1984; Crossey, 1985; Kawamura et al., 1986; Kawamura and Kaplan, 1987; Surdam and MacGowan, 1987; Lundegard and Senftle, 1987; MacGowan et al., 1988; Kharaka et al., 1989; Lewan and Fisher, 1994). Therefore, the general absence of petroleum source rocks in geothermal terrains indicates that CAAs would not be present in these waters, and thus the assumption is valid.

Water analyses originally reported in molality were converted to parts-per-million (ppm) before the calculation of temperature since the original equations were developed for concentration expressed in ppm. For analyses reported in mg/l, it was assumed that the specific gravity of the water was equal to one (and therefore that mg/l equals ppm) for purposes of the temperature calculations. To test the validity of this assumption, temperatures were calculated on three highly saline brines using both units for the Na-K-Ca equation (Fournier and Truesdell, 1973). The results are on Table 3 which shows that calculated temperatures differed by less than 2%, even for the most dense sample.

RESULTS

Measured vs. Predicted Temperatures for Oilfield and Geothermal Waters

A temperature was calculated for each of the 334 water analyses using each equation given on Table 1. These are referred to as the predicted temperatures, and they are referenced to the type of equation used (i.e., the Na-K predicted temperature or the Na-K-Ca predicted temperature).

Plots were made of measured temperature (Tm) vs. predicted temperature (Tp) for each of the geothermometer equations to confirm our initial observation that cation ratio geothermometer equations give consistently less reliable temperature predictions in oilfield waters than in geothermal waters. One example from each of the four types of cation ratio geothermometer equations is presented in Figure 1 with geothermal data plotted separately from oilfield data. The 1:1 line is shown on Figure 1 to provide visual evidence of the accuracy of the geothermometer equations. Lack of *accuracy* of the temperature prediction is indicated by how far a point lies away from the 1:1 line. The scatter in the geothermal data sets can be visually compared with the scatter in the oilfield data set.

The measure of *precision* of the temperature predicted by a geothermometer equation is given by the correlation coefficient for each plot. A value near one indicates a strong relationship between Tm and Tp, whereas a value nearer zero would indicate that there is little correlation between the two. It is important to note that the correlation coefficient is unrelated to the 1:1 line shown on Figure 1. The correlation coefficient indicates how well Tp correlates with Tm, yet it is possible for the two variables to correlate well without plotting along the 1:1 line. Correlation coefficients are relative and can be compared within a given data set. However, the strength of comparison can break

TABLE 1.—CATION RATIO GEOTHERMOMETER EQUATIONS USED IN THIS STUDY

Type	Geothermometer Equation	Reference
1. Na-K	$T°C = \dfrac{1217}{\log(Na/K) + 1.483} - 273.15$	Fournier, 1981
2. Na-K	$T°C = \dfrac{855.6}{\log(Na/K) + 0.857} - 273.15$	Truesdell, 1976
3. Na-K	$T°C = \dfrac{933}{\log(Na/K) + 0.933} - 273.15$	Arnorsson, 1983
4. Na-K	$T°C = \dfrac{1180}{\log(Na/K) + 1.31} - 273.15$	Kharaka & Mariner, 1989
5. Na-K-Ca*	$T°C = \dfrac{1647}{\log(Na/K) + \beta[\log(\sqrt{Ca}/Na) + 2.06] + 2.47} - 273.15$	Fournier & Truesdell, 1973
6. Na-K-Ca*	$T°C = \dfrac{699}{\log(Na/K) + \beta[\log(\sqrt{Ca}/Na) + 2.06] + 0.489} - 273.15$	Kharaka & Mariner, 1989
7. Na-K-Ca (no Mg avail)	$T°C = \dfrac{1120}{\log(Na/K) + 1/3[\log(\sqrt{Ca}/Na) + 2.06] + 1.32} - 273.15$	Kharaka & Mariner, 1989
8. Mg-Li	$T°C = \dfrac{2200}{\log(\sqrt{Mg}/Li) + 5.47} - 273.15$	Kharaka & Mariner, 1989
9. Mg Correction*†	$T_{MgCorr} = T_{NaKCa} - \Delta T_{Mg}$	Fournier & Potter, 1979

$R = \dfrac{([Mg] \times 100)}{[Mg] + (0.61 \times [Ca]) + (0.31 \times [K])}$ for $5 < R < 50$: $\Delta T_{Mg} = 10.7 - 4.74R + 326(\log R)^2 - 1.03 \times 10^5(\log R)^2/T$

$\qquad\qquad\qquad\qquad - 1.97 \times 10^7(\log R)^2/T^2 + 1.61 \times 10^7(\log R)^3/T^2$

$\qquad\qquad$ for $.5 < R < 5$: $\Delta T_{Mg} = 1.03 + 60\log R + 145(\log R)^2$

$\qquad\qquad\qquad\qquad - 3.67 \times 10^5(\log R)^2/T - 1.67 \times 10^7\log R/T^2$

*see original reference for use of this equation.
†numbers here have been rounded off.

TABLE 2.—SOURCES OF WATER ANALYSES USED IN THIS STUDY

Reference: Region(s)

Arnorsson, 1975: Iceland
Carothers and Kharaka, 1978: Gulf Coast, TX; Kettleman Dome, CA
Carpenter et al., 1974: central Mississippi
Fisher and Boles, 1990: San Joaquin Basin, CA
Fournier and Truesdell, 1973: Iceland; Kettleman Dome, CA; Steamboat Spr., NV; Yellowstone, WY
Kharaka and Berry, 1976: Kettleman Dome, CA
Kharaka and Mariner, 1989: North Slope, AK; Gulf Coast, LA; Gulf Coast, TX; San Joaquin Basin, CA; Paris Basin, France; geothermal areas of CA, UT, ID, NV, OR, WY, Japan, Chile, Mexico, New Zealand, Turkey
Kharaka et al., 1977a,b: Gulf Coast, TX
Land and Macpherson, 1989: Gulf Coast, LA
Land and Macpherson, 1992b: Gulf Coast, LA & TX
Meshri and Walker, 1990: Washakie Basin, WY
Surdam and Crossey, 1985: San Joaquin Basin, CA
White et al., 1963: Gulf Coast, LA

TABLE 3.—COMPARISON OF PREDICTED TEMPERATURES FOR THREE WATER SAMPLES CALCULATED USING PPM AND USING MG/L

	Sample One		Sample Two		Sample Three	
	ppm	mg/l	ppm	mg/l	ppm	mg/l
Na	63,900	55,517	51,500	42,422	22,500	17,415
K	869	755	2,460	2,026	9,120	7,059
Ca	9,210	8,002	32,800	27,018	74,800	57,895
TDS	200,000		256,000		331,000	
Density	1.151		1.214		1.292	
Na-K-Ca (°C)*	130.0	129.0	177.0	174.1	292.3	288.9
difference (ppm—mg/l)		1.0°C		1.9°C		3.4°C
percent difference		0.7%		1.1%		1.2%

*Temperature calculated using Equation 5, Table 1

down when applied between data sets (Mann, 1987). Hence, r values of a given geothermometer for oilfield vs. geothermal data sets are not explicitly compared herein.

Na-K Equations.—

Measured vs. predicted temperatures for the Na-K equation (Eq. 2 on Table 1) of Truesdell (1976) are shown in Figures 1A-C. This equation is a fairly reliable predictor of temperature for the geothermal samples (Fig. 1A). The geothermometry equation is fairly accurate because the data tend to cluster around the 1:1 line. This is for the entire geothermal data set, including samples with temperatures up to 400°C.

Temperature prediction by this Na-K geothermometer equation for the oilfield data set (Fig. 1B) is much less consistent. The amount of scatter in the data is high. However, when comparing geothermometers, the two datasets must be from the same measured temperature interval. Therefore, Figure 1C shows a subset of the geothermal data with only the data below 250°C measured temperature — the same temperature range of the oilfield samples. Even when comparing the same temperature interval, this Na-K geothermometer is much less reliable

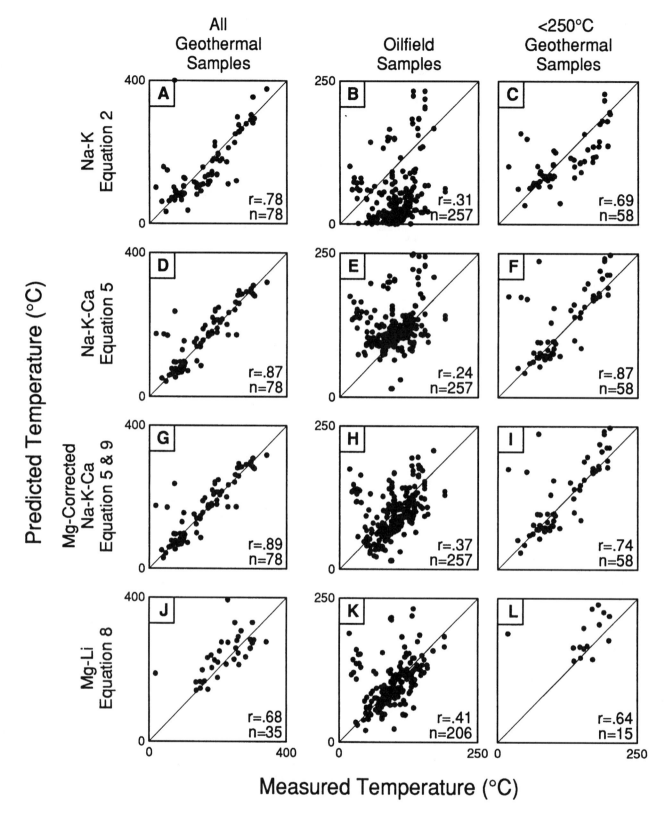

FIG. 1.—Plots of measured vs. geothermometry-predicted temperature. Each row of plots uses a different geothermometry equation, identified on the far left by name and by the equation number on Table 1. Each column represents a different data set, from left to right: all geothermal well samples (0–400°C), oilfield well samples (0–250°C), and a subset of the geothermal samples in a lower temperature range (0–250°C). The solid line is the 1:1 line and is shown for a visual representation of the accuracy of geothermometry prediction. The correlation coefficient for each plot is in the lower-right corner of each plot. Note that the scale for the plots using all geothermal samples is 0–400°C, whereas the scale for the plots using low temperature geothermal waters and oilfield waters is 0–250°C. A few of the predicted temperatures were negative and thus do no appear on the plots.

in the oilfield waters. The amount of scatter in the oilfield waters is more than in the low-temperature geothermal waters.

The other Na-K equations give similar results. In each case, there is more scatter in the measured vs. predicted data for the oilfield samples than for the low-temperature geothermal samples. Even the empirically derived Na-K equation of Kharaka and Mariner (1989, Eq. 4, Table 1), which included oilfield data in the empirical derivation of the equation, has more scatter in the oilfield dataset.

Na-K-Ca Equations.—

The Na-K-Ca geothermometer equation yields results that are similar to the Na-K equations. This equation was originally developed by Fournier and Truesdell (1973) to correct for the spurious results of the Na-K geothermometers in high-Ca waters, some of which were from oilfields. For the geothermal waters, Fournier and Truesdell's (1973) equation gives better temperature predictions than the simple Na-K equations. Both the entire geothermal dataset (Fig. 1D) and the low-temperature geothermal dataset (Fig. 1F) have relatively small scatter in the measured vs. predicted temperature relationship. However, when applied to the oilfield data set (Fig. 1E), this equation yields worse temperature predictions than do those of the simple Na-K equations.

The other two Na-K-Ca equations (Eqs. 6, 7, Table 1), not presented in Figure 1, give results similar to those of the Fournier and Truesdell (1973) equation. Clearly, there is considerably more scatter in the oilfield data than in the geothermal data.

Mg-Correction for the Na-K-Ca Equation.—

The magnesium correction (Fournier and Potter, 1979) applied to the Na-K-Ca equations was meant to improve the results for low-temperature Mg-rich waters, most of which are oilfield waters. The plots of measured vs. predicted temperature with the Mg correction applied to Fournier and Truesdell's (1973) equation are given in Figure 1G-I. For both the oilfield waters (Fig. 1H) and the low-temperature geothermal waters (Fig. 1I), the Mg correction does improve temperature prediction. However, the Mg correction, which was developed primarily for oilfield waters, clearly gives a poorer fit for the oilfield waters than for the low-temperature geothermal waters. The Mg correction applied to Kharaka and Mariner's (1989) equation, not shown on Figure 1, also gives similar results.

Mg-Li Equation.—

The Mg-Li geothermometer of Kharaka and Mariner (1989) was also developed primarily for oilfield waters. The plots of measured vs. predicted temperature using this equation (Eq. 8, Table 1) on the two datasets are shown as Figures 1J-L. This equation does predict temperature in the oilfield waters better than the other cation ratio geothermometers, as shown previously by Land and Macpherson (1992a). However, there is more scatter in the oilfield data than in the low-temperature geothermal data, even though there is considerably less lithium data in the low-temperature geothermal data set.

Summary of Measured vs. Predicted Plots.—

Regardless of which type of cation ratio geothermometer equation is used to predict temperature, the results are similar. The fit between measured temperature and predicted tempera-

ture is always better in the geothermal water dataset than in the oilfield water dataset. Having established the consistent discrepancy in the reliability of predicted temperature between the geothermal and the oilfield waters, the oilfield water dataset was examined in more detail.

Predicted Temperature and CAAs in Oilfield Waters

Plots were constructed for both data sets of measured temperature minus predicted temperature vs. measured temperature (Tm-Tp vs. Tm) to assess which measured temperature interval had the greatest deviations in predicted temperature. A point which has a value of Tm-Tp = 0 on the x-axis of this plot appears on the 1:1 line on a plot of Tm vs. Tp. The larger the absolute value of Tm-Tp, the farther the predicted temperature is from the measured temperature. Therefore, a plot of Tm-Tp vs. Tm shows where the greatest data scatter occurs. Additional plots were constructed of Tm-Tp vs. percent carboxylic acid anions in the total alkalinity.

The equation that yielded one of the best correlations for the oilfield waters, the Mg-corrected Na-K-Ca equation (Eqs. 5, 9, Table 1), was chosen for examination. Although the Mg-Li equation yielded a better correlation, this equation was chosen for the plots on Figure 2 due to there being a paucity of lithium data in the geothermal water dataset. Figure 2A shows a plot of the measured temperature minus the predicted temperature vs. the measured temperature for only those oilfield samples with analyses of carboxylic acid anions. Measured temperature on the y-axis is plotted with temperature increasing downward, as it would with depth. There is increased scatter in the oilfield data in the temperature interval between 80 and 120°C, highlighted in Figure 2. The maximum scatter in the oilfield data corresponds to the temperature interval in which samples have the highest percentage of CAAs in the total alkalinity (Fig. 2C). No similar trends are evident for the geothermal water samples (Fig. 2B). These same relationships exist for each of the other cation geothermometers as expected since the others also had worse correlation coefficients in the geothermal waters.

The relationship of increased variability in predicted temperature to the percent CAAs present is shown also in Figure 3. This plot of measured temperature minus Mg-corrected Na-K-Ca predicted temperature vs. percentage organic acid anions in the total alkalinity shows a clear trend: where CAAs are present, the maximum data scatter occurs in samples where CAAs comprise more than 80% of the total alkalinity.

This same relationship holds for the other cation ratio geothermometers which yielded correlation coefficients lower than those of the Mg-corrected Na-K-Ca equation. For example, Figure 4 shows the measured minus predicted Mg-Li temperature vs. percentage of organic acid anions in the total alkalinity. Using Equation 8 (Table 1) which was developed primarily for oilfield waters, a strong correlation exists between percent CAAs and maximum scatter in the data where CAAs contribute to the alkalinity.

DISCUSSION

Several authors have attempted to accommodate oilfield waters in their geothermometry equations. Fournier and Truesdell (1973) developed the Na-K-Ca geothermometer in part to accommodate low-temperature, high-Ca oilfield waters. Also,

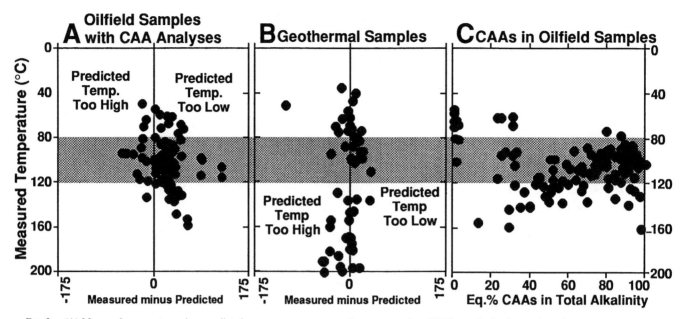

FIG. 2.—(A) Measured temperature minus predicted temperature vs. measured temperature for oilfield samples having analyses for carboxylic acid anions. The predicted temperature is that which is given by the Mg-corrected Na-K-Ca equation (Eq. 5 with Eq. 9, Table 1). Note the increased scatter in the 80–120°C temperature range. (B) Same as (A) with the exception that the data set is for the <250°C geothermal samples. Note that there is no increase in scatter in the 80–120°C temperature interval. (C) Plot of percent CAAs in the total alkalinity of the oilfield samples vs. the measured temperature. CAAs dominate the alkalinity in the 80–120°C temperature interval, where there is also increased scatter in the predicted temperature data in (A).

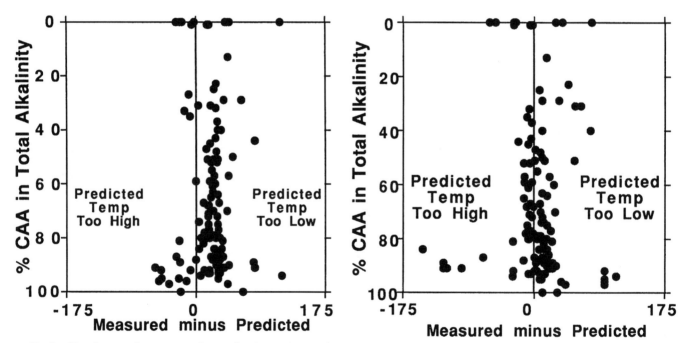

FIG. 3.—Plot of measured temperature minus predicted temperature vs. the percent CAAs in the total alkalinity in the oilfield samples. The predicted temperature was calculated using the Mg-corrected Na-K-Ca.

FIG. 4.—Same as Figure 3, except that the temperature was predicted by the Mg-Li equation (Eq. 8, Table 1).

Fournier and Potter (1979) developed their Mg correction primarily around oilfield waters having low temperatures and high magnesium concentrations. Kharaka and Mariner (1989) developed their Mg-Li geothermometer for oilfield waters on the basis that Mg and Li tend to substitute for each other in clay

minerals and that Mg concentration generally decreases with increasing temperature.

Several reasons have been proposed for the observed differences in temperature predictability, by cation geothermometry, between geothermal and oilfield waters. Kharaka and Mariner (1989) suggested that the lower temperature, generally higher pressure, and higher salinity of oilfield waters contribute to the observed differences in cation geothermometry behavior between oilfield and geothermal systems. Land and Macpherson (1992a) attributed the poor temperature predictions in oilfield waters to kinetic rather than equilibrium considerations.

High salinity cannot account for the observed differences. First, the Na-K geothermometer should work better in highly saline waters, such as those commonly encountered in oilfields. High salinity reduces the activity of water. Reduced water activity, in turn, favors anhydrous mineral phases such as microcline and albite (minerals commonly assumed to control the Na-K ratios) over hydrous phases such as illite or chlorite. Second, if salinity were the problem, the use of activity in the geothermometer equations rather than total analytical concentration should correct the problem. However, it does not. The re-calculated temperatures are either the same as for total analytical concentration or they are worse. This was checked using both the B-dot and the Pitzer equations in SOLMINEQ.88 (Kharaka et al., 1988) to calculate activities for highly saline (>100,000 ppm total dissolved solids) samples where a full suite of analyses were available (Na, K, Ca, Mg, Cl, SO_4, and HCO_3). Concentrations of sodium, potassium, and calcium, corrected for speciation, were then used to calculate temperature using two of the geothermometer equations. Table 4 shows an example of these two "activity-corrected" temperatures compared with the temperature calculated using total analytical concentration. Speciated at both 25 and 75°C, the calculated temperatures are very nearly the same as those calculated using analytical concentration. The only exception is for the Na-K-Ca temperature calculated using the concentration corrected by the Pitzer equation at 75°C. That calculated temperature is considerably higher (almost 17°C) than the total analytical concentration temperature. However, both of them are wrong; the actual temperature for the example is 87°C. The effect of speciation corrections on the calculated temperature is either negligible or worse than the temperature calculated using total analytical concentration.

The effect of higher pressure could not be evaluated due to the lack of pressure data reported with the published chemical analyses used in this study. Among all the references used for published water analyses, only Kharaka et al. (1977a, b) report overpressuring in their samples. Kharaka and Mariner (1989) indicate that high pressure in gas wells complicates the use of geothermometers by dilution of the water sample. This occurs by condensation of water vapor that produces with the natural gas. However, this phenomenon should not affect the Na-K geothermometers because both constituents would be diluted equally. Use of the Na-K-Ca geothermometer might be less straightforward in high-pressure natural gas wells that also pro-

duce a large amount of CO_2. The complication would occur if significant amounts of $CaCO_3$ precipitated due to CO_2 degassing causing a significant loss of calcium from the water.

Higher temperatures and the faster kinetics associated with higher temperatures also cannot account for the observed differences in the behavior of the geothermometers between oilfield and geothermal systems. First, this study compared water analyses from both systems over the same temperature range (0–250°C). Second, if kinetics related to higher temperature were the key factor, then the predictability of the geothermometers would steadily improve with increasing temperature in both systems; less and less scatter would be observed in the data as temperature increased. That pattern is not present in the oilfield water data. There was less scatter in the predicted temperature data at low temperatures, increased scatter in the 80–120°C temperature range, and less scatter again above 120°C (Fig. 2A).

Differences in lithology also cannot, alone, account for the observed differences in geothermometry behavior. The geothermal waters were derived from plutonic, volcanic, and sedimentary rocks, mostly devoid of organic matter which is known to be the source of CAAs (Vandenbroucke, 1980; Rouxhet et al., 1980; Vandergrift et al., 1980; Surdam et al., 1984; Crossey, 1985; Kawamura et al., 1986; Kawamura and Kaplan, 1987; Surdam and MacGowan, 1987; Lundegard and Senftle, 1987; MacGowan et al., 1988; Kharaka et al., 1989; Lewan and Fisher, 1994). The oilfield waters were derived from hydrocarbon-bearing sandstone/shale sequences, several of which include abundant volcanogenic detritus. Thus, both sets of waters produce from rocks containing various amounts of feldspar, quartz, and other common rock-forming minerals. Both host suites alter to form the same suite of authigenic phases (i.e., clays, feldspars, carbonates, zeolites, etc.). The presence/absence of hydrocarbons is clearly a separating factor.

Thus, neither salinity, pressure, temperature, nor lithology can fully account for the observed differences between geothermal and oilfield waters. Furthermore, the effect that CAAs have on alkalinity and mineral equilibria invalidates some of the assumptions upon which the geothermometer equations are based. These assumptions are that cation exchange between solid phases is the sole control on cation ratios, that aluminum is conserved in solid phases and that hydrogen ions involved in hydrolysis cancel out of the net chemical reaction equations.

The assumptions that cation ratios are controlled only by exchange between solid phases and that aluminum is conserved in solid phases are both invalidated where carboxylic acid anions exert a strong influence on solubility. Recent experimental work by Reed and Hajash (1992) shows that Na-bearing aluminosilicates are more soluble than K-bearing ones in pH-buffered oxalic acid and in buffered acetate solutions. This would lead to Na/K ratios that are higher than would be expected if simple cation exchange between solid phases were controlling the ratio. In turn, that would lead to a lower predicted temperature. In the present study, the simple Na-K geothermometers predicted overall lower temperatures in the oilfield waters than in the geothermal waters (compare Figs. 1B and 1C, for example) thus supporting the conclusion of Reed and Hajash (1992). Also, Fein (1991a) demonstrated that acetate forms moderately stable complexes with Ca and Mg, two cations for which geothermometry equations were adapted with oilfield

TABLE 4.—EFFECT OF SPECIATION ON CALCULATED TEMPERATURES

	Total Analyzed	Speciated at 25°C		Speciated at 75°C	
	ppm	B-dot	Pitzer	B-dot	Pitzer
Na	63,900	57,280	53,080	54,490	45,130
K	869	795	770	751	671
Ca	9,210	8,375	4,254	7,853	842
Na-K (°C)*	90.2	91.2	93.3	90.8	94.4
Na-K-Ca (°C)**	130.3	130.3	129.5	129.7	146.9

*Calculated using Equation 1, Table 1
**Calculated using Equation 5, Table 1

waters in mind. Therefore the presence of abundant CAAs exerts a strong influence not only on the amount of cations but their relative abundances as well.

Carboxylic acid anions also have been shown to be strong complexers of Al. In Reed and Hajash's (1992) experiments, for example, molar ratios of Al/Si, Al/K and Al/Na all exceeded the stoichiometric ratios for feldspars; Reed and Hajash (1992) concluded that the feldspars dissolve incongruently "by preferential formation of Al-oxalate complexes." Franklin et al. (1994) found in their experiments that Al-acetate complexes were more effective than previously recorded at maintaining high concentrations of Al in solution. Further, Fein (1991a, b) demonstrated that Al-acetate complexes are stable and enhance the solubility of gibbsite at 80°C. In similar experiments that expanded the range of temperatures, Benezeth et al. (1994) found Al-acetate association constants in close agreement with Fein. They also found that Al-hydroxide-acetate complexes enhanced Al concentrations five to seven times over that expected from an equilibrium assemblage of albite, quartz, and kaolinite. Fein (1991a, b) also demonstrated that Al-oxalate complexes increase the solubility of gibbsite by 2.3 to 3.1 orders of magnitude. Finally, solutions containing carboxylic acids accelerate dissolution rates of feldspar (Manning et al., 1992; Knauss and Copenhaver, 1995) and plagioclase (Welch and Ullman, 1993). The implication of all this for geothermometry is that solid products in the assumed chemical reactions are not the only sinks for Al in systems where aqueous aluminum-CAA complexes are present. Aluminum is not necessarily conserved in solid phases.

The assumption that hydrogen ions involved in hydrolysis cancel out of the net chemical reaction equation is also invalidated by the presence of abundant organic components. This assumption implies that neither H^+ nor CO_2 enters into the net chemical reaction and that only silicate hydrolysis buffers the pH of the water (Fournier and Truesdell, 1973). However, carboxylic acid anions have been found to be important pH buffers even in feldspar-rich systems, thereby allowing the dissolution of aluminosilicate phases (Barth et al., 1990; Huang and Longo, 1992; Reed and Hajash, 1992). Thus for the Na-K geothermometers, H^+ does in fact enter into the equation when CAAs are present. CAAs' buffering capacity also affects the geothermometer equations involving calcium. Surdam et al. (1984) theorized that a pH buffered by acetate would allow carbonate cement precipitation by accommodating increased P_{CO2} without a concomitant lowering of pH. Lundegard and Land (1989) found that carbonate cement did not precipitate. However, they did find that less carbonate cement dissolution took place at elevated P_{CO2} in the presence of acetate. Whether carbonates precipitate or simply less dissolution takes place, calcium concentrations are altered from those anticipated in a system devoid of CAAs and dominated by silicate hydrolysis.

The science of cation ratio geothermometry has developed by progressively attempting to accommodate "anomalous" oilfield waters by making corrections and additions to the original Na-K equations. Fournier and Truesdell (1973) added calcium to the equation because they observed that high Ca-concentration waters, such as those producing from the Kettleman Hills oilfield of California, yielded high Na/K ratios. Fournier and Potter (1979) formulated the magnesium correction to the Na-K-Ca equation (Fournier and Truesdell, 1973) after their observation that many low temperature, Mg-rich waters yielded estimated Na-K-Ca temperatures that were too high. To develop their Mg correction, their database of Mg-rich waters consisted largely of brines produced from oil fields.

These equations and corrections developed to date were done so empirically as described in the section on Cation Ratio Geothermometer Equations. The original Na-K empirical correlations were developed using water samples from geothermal water-rock systems. Those systems appeared to behave as predicted by the theoretical thermodynamic considerations of cation exchange between specific sodium-bearing and potassium-bearing feldspars. Therefore, the empirical correlations for temperature worked well, and the assumptions upon which they were based, as well as the theory, were validated. Only small adjustments have been made to the Na-K geothermometer to reflect differences in the particular feldspar compositions controlling the equilibria (Truedell, 1976; Arnorrson et al., 1983).

However, departures from these empirical correlations in hydrocarbon-bearing sedimentary sections have been clearly described above. Merely making correction to the original empirical correlations may not offer the most satisfactory approach. In this aspect, Kharaka and Mariner (1989) provided a new step in improving cation ratio geothermometry. The next generation of cation ratio geothermometer equations needs to deal explicitly with the now-recognized importance of organic-inorganic interactions. Merely ascribing the loss of fit of the geothermometers to the vagaries of the sedimentary system is unsatisfactory.

CONCLUSIONS

Carboxylic acid anions exert a strong control on the cation ratios in those oilfield waters where they are present and make up more than 80% of the total alkalinity. Their presence contributes significantly to the observed differences in cation ratio geothermometry behavior between oilfield and geothermal systems.

The cations used in these cation ratio equations, Ca and Mg as well as Na and K, are strongly affected by the presence of carboxylic acid anions. No information exists on Li-acetate complexes. However, the Mg-Li geothermometer is most likely affected by Mg-acetate complexes or by pH-buffering effects of the CAAs which affects the solubilities of common sedimentary rock-forming minerals containing Mg.

The strong complexing of acetate and oxalate with Al, the ability of CAAs to dissolve aluminosilicate minerals, particularly Na-bearing ones preferentially over K-bearing ones, and the buffering capacity of CAAs indicate that where carboxylic acid anions dominate the alkalinity, they play an important role in diagenetic processes. This, coupled with the fact that cation ratios are poor predictors of temperature where CAAs are abundant, indicates that CAAs should be included in the analysis if the use of cation ratio geothermometers in oilfields is to be improved.

ACKNOWLEDGMENTS

An earlier version of this manuscript benefitted from input by J. I. Drever. We also thank Dave Copeland for editing and Al Deiss for drafting. Reviews by R. Loucks, S. G. Franks, and an anonymous reviewer improved the final manuscript. Funding for this research was provided by NSF/EPSCoR-ADP grant #EHR-910–8774.

REFERENCES

ARNORSSON, S., 1975, Application of the silica geothermometer in low temperature hydrothermal areas in Iceland: American Journal of Science, v. 275, p. 763–784.

ARNORSSON, S., GUNLAUGSSON, E., AND SVARVARSSON, H., 1983, The chemistry of geothermal waters in Iceland. III. Chemical geothermometry in geothermal investigation: Geochimica et Cosmochimica Acta, v. 47, p. 567–577.

BARTH, T., BORGUND, A. E., AND RIIS, M., 1990, Organic acids in reservoir waters — relationship with inorganic ion composition and interactions with oil and rock: Organic Geochemistry, v. 16, p. 489–496.

BENEZETH, P., CASTET, S., DANDURAND, J-L., GOUT, R., AND SCHOTT, J., 1994, Experimental study of aluminum-acetate complexing between 60 and 200°C: Geochimica et Cosmochimica Acta, v. 58, p. 4561–4571.

CAROTHERS, W. W. AND KHARAKA, Y. K., 1978, Aliphatic acid anions in oilfield waters—implications for origin of natural gas: American Association of Petroleum Geologists Bulletin, v. 62, p. 2441–2453.

CARPENTER, A. B., TROUT, M. L., AND PICKETT, E. E., 1974, Preliminary report on the origin and chemical evolution of lead- and zinc-rich brines in central Mississippi: Economic Geology, v. 69, p. 1191–1206.

CROSSEY, L. J., 1985, The origin and role of water soluble organic compounds in clastic diagenetic systems: Unpublished Ph.D. Dissertation, University of Wyoming, Laramie, 135 p.

CURTIS, C. D., 1978, Possible links between sandstone diagenesis and depth related geochemical reactions occurring in enclosing mudstones: Quarterly Journal of the Geological Society of London, v. 135, p. 107–117.

FEIN, J. B., 1991a, Experimental study of aluminum-, calcium-, and magnesium-acetate complexing at 80°C: Geochimica et Cosmochimica Acta, v. 55, p. 955–964.

FEIN, J. B., 1991b, Experimental study of aluminum-oxalate complexing at 80°C: Implication for aluminum mobility in sedimentary basin fluids: Geology, v. 19, p. 1037–1040.

FERTL, W. H. AND OVERTON, H., 1982, Formation evaluation, in Edwards, L. M., Chilingar, G. V., Rieke III, H. H., AND Fertl, W. H., eds., Handbook of Geothermal Energy: Houston, Gulf Publishing Co., p. 326–387.

FISHER, J. B. AND BOLES, J. R., 1990, Water-rock interaction in Tertiary sandstones, San Joaquin Basin, California, U.S.A.: Diagenetic controls on water composition: Chemical Geology, v. 82, p. 83–101.

FOURNIER, R. O., 1981, Application of water geochemistry to geothermal exploration and reservoir engineering, in Rybach, L. and Muffler, L. J. P., eds., Geothermal Systems: Principles and Case Histories: New York, John Wiley & Sons Ltd., p. 109–143.

FOURNIER, R. O. AND POTTER, R. W., III, 1979, Magnesium correction to the Na-K-Ca chemical geothermometer: Geochimica et Cosmochimica Acta, v. 43, p. 1543–1550.

FOURNIER, R. O. AND TRUESDELL, A. H., 1973, An empirical Na-K-Ca geothermometer for natural waters: Geochimica et Cosmochimica Acta, v. 37, p. 1255–1275.

FOURNIER, R. O., WHITE, D. E., AND TRUESDELL, A. H., 1974, Geochemical indicators of subsurface temperature. Part 1. Basic assumptions: Washington, D.C., United States Geological Survey Journal of Research, v. 2, p. 259–262.

FRANKLIN, S. P., HAJASH, A., Jr., DEWERS, T. A., AND TIEH, T. T., 1994, The role of carboxylic acids in albite and quartz dissolution: An experimental study under diagenetic conditions: Geochimica et Cosmochimica Acta, v. 58, p. 4259–4279.

HUANG, W-L. AND LONGO, J. M., 1992, The effect of organics on feldspar dissolution and the development of secondary porosity: Chemical Geology, v. 98, p. 271–292.

KAWAMURA, K. AND KAPLAN, I. R., 1987, Dicarboxylic acids generated by thermal alteration of kerogen and humic acids: Geochimica et Cosmochimica Acta, v. 81, p. 3201–3207.

KAWAMURA, K., TANNEBAUM, E., HUIZINGA, B. J., AND KAPLAN, I. R., 1986, Volatile organic acids generated from kerogen during laboratory heating: Geochemical Journal, v. 20, p. 51–59.

KHARAKA, Y. K., AMBATS, G., AND LUNDEGARD, P. D., 1989, Origin and inorganic interactions of organic species in formation waters from sedimentary basins: Washington, D.C., Proceedings of the 28th International Geological Congress 2, p. 184.

KHARAKA, Y. K. AND BERRY, E. A. F., 1976, The influence of geological membranes on the geochemistry of subsurface waters from Eocene sediments at Kettleman North Dome, California: An example of effluent-type waters, in Cadik, J. and Paces, T., eds., Proceedings of the 1st International Symposium on Water-Rock Interaction: Prague, The Geological Survey of Czechoslovakia, p. 268–277.

KHARAKA, Y. K., CALLENDER, E., AND CAROTHERS, W. W., 1977a, Geochemistry of geopressured geothermal waters from the Texas Gulf Coast: Lafayette, Proceedings of the 3rd Geopressured-Geothermal Energy Conference, University of Southwestern Louisiana, v. 1, p. G1121–G1165.

KHARAKA, Y. K., CALLENDER, E., AND CAROTHERS, W. W., 1977b, Geochemistry of geopressured geothermal waters of the northern Gulf of Mexico basin, 1. Brazoria and Galveston counties, Texas, in Paquet, H. and Tardy, Y., eds., Proceedings of the Second International Symposium on Water-Rock Interaction: Strasbourg, Université Louis Pasteur, Centre National de la Recherche Scientifique, Institut de Géologie, p. II 32–II 41.

KHARAKA, Y. K., GUNTER, W. D., AGARWALL, P. K., PERKINS, E. H., AND DEBRAAL, J. D., 1988, SOLMINEQ.88: A computer program code for geochemical modeling of water-rock interactions: Washington, D.C., United States Geological Survey Water Investigations Report 88–4227, 420 p.

KHARAKA, Y. K. AND MARINER, R. H., 1989, Chemical geothermometers and their application to formation waters from sedimentary basins, in Naeser, N. and McCulloh, T., eds., Thermal History of Sedimentary Basins: New York, Springer Verlag, p. 99–117.

KNAUSS, K. G. AND COPENHAVER, S. A., 1995, The effect of malonate on the dissolution kinetics of albite, quartz, and microcline as a function of pH at 70°C: Applied Geochemistry, v. 10, p. 17–33.

LAND, L. S. AND MACPHERSON, G. L., 1989, Geochemistry of formation water, Plio-Pleistocene reservoirs, offshore Louisiana: Transactions, Gulf Coast Association of Geological Societies, v. 39, p. 421–430.

LAND, L. S. AND MACPHERSON, G. L., 1992a, Geothermometry from brine analyses: Lessons from the Gulf Coast, U.S.A.: Applied Geochemistry, v. 7, p. 333–340.

LAND, L. S. AND MACPHERSON, G. L., 1992b, Origin of saline formation waters, Cenozoic section, Gulf of Mexico sedimentary basin: American Association of Petroleum Geologists Bulletin, v. 76, p. 1344–1362.

LAND, L. S., MACPHERSON, G. L., AND MACK, L. E., 1988, The geochemistry of saline formation waters, Miocene, offshore Louisiana: New Orleans, Transactions, Gulf Coast Association of Geological Societies, v. 38, p. 503–511.

LEWAN, M. D. AND FISHER, J. B., 1994, Organic acids from petroleum source rocks, in Pittman, E. D. AND Lewan, M. D., eds., Organic Acids in Geological Processes: Berlin, Springer-Verlag, p. 70–114.

LUNDEGARD, P. D. AND LAND, L. S., 1989, Carbonate equilibria and pH buffering by organic acids — response to changes in P_{CO2}: Chemical Geology, v. 74, p. 277–287.

LUNDEGARD, P. D. AND SENFTLE, J. T., 1987, Hydrous pyrolysis—a tool for the study of organic acid synthesis: Applied Geochemistry, v. 2, p. 605–612.

MACGOWAN, D. B., THYNE, G. D., AND SURDAM, R. C., 1988, Prediction of changes in porosity due to feldspar dissolution using hydrous pyrolysis experiments: An example from the Bighorn Basin: Geological Society of America Program with Abstracts, v. 20, p. 347.

MANN, C. J., 1987, Misuses of linear regression in earth sciences, in Size, W. B., ed., Use and Abuse of Statistical Methods in the Earth Sciences: International Association for Mathematical Geology: Studies in Mathematical Geology No. 1: New York, Oxford University Press, p. 74–106.

MANNING, D. A. C., GESTSDOTTIR, K., AND RAE, E. I. C., 1992, Feldspar dissolution in the presence of organic acid anions under diagenetic conditions: An experimental study: Organic Geochemistry, v. 19, p. 483–492.

MESHRI, I. AND WALKER, J., 1990, A study of rock-water interaction and simulation of diagenesis in the Upper Almond Sandstones of the Red Desert and Washakie basins, Wyoming, in Meshri, I. D. and Ortoleva, P. J., eds., Prediction of Reservoir Quality Through Chemical Modeling: Tulsa, American Association of Petroleum Geologists Memoir 49, p. 55–70.

REED, C. L. AND HAJASH, A., Jr., 1992, Dissolution of granitic sand by pH-buffered carboxylic acids: A flow-through experimental study at 100°C and 345 bars: American Association of Petroleum Geologists Bulletin, v. 76, p. 1402–1416.

ROUXHET, P. G., ROBIN, P. L., AND NICAISE, G., 1980, Characterization of kerogens and of their evolution by infrared spectroscopy, *in* Durand, B., ed., Kerogen: Paris, Editions Technip, p. 163–190.

SURDAM, R. C., BOESE, S. W., AND CROSSEY, L. J., 1984, The chemistry of secondary porosity, *in* McDonald, D. A. and Surdam, R. C., eds., Clastic Diagenesis: Tulsa, American Association of Petroleum Geologists Memoir 37, p. 127–149.

SURDAM, R. C. AND CROSSEY, L. J., 1985, Organic-inorganic reactions during progressive burial: Key to porosity and permeability enhancement and preservation: Philosophical Transactions of the Royal Society of London, v. 315, p. 135–156.

SURDAM, R. C. AND MACGOWAN, D. B., 1987, Oilfield waters and sandstone diagenesis: Applied Geochemistry, v. 2, p. 613–619.

TRUESDELL, A. H., 1976, Summary of section III, geochemical techniques in exploration, *in* Fournier, R. O., ed., Proceedings of the Second United Nations Symposium on the Development and Use of Geothermal Resources: United Nations, v. 1 p. 1iii–1xxix.

VANDENBROUCKE, M., 1980, Structure of kerogen as seen by investigation on soluble extracts, *in* Durand, B., ed., Kerogen: Paris, Editions Technip, p. 415–444.

VANDERGRIFT, G. F., WINANS, R. E., SCOTT, R. G., AND HORWITZ, E. P., 1980, Quantitative study of the carboxylic acids in the Green River oil shale bitumen: Fuel, v. 59, p. 627–633.

WELCH, S. A. AND ULLMAN, W. J., 1993, The effect of organic acids on plagioclase dissolution rates and stoichiometry: Geochimica et Cosmochimica Acta, v. 57, p. 2725–2736.

WHITE, D. E., 1963, Saline waters of sedimentary rocks, *in* Young, A. and Galley, J. E., eds., Fluids in Subsurface Environments: Tulsa, American Association of Petroleum Geologists Memoir 4, p. 342–366.

WILLEY, L. M., KHARAKA, Y. K., PRESSER, T. S., RAPP, J. B., AND BARNES, I., 1975, Short chain aliphatic acid anions in oil field waters and their contribution to the measured alkalinity: Geochemica et Cosmochimica Acta, v. 39, p. 1707–1711.

FLUID-FLOW REGIMES AND SANDSTONE/SHALE DIAGENESIS IN THE POWDER RIVER BASIN, WYOMING

RONALD C. SURDAM, ZUN SHENG JIAO, AND PEIGUI YIN

Institute for Energy Research, University of Wyoming, P.O. Box 4068, Laramie WY 82071–4068

ABSTRACT: Many of the most productive Cretaceous hydrocarbon reservoir rocks in Rocky Mountain Laramide basins are in close stratigraphic proximity to organic-rich source rocks; an example is the Muddy Sandstone/Mowry Shale system in the Powder River Basin. The spatial attributes of mass-transfer processes characterizing the diagenesis and maturation of these reservoir rocks and source rocks during basin evolution are interrelated.

As is typical in Wyoming Laramide basins, that portion of the Cretaceous shale section in the Powder River Basin below a present-day depth of approximately 2700 m (9000 ft) constitutes one anomalously pressured, gas-saturated, basinwide, dynamic fluid-flow compartment. The driving mechanism for the compartmentalization of these shales was the generation, storage, and subsequent reaction of hydrocarbons. As these processes proceeded in the shales, the fluid-flow regime was converted to a multiphase regime; low-permeability fluid-flow barriers were converted to capillary seals; and three-dimensional closure of these seals created the compartment. Above these anomalously pressured, gas-saturated shales, the fluid-flow regime remains single-phase and typically under water drive (influenced by the meteoric water regime).

The vertical compartmentalization of the fluid-flow regime has had a pronounced effect on the different mass-transfer characteristics in the sandstone bodies within and above the basinwide pressure compartment in the shales. Mass transfer in those portions of the basin characterized by a single-phase fluid-flow regime is large-scale and regionally significant, as reflected in the alteration of framework grains and cementation/decementation reactions. In contrast, in those portions of the basin characterized by a multiphase fluid-flow regime, mass transfer in the sandstones, while commonly intense, is on a much smaller scale and typically is confined to relatively small, isolated fluid-flow compartments. These compartments in the sandstone have trapped the hydrocarbon whose presence led to their formation.

INTRODUCTION

The purposes of this paper are (1) to review the fluid-flow regime in the Powder River Basin, Wyoming and (2) to determine what effect the fluid-flow regime has had on progressive sandstone and shale diagenesis and on the sealing capacity of the pressure seals associated with hydrocarbon accumulations. Ultimately, by integrating the fluid-flow regime with diagenesis, it should be possible to lessen uncertainty about porosity prediction (i.e., reservoir quality) and about hydrocarbon accumulation and pressure-seal distribution and character.

We concentrate on the Muddy Sandstone/Mowry Shale reservoir/source-rock system because we have established a large mineralogic, petrographic, and petrophysical database for these rocks (Surdam et al., 1995b; Jiao and Surdam, 1993). There also exists a well documented depositional model for the Muddy/Mowry system (Odland et al., 1988; Martinsen, 1995).

FLUID-FLOW/PRESSURE REGIME

The Powder River Basin of Wyoming (Fig. 1), a highly productive hydrocarbon province, is a typical asymmetrical Laramide basin. The basin contains a thick section of Cretaceous shales with associated relatively thin sandstones (Fig. 2). This shale/sandstone system was deposited in the large inland sea that occupied the western North American interior during most of the Cretaceous period. Some of the shales are organic-rich and can be significant hydrocarbon sources (e.g., the Mowry Shale). The sandstone/shale system contains several types of low-permeability fluid barriers; hydrocarbons are trapped in a variety of ways in the sandstones. Thus, the Cretaceous shale/sandstone system also is a hydrocarbon source/seal/reservoir system (a "petroleum system" in the sense of W. G. Dow and L. B. Magoon; see discussion in Magoon and Dow, 1994).

Shales

The Cretaceous shales in the Powder River Basin below a present-day depth of approximately 2400 to 2700 m (8000 to 9000 ft) are typically overpressured (subject to a pressure gra-

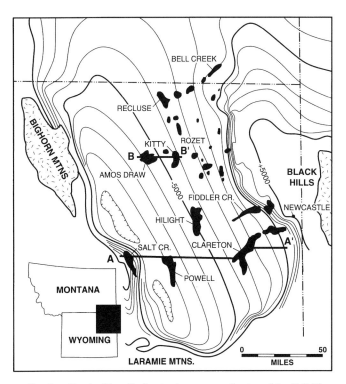

FIG. 1.—Powder River Basin structure map on the top of the Fall River Sandstone, with a contour interval of 1000 ft. Major oil and gas fields and pressure-panel lines of section are shown. Modified from Waring (1976).

dient greater than 0.433 psi/ft; Surdam et al., 1995a, b). Based on mud-weight profiles and pressure measurements from DSTs, the top of the overpressuring in the basin center is marked by a transitional zone, 300–600 m (1000–2000 ft) thick, that most often occurs within the Upper Cretaceous Steele Shale. The overpressured zone, ~600 m (~2000 ft) thick, is located below the transitional zone and persists down to the lowermost organic-rich shale in the Lower Cretaceous Fuson Shale. Below

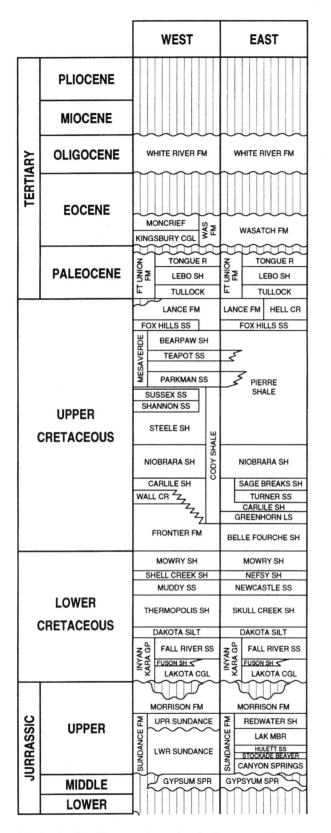

FIG. 2.—Stratigraphic nomenclature chart for Cretaceous and contiguous units in the Powder River Basin. The basin is characterized by a thick section of Cretaceous shales with associated relatively thin sandstones. Modified from Wyoming Geological Association (1988).

the Fuson Shale, the fluid-flow regime is normally pressured (Fig. 3). Both the transitional and overpressured zones, together termed the anomalously pressured zone in this paper, are characterized by marked increases in sonic transit time, hydrocarbon production index, clay diagenesis (smectite → illite), and vitrinite reflectance (Surdam et al., 1995a, b). The produced water in the sandstones associated with the Cretaceous shales changes relatively abruptly from fresh to nearly marine as the top of the anomalously pressured zone is entered from above. These anomalously pressured, organic-rich shales typically contain bitumen-filled microfractures.

The regional geometry of the overpressured shale section can be determined by reconstructing the pressure regime of the Cretaceous shale section using the techniques described in Surdam et al. (1995a, b). Briefly, these techniques involve generating

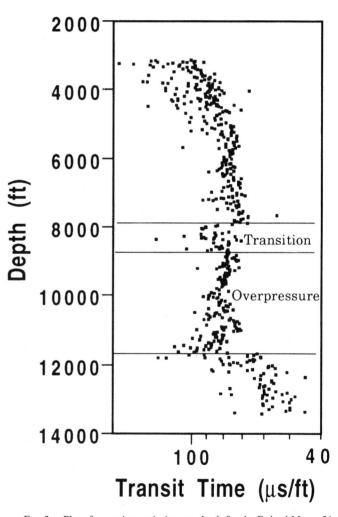

FIG. 3.—Plot of acoustic transit time vs. depth for the Federal Means 21 well, Sec. 13, T46N, R75W. Transit time may be associated with pressure. The top of the overpressuring is marked by a transitional zone (~1000 ft thick). The overpressured zone (here ~2500 ft thick) is located below the pressure transitional zone and persists down to the lowermost organic-rich shale in the Cretaceous.

sonic velocity panels, or cross-sections, such as the one shown in Figure 4 (see p. 94), filtered for lithology, that are first de-compacted to isolate sonic velocities deviating from normal compaction trends. These sonic anomalies are then scaled, as-suming that any particular anomaly can be attributed to gas and pressure effects. The result of this process is a gas-pressure panel showing relative gas-pressure values. Figure 4 is a west-east gas-pressure panel approximately 110 miles long through the Powder River Basin (A-A' in Fig. 1).

Figure 4 exemplifies our findings throughout the Powder River Basin (Surdam et al., 1995a). The top of the anomalously pressured shale, the transitional zone, is approximately hori-zontal in the basin center. Where observed, the bottom of the overpressured shale section is regionally coincident with the lowermost organic-rich shale in the Cretaceous section. Typi-cally, the highest abnormal pressures in the shale section are found in the lower half of the overpressured column. These observations are not repeated along the margins of the basin, where the Cretaceous section approaches the surface and uplift and erosion have been greatest. Along the margins of the Pow-der River Basin, the overpressured shale volume is wedge-shaped (Fig. 5). There exists ample evidence that the Powder River Basin, like most Laramide basins, has undergone differ-ential unroofing during the last 10–20 million years (Heasler et al., 1995).

It has been observed that hydrocarbons produced from sand-stones within the overpressured volume are from gas depletion drive accumulations. It has been concluded that the overpres-sured Cretaceous shale section is a basinwide, gas-saturated pressure compartment (Fig. 4; Jiao and Surdam, 1993).

Surdam et al. (1995a, b), on the basis of maturation indica-tors, thermal modeling, and analytical work, have concluded that the dominant mechanism driving the overpressuring of the Cretaceous shale system is the generation of liquid hydrocar-bons and the subsequent reaction of oil to gas in the shale. This process of generation, storage, and reaction converts the fluid-flow regime from a water-dominated single-phase regime to a hydrocarbon-dominated multiphase regime. As generated but stored liquid hydrocarbons react progressively to gas, the fluid-flow regime evolves to a gas/oil/water regime. Consequently, capillarity becomes increasingly important over time; the dis-placement pressure in the shale section increases substantially; and abnormally high pressure results.

In summary, the overpressured Cretaceous shale section in the Powder River Basin is a regional, gas-saturated pressure compartment. The driving mechanism is the reaction of stored liquid hydrocarbons to gas. This reaction ensures the conversion of the fluid-flow regime from a single-phase water-dominated regime to a multiphase hydrocarbon-dominated regime. As a consequence of capillarity, the displacement pressure within the fine-grained rocks increases significantly, as fluid-flow barriers are converted to capillary seals. As a result of the progress of the oil-to-gas reaction, the conversion of the fluid-flow regime from single-phase to multiphase, and the resultant capillary ef-fects, the shale section becomes a regional pressure compart-ment, probably at about the time of maximum burial (Surdam et al., 1995a, b). In contrast, above and below the regional ov-erpressured Cretaceous shale section, where there is no asso-ciation with hydrocarbon-saturated source rocks, the fluid-flow regime is a water-dominated, single-phase fluid-flow regime at normal hydrostatic pressure.

Sandstones

The pressure and fluid-flow regime of Cretaceous sandstones in the Powder River Basin is different from the pressure regime characterizing the shales. The sandstones are compartmental-ized on a much smaller, or local, scale (Heasler et al., 1995). Individual sandstone intervals are characterized by fluid-flow and pressure regimes that are separated spatially, both vertically and horizontally (i.e., they are not stacked). Individual sand-stone intervals (e.g., the Dakota, Muddy, and Frontier Sand-stones) are subdivided into relatively small, isolated compart-ments—typically, the longest dimension is 1.6 to 16 km (1 to 10 mi). This fluid/pressure compartmentalization is the result of the closure of internal seals, which are stratigraphic elements within the individual sandstone intervals (e.g., lowstand uncon-formities, transgressive shales, and paleosols) that have been modified diagenetically (Fig. 6; Martinsen, 1995; Jiao and Sur-dam, 1993). These diagenetically modified internal strati-graphic elements are low-permeability rocks in a single-phase fluid-flow regime, with finite leak rates; as the fluid-flow regime evolves into a multiphase fluid-flow regime (see especially Berg, 1975; Iverson et al., 1995), these elements evolve into capillary seals. If three-dimensional closure of the capillary seals occurs, the sandstone is compartmentalized; the fluids within the compartment are isolated from surrounding fluids (Fig. 6C). With the exception of sandstones at the top of the anomalously pressured shale section, the sandstones that occur within the overpressured shale section are compartmentalized and overpressured.

Figure 7 illustrates the details of compartmentalization within an individual reservoir facies (in this case, the Muddy Sandstone along an east-west cross section from the Amos Draw to the Kitty to the Ryan field in the Powder River Basin,

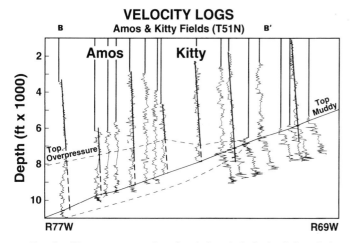

FIG. 5.—West-east cross section of sonic logs (velocity logs) through the Amos Draw and Kitty fields, line of section B-B' in Figure 1 (T51N). The vertical solid lines show well locations; the tilted solid lines indicate local normal compaction trend above the overpressured volume, and are extrapolated as dashed lines into the overpressured volume. The cross section shows how the boundary of the overpressured shale section wedges out at the edge of the Powder River Basin. Too little data exists at the western edges of the basin to model the shape of the section there. From Maucione et al. (1995).

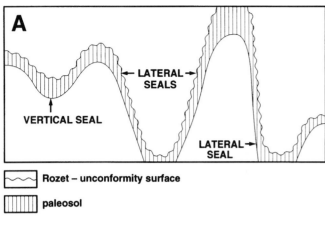

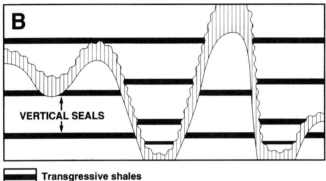

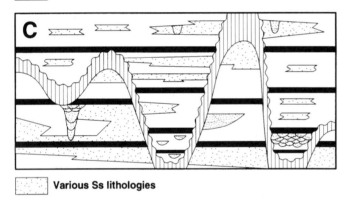

FIG. 6.—Hierarchy of stratigraphic compartmentalization observed in the Muddy Sandstone in the Powder River Basin, associated with a prominent unconformity. (A) First-level compartment formed by intersection of a lowstand surface erosion (LSE) with low-permeability shales that bound the Muddy Sandstone. Note that both vertical and lateral seals to fluid flow exist. Geometry is locally modified by diagenesis within the paleosol zone. (B) Second-level compartmentalization is the product of transgressive shale. (C) Third-level compartmentalization is defined by the distribution of lithofacies. From Martinsen (1995).

B-B′ in Fig. 1; also see Fig. 5). Along this section, the Muddy Sandstone is part of a shoreface/valley-fill depositional system in which the younger, valley-fill elements are separated from older, shoreface sandstones by a paleosol developed along a regionally prominent lowstand unconformity, with intensive clay infiltration into the underlying sandstone (Martinsen, 1995). As described by Jiao and Surdam (1993), the paleosol horizons are low-permeability horizons that in a multiphase

fluid-flow regime are impermeable to fluid flow until a finite displacement pressure is exceeded.

As a consequence, the Muddy Sandstone (shown in Figure 7 in an oil/water or gas/oil dominated fluid-flow regime) is externally sealed above and below by the Mowry and Skull Creek Shales (transgressive shales) and is internally separated into fluid-flow compartments by impermeable horizons within the sandstone (Fig. 7). The network of sealing boundaries above, below, and within the sandstone determines the three-dimensional closure of seals in a multiphase fluid-flow system — in a single-phase (water) system, the low-permeability horizons would have finite leak rates, and fluid compartmentalization (differential pressure) would be short-lived (Iverson et al., 1995). A result of the three-dimensional closure of the seals is the formation of pressure/fluid compartments completely isolated from the shales above and below, and from the sandstone updip and downdip. It should be noted that along the cross section shown in Figure 7, the inclination of the rocks is 2° regionally and there is no structural closure along the section.

Figure 8 (see p. 94) summarizes the relationship between sandstone pressure compartments and the regional overpressuring of the Cretaceous shales. The distribution of sandstones and shales in the Powder River Basin as determined from gammaray logs is shown in Figure 8. Also shown are the volume of regionally overpressured shales and the distribution of anomalously pressured sandstones. It should be noted that sandstones close to the upper boundary of the overpressured shale section may be underpressured; see Surdam et al. (1995b) for a discussion of underpressured compartments.

Differential Uplift and Erosion

Other sources have described the existence of differential uplift across the Powder River Basin (Jiao, 1992). It has been suggested that in the basin center, there has been 300 to 600 m (1000 to 2000 ft) of unroofing, whereas on the basin margin, the amount of unroofing may be more than 1500 m (5000 ft). These estimates are based on reconstructions using vitrinite reflectance profiles and sonic logs (Jiao, 1992). The underpressuring that occurs at the top of the anomalously pressured shales (Fig. 8) is thought to be a product of the uplift and erosion of the basin (see Surdam et al., 1995b).

Burial Histories/Fluid-Flow Variations

A range of burial history scenarios is possible for the Muddy/Mowry system, depending on position in the basin. We present three potential burial history scenarios (Figs. 9A, 9B, 9C), corresponding to positions at the basin center, halfway to the margin, and at the basin margin. Significant differences exist between the three scenarios and in the evolution of the corresponding fluid-flow regimes. In the first scenario (Fig. 9A), the Muddy Sandstone has contained only marine connate waters. Its fluid-flow regime has evolved from single-phase to multiphase and from normally pressured to overpressured. In the second (Fig. 9B), the rocks also have contained only marine connate waters, and the fluid-flow regime has evolved from single-phase to multiphase; but initial normal pressuring has been followed by overpressuring, and finally by underpressuring (see Surdam et al., 1995b). The third scenario (Fig. 9C) includes rocks that have contained early connate water and later have

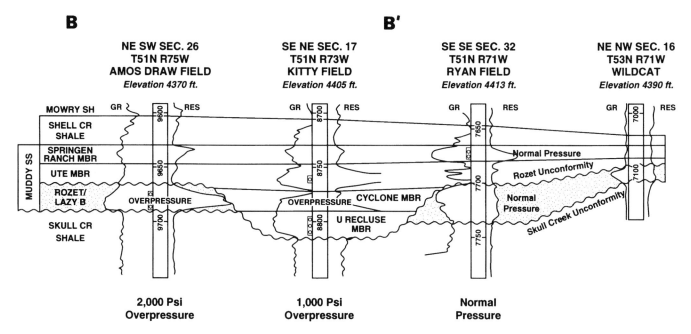

FIG. 7.—West-east cross section of the lower Cretaceous stratigraphic sequence through the Amos Draw, Kitty, and Ryan fields and a wildcat well in the Powder River Basin. The line of section continues line B-B' in Figure 1 eastward. The Muddy Sandstone reservoirs along this cross section are oil/water or gas/water dominated fluid-flow systems and are externally sealed above and below by the Mowry and Skull Creek shales. The Muddy Sandstone is internally separated into flow compartments by a fluid-flow barrier (here the Rozet unconformity) within the sandstone. The Rozet and Skull Creek unconformities are shown by wavy lines on the cross section. The stippling represents shoreface sandstone.

been flushed by meteoric water, but always in a single-phase, normally pressured fluid-flow regime. Figures 9A, 9B, and 9C show the differences in fluid-flow regime that characterize the three burial histories shown in Figures 10A, 10B, and 10C determined for these three positions in the basin. If fluid-flow regime plays an important role during diagenesis, then these three burial scenarios should be characterized by significant variations in diagenetic history and thus in porosity/permeability character. The remainder of this paper will be devoted to determining the effect of fluid-flow variation on the diagenesis of sandstones and shales and on the sealing capacity of hydrocarbon pressure seals.

EFFECT OF FLUID-FLOW REGIME ON SANDSTONE DIAGENESIS

Porosity and Permeability Distribution

In this study, a total of 1533 conventional porosity and permeability measurements of Muddy Sandstone core samples collected from 48 wells in the Powder River Basin were retrieved from the Wyoming Geological Survey in Laramie. Porosity and permeability for each depth interval are displayed as bar graphs in Figures 11 and 12. The range of values is widely scattered in each depth interval, perhaps due to variations in sandstone depositional setting, mineralogic composition, and diagenetic evolution. The porosity and permeability distribution with depth in the Muddy sandstones generally can be divided into two trends. From 800 to 2300 m (2500 to 7500 ft) deep, the sandstones exhibit a relatively normal compaction trend with increasing depth. In contrast, in the Muddy sandstones from 2300 to 3900 m (7500 to 12,800 ft) deep (i.e., in the anomalously pressured shale volume), the median populations of porosity show little change with depth. The porosity-loss trend in these

sandstones changes across the top overpressured shale boundary, and both porosity and permeability are greater above the boundary than below it. Porosity generally decreases with depth in the Muddy sandstones above the pressure boundary; however, the median porosity and permeability in the overpressured zone remain relatively constant over a depth interval of 1600 m (5300 ft).

The Muddy sandstones from 2300 to 3900 m (7500 to 12,800 ft) deep are abnormally pressured and petroliferous. The relative constancy of porosity and permeability in this depth interval has probably resulted from a combination of overpressuring and hydrocarbon accumulation. Abnormal pressure development in the Muddy Sandstone probably coincided with either maximum burial or depth of burial in excess of approximately 3000 m (10,000 ft). With late tectonic activity, the basin was subject to subsidence and uplift, the extent of which varied across the basin. Our burial histories show that the section of the Muddy Sandstone presently located 2700 m (9000 ft) deep (halfway to basin margin; Fig. 10B) was once buried to a maximum depth of 3700 m (12,000 ft), and the section presently 4300 m (14,100 ft) deep (basin center; Fig. 10A) had a maximum burial of 5000 m (16,400 ft).

Figure 13 is a plot of permeability versus porosity in the Muddy sandstones. The points are widely scattered, indicating large variations in pore configuration in the Muddy sandstones. Sandstones with the same porosity values may have permeability variations up to four orders of magnitude. The highest permeability values are measured in clean sandstones with well-connected pores. Low permeability is generally characteristic of dirty sandstones with abundant lithic fragments and clay matrix. The lithic fragments and clay matrix may have abundant

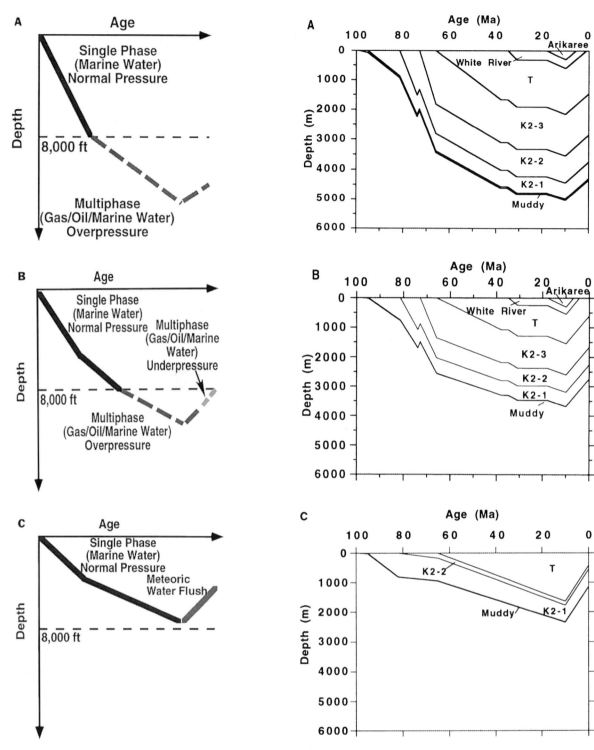

FIG. 9.—Schematic diagram showing three potential burial history scenarios, in the center, halfway to the margin, and at the margin of the Powder River Basin. In the center of the basin, (A) the Muddy sandstone has seen only marine connate waters. Its fluid-flow system has evolved from single-phase to multiphase and from normally pressured to overpressured. Halfway to the basin margin, (B) the rocks also have seen only marine connate waters, and the fluid-flow system has evolved from single-phase to multiphase, but the initial normally pressured regime has been followed by overpressuring and finally by underpressuring. At the basin margin, (C) the Muddy Sandstone has seen early connate waters, later flushed by meteoric water, but always in a single-phase, normally pressured fluid-flow regime.

FIG. 10.—Reconstructed burial history diagrams for the center, half way to the margin, and the margin of the Powder River Basin. (A, B) In the basin center and half way to the basin margin, K_{2-1} represents the Lower Cretaceous Mowry Shale and the Upper Cretaceous Frontier and Niobrara formations. K_{2-2} represents the Upper Cretaceous Steel Shale. K_{2-3} represents the Upper Cretaceous Mesaverde and Lance formations. T represents the Tertiary Fort Union and Wasatch formations. (C) At the basin margin, K_{2-1} represents the Lower Cretaceous Mowry Shale and the Upper Cretaceous Frontier and Niobrara formations. K_{2-2} represents the Upper Cretaceous Steel Shale, Mesaverde, and Lance formations. T represents the Tertiary Fort Union and Wasatch formations.

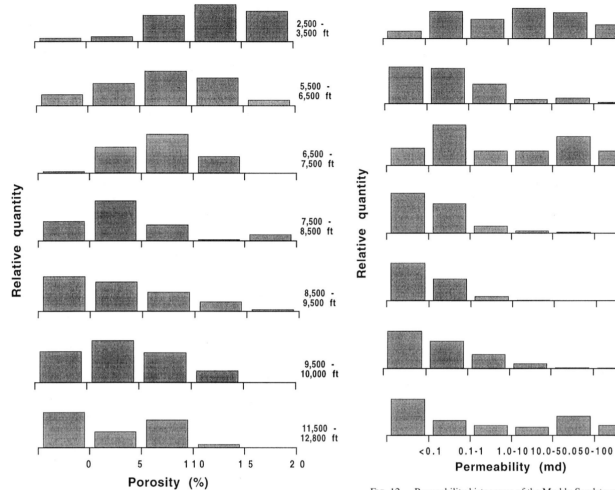

FIG. 11.—Porosity histograms of the Muddy Sandstone in the Powder River Basin. There is no apparent decrease in porosity with increasing burial within the pressure compartment between depths of 7500 and 11,500 ft.

FIG. 12.—Permeability histograms of the Muddy Sandstone in the Powder River Basin. There is no apparent decrease in permeability with increasing burial within the pressure compartment between depths of 7500 and 11,500 ft.

micropores, but these pores are usually poorly connected, and permeability in these rocks is thus very low.

Diagenesis of the Muddy Sandstone

The majority of Muddy sandstones can be classified into two major groups: quartzarenites and litharenites. The quartz-rich arenites are composed predominantly of quartz and chert grains. The lithic-rich arenites contain abundant rock fragments and feldspar grains. A portion of the clay matrix in these lithic-rich sandstones was derived from clay infiltration related to pedogenic processes, and some of the matrix is crushed lithic fragments deformed during compaction.

Because the Muddy Sandstone in the Powder River Basin has been subject to differential subsidence and uplift, the reservoir sandstones in the formation have experienced various chemical, thermal, and hydrologic conditions. These post-depositional changes have complicated the diagenetic evolution of the Muddy Sandstone. Pressure change with depth and hydrocarbon accumulation in the deep sandstones have also affected diagenesis.

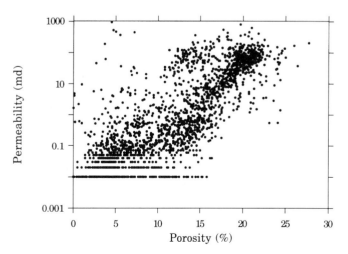

FIG. 13.—Plot of core porosity versus permeability in the Muddy Sandstone in the Powder River Basin.

Most of the quartzarenites in the Muddy Sandstone are porous, and those at shallow depth show high permeability (Fig. 14A, see p. 95). Most of the well-connected pores in these sandstones are secondary pores obtained from the dissolution of early carbonate cement. Scattered carbonate cement patches with irregular, embayed margins observed in the sandstones are evidence of this carbonate dissolution (Fig. 14B, see p. 95). The high intergranular volume filled by carbonate cement in the cemented patches indicates that the precipitation of the carbonate cement occurred at shallow burial depth. The early carbonate cement protected the sandstones from subsequent mechanical compaction. Dissolution of the cement after uplift is responsible for some of the high porosity values in the present-day 0 to 2400 m (8 to 8000 ft) depth interval.

Dissolution of lithic fragments and clay matrix are observed in some of the lithic-rich sandstones at a depth of around 900 m (3000 ft). Authigenic kaolinite was precipitated in some of the dissolution pores, as well as in intergranular pores (Fig. 14C, see p. 95). Dissolution of the solid phases has enhanced (moldic) porosity in these sandstones. However, these moldic pores can also be poorly connected, and permeability in the lithic-rich sandstones is usually low.

Some of the quartz-rich sandstones at shallow depth are intensively cemented by quartz overgrowths (Fig. 14D, see p. 95). The quartz overgrowth cement interlocks the detrital grains and fills all the pore spaces in the sandstones, destroying all effective porosity. Burial modeling of the Powder River Basin suggests that at least 900 m (3000 ft) of overburden have been eroded from above the quartz-cemented sandstones on the east flank of the basin (Jiao, 1992).

Below a depth of 2400 m (8000 ft), intergranular pores are still abundant and well connected in some of the quartz-rich sandstones. Porosity in some of these sandstones is greater than 10% at a depth of 4300 m (14,000 ft) (Fig. 15A, see p. 96). These clean pores have probably been created by the dissolution of carbonate cement. The existence of irregular carbonate cement patches in these sandstones supports this conclusion (Fig. 15B, see p. 96). These quartz-rich sandstones were probably cemented by carbonate at shallow burial, and this early cement partially protected the sandstones from crushing by mechanical compaction with burial. Dissolution of the early carbonate cement under favorable geochemical conditions subsequently enhanced porosity and permeability in the sandstones. Alternatively, these sandstones may have resisted compaction and retained some degree of primary porosity.

Some of the Muddy sandstones below 2400 m (8000 ft) in depth are still cemented by carbonate (Fig. 15C, see p. 96). Also, some of the quartz-rich sandstones in the deeply-buried section of the Muddy Sandstone are intensively cemented by quartz overgrowths (Fig. 15D, see p. 96).

Deep in the section, most of the litharenites are tightly compacted. Deformed rock fragments and infiltrated clay matrix occlude the intergranular pores in these sandstones. Dissolution of lithic and feldspar grains has led to the development of a few isolated intragranular pores, or dissolution pores; these sandstones are nearly impermeable.

As mentioned above, the Muddy Sandstone probably has been characterized by overpressuring since it was buried to 2750+ m (9000+ ft) deep (Jiao and Surdam, 1993). The boundary for abnormal pressure development is presently located at a depth of around 2400 to 2700 m (8000 to 9000 ft), and separates two fluid-flow regimes. The formation fluids above the top pressure boundary are in hydrologic communication with the surface water system, and the formation fluids below the seal are restricted to overpressured compartments at the various scales discussed above. MacGowan et al. (1995) observed that the total dissolved solids values in Muddy Sandstone formation waters are less than 1000 ppm above the pressure boundary, but abruptly increase to 3000–4000 ppm or more below the boundary. Their investigations indicate that formation fluids presently found below the pressure boundary consist of marine connate water that was trapped at, or close to, the sediment-water interface. The shallow sandstones above the top pressure seal have been thoroughly flushed by meteoric waters since uplift. The early carbonate cementation and some subsequent dissolution probably occurred as the formation subsided. However, after uplift, the meteoric water appears to have continuously dissolved the early carbonate cement, creating porosity and permeability in the shallow sandstones. Where the Muddy Sandstone entered the overpressured zone, the restriction of formation-fluid circulation and replacement of water by hydrocarbons retarded diagenetic changes in these sandstones.

The porosity and permeability of Muddy sandstones were greatly modified following deposition, during basin subsidence and uplift. Negative diagenetic changes affecting sandstone porosity include mechanical compaction and carbonate and quartz cementation. Positive changes include dissolution of some solid phases and retardation of compaction and cementation by overpressuring and hydrocarbon accumulation. Therefore, the sandstones with the greatest porosity and permeability include those quartz-rich sandstones above and below the pressure seal that have experienced carbonate cementation and subsequent dissolution and minimum quartz cementation. Again, although there is a progressive downward loss of porosity in the upper 2400 m (8000 ft) of the section, once the abnormally pressured section is entered, porosity loss is significantly inhibited (Fig. 11).

EFFECT OF THE FLUID-FLOW REGIME ON THE SEALING CAPACITY OF
FLUID-FLOW BARRIERS AND ON SHALE DIAGENESIS

Capillary seals resulting from the evolution of a single-phase fluid-flow regime to a multiphase regime are important components of potential hydrocarbon accumulation, because these seals may lead to pressure compartmentalization. Without effective capillary sealing, any pressure across a flow barrier is dispelled in a (geologically) short time (Iverson et al., 1995). Low-permeability flow barriers may result from the original depositional environment, compaction, clay diagenesis, diagenetic cementation, or a combination of these elements.

According to the relationship $P_c = 2\gamma \cos\theta/R$ (Purcell, 1949), the capillary pressure P_c in a fluid-filled rock is a function of hydrocarbon-water interfacial tension γ, wettability θ, and pore-throat radius R. This relationship shows that P_c increases as pore-throat radius and wettability decrease and as hydrocarbon-water interfacial tension increases. Typically, the most effective seals to fluid flow are low-permeability, laterally continuous, ductile rocks with high capillary displacement pressure (e.g., evaporite, organic-rich shale, and regionally developed paleosol). The sealing capacity of a seal is defined as the minimum pressure required to displace formation fluids through or across it, and is expressed as the height H of the fluid column that the seal would support (Sneider et al., 1991).

In the Powder River Basin, the most common low-permeability rock associated with hydrocarbon accumulations or pressure compartment boundaries is transgressive shale. This situation is typical of many hydrocarbon accumulations, for more than 60% of the known giant oilfields worldwide have shale seals on sandstone reservoirs (North, 1985). However, it should be noted that burial diagenesis, recrystallization, pressure solution and reprecipitation, deformation of ductile grains, or pedogenic processes can convert a normally porous sedimentary rock into a fluid-flow barrier.

Thus, rocks of many types serve as potential capillary seals and pressure compartment boundaries in the Powder River Basin, including paleosols associated with lowstand unconformities (lowstand surfaces of erosion, or LSEs), clay infiltrated sandstones beneath these unconformities, transgressive shales, and carbonate and quartz cemented sandstones associated with fractures and perhaps faults. However, even though most potential seals have a measurable permeability of less than 0.1 md, no sedimentary rock is totally impermeable. A low-permeability rock in a multiphase fluid-flow system will become a pressure compartment or hydrocarbon seal only insofar as the differential pressure in the compartment is less than the sealing rock's displacement pressure. Capillary pressure is thus a key element in the sealing capacity of a pressure compartment or hydrocarbon seal.

Paleosol and Sandstone with Infiltrated Clay

Porosity and Permeability.—

In the Powder River Basin, paleosols and clay-infiltrated sandstones are generally associated with several regional lowstand unconformities (LSEs; Dolson et al., 1991; Martinsen, 1995). Petrographic evidence shows these sandstones to be characterized by abundant clay matrix and an absence of intergranular pore space (Fig. 16A, see p. 97). The clay matrix constitutes about 20% of the rock volume, on average, and in some samples up to 55% of the rock volume. The clay matrix percentage rapidly increases toward a lowstand unconformity, and porosity decreases from approximately 18% in the reservoir sandstone (Fig. 16B, see p. 97) to 5% in the sandstone in the unconformity zone (Fig. 16A; Jiao and Surdam, 1993).

Scanning electron microscope photomicrographs show that intergranular pore space is absent in the sandstone associated with the unconformity zones, but that microporosity may be common in the matrix. An example of microporosity associated with abundant clay is found in core samples from the Amos Draw 3 well (Samples 1 and 2, Table 1 and Fig. 17); another example is a soil-zone sandstone sample (not listed in Table 1) taken from the Federal 1–33 well, which has an average 10.4% porosity but permeability below 0.01 md. This low permeability results from an almost complete lack of interconnected pores in the clay matrix of the pedogenically modified sandstone.

A marked decrease in pore-throat radii in the unconformity zone is shown in Figure 18. Samples 1 and 2 are from the unconformity zone, and pore-throat radii are primarily in the subnano and nano categories (<0.01 to 0.05 microns); permeability ranges from 0.02 to 0.08 md. Sample 3, located 12 ft

Sample No.	Depth (ft)	Porosity (%)	Perm. (md)	Pd(psi)* Gas/Brine	Sealing Capacity H (ft, gas)
Paleosol					
4	5541	9.0	0.2	60	160
5	8814	7.3	0.1	130	350
1	10,010	7.4	0.08	1800	4600 (reservoir
3	10,022	17.1	12	3	sandstone)
Clay Infiltrated Sandstone					
2	10,014	7.0	0.02	1000	2770

*Pd is displacement pressure
1–3, #3, Amos Draw Field
4, Daly 17–1, Kitty Field
5, Government #1, Wildcat Field

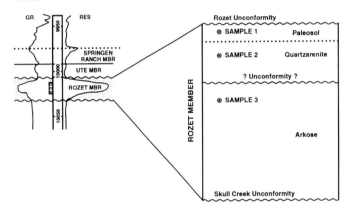

NESE SEC. 29 T51N R75W
AMOS DRAW FEDERAL #3

FIG. 17.—Location of samples for high-pressure mercury injection tests: three Muddy sandstone samples from the Amos Draw Federal #3 well. Samples 1 and 2 were taken from the Rozet unconformity zone, Sample 3 from reservoir sandstone.

below Sample 1, has pore throats primarily in the meso and macro categories (0.25 to 2.5 microns), and permeability is three orders of magnitude higher at 12 md. Sample 3 was taken from the hydrocarbon producing interval at the Amos Draw field.

Sealing Capacity.—

The sealing capacity of a pressure seal is a function of pore-throat radii in the rock and fluid characteristics within the pore space. The sealing capacity of a seal is expressed as the height in feet of the hydrocarbon column that can be supported by the seal. The equation developed by Smith (1966) was used in this study for calculating the sealing capacity H, in terms of minimum displacement pressure Pd of an unconformity seal.

In our investigation, three bulk sandstone samples (Sample 1, 2, and 3, Table 1 and Fig. 17) associated with unconformities were subjected to high-pressure mercury injection tests (Fig. 19). Sample 1 was taken from the paleosol zone, Sample 2 from the clay infiltration and diagenetic alteration zone, and Sample 3 from the reservoir sandstone. Data on two additional pedogenic samples (Samples 4 and 5, Table 1) were assembled from the USGS core library. For each sample, the displacement pressure Pd, the minimum pressure required to start mercury mov-

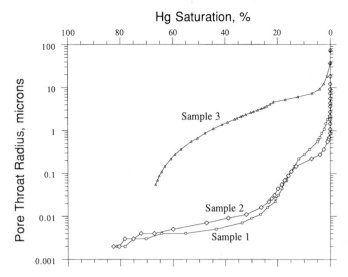

FIG. 18.—Pore-throat radius distribution as determined from the results of mercury injection tests on samples of paleosol (Sample 1), clay-infiltrated sandstone (Sample 2), and reservoir sandstone (Sample 3). Samples 1 and 2 contain pore-throats primarily in the sub-nano and nano categories (<0.01 to 0.5 μm), while Sample 3 contains pore-throats primarily in the meso and macro categories (0.25 to 2.5 μm), as determined for histograms.

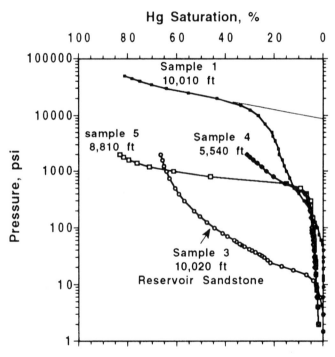

FIG. 19.—Mercury injection pressure curves. Data on Samples 1, 2, and 3 (located in Fig. 17) are from this study. Data on Samples 4 and 5 were assembled from the USGS Core Library. Samples 1–5 correspond to Samples 1–5 in Table 1. Sample 2 curve is close to Scale 1 curve.

ing through the seal, was determined by extrapolating the injection pressure curve (plateau portion) to 0% mercury saturation (as for Sample #1 in Fig. 19). This pressure was converted to the displacement pressure Pd for the subsurface

gas/water and oil/water systems by using Schowalter's (1979) nomograms. Results are shown in Table 1.

According to the Sneider classification (Sneider et al., 1991), our Samples 1 and 2 are A-type seals, and Samples 4 and 5 are C-type seals (Sample 3 is a reservoir sand, not a seal). Thus, the sealing capacity of the unconformity zone (in this case the Rozet unconformity zone) increases substantially with increasing depth of burial.

It is interesting that the increased sealing capacity of the Rozet unconformity coincides with progressive clay diagenesis as the structure of mixed-layer illite/smectite (I/S) changes from random to ordered at a depth of approximately 2700 m (9000 ft); on the basis of these few samples, the sealing capacity of the Rozet unconformity appears to jump from C-type to A-type at the top of the anomalously pressured zone (Fig. 20). It appears that there is a marked coincidence between clay diagenesis, displacement pressure, sealing capacity, and the fluid-flow regime in rocks associated with the lowstand unconformities in the basin.

Transgressive Shales

Porosity and Permeability.—

The most important Cretaceous transgressive shales in the Powder River Basin are the Skull Creek, Mowry, Frontier, Niobrara, and Steele shales. The measured porosity in these shales is 12% at a depth of 1500 m (5000 ft) and rapidly decreases to 2% at a depth of 3000 m (10,000 ft). The measured permeability is 0.012 md at 1500 m (5000 ft) depth and 0.002 md at 3000 m (10,000 ft) (Table 2). These low-permeability shales provide potential seals for hydrocarbon accumulations in pressure compartments.

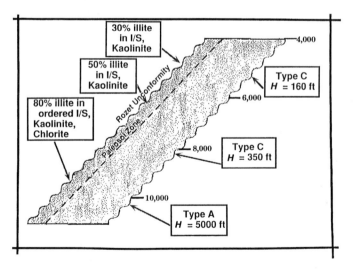

FIG. 20.—Schematic diagram showing the sealing capacity H of the Rozet unconformity pressure seal progressively increasing with increasing burial depth. The proportion of illite in the mixed-layer smectite/illite clays increases from 30% to 80% between 160 and 5000 ft depth, and the sealing capacity of the low-permeability rocks associated with the unconformity increases by an order of magnitude, from 350 ft (gas column) at 8000 ft deep depth to 5000 ft at 9000 ft depth. Type A seal was formed when the unconformity zone was buried more than 9000 ft deep.

TABLE 2.— DISPLACEMENT PRESSURE *Pd* AND SEALING CAPACITY *H* OF TRANSGRESSIVE SHALES, POWDER RIVER BASIN

Depth (ft)	Porosity Ø (%)	Permeability K (md)	*Pd* (psi, gas/brine)	Sealing capacity *H* (ft, gas column)
5500	11.7	0.012	400	1000
7900	5.6	0.013	1500	3800
7900	5.8	0.059	1300	3300
8800	4.9	0.012	1800	4600
9900	2.7	0.006	2800	7100
10,000	2.2	0.002	3000	7600
13,000	2.2	0.004	4000	10,000

Sealing Capacity.—

In our investigation, seven bulk transgressive shale samples (Table 2) from the Cretaceous section were subjected to high-pressure mercury injection tests; results from three of these tests are shown in Figure 21. The displacement pressure and sealing capacity were determined using the methods described by Schowalter (1979) and Smith (1966). Results are shown in Table 2 and Figure 22; the displacement pressure and thus the sealing capacity of the transgressive shale increase steadily with burial depth (Fig. 22 shows sealing capacity vs. depth; the plot of displacement pressure vs. depth is identical). For the gas/water system, the sealing capacity of the Cretaceous transgressive shales is 300 m (1000 ft; gas column) at a present-day depth of 1700 m (5500 ft) and rapidly increases to 1200 m (4000 ft) at 2400 m (8000 ft) and to 3000 m (10,000 ft) at 4000 m (13,000 ft). The displacement pressure also shows significant increase with burial. At a present-day depth of 4000 m (13,000 ft), the

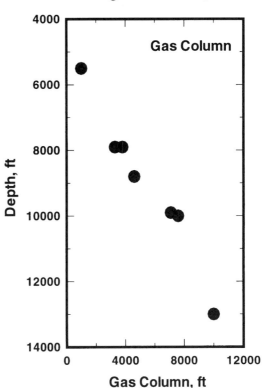

FIG. 22.—Plot of sealing capacity versus depth for transgressive shales, Powder River Basin, showing significant increase in sealing capacity below 8000 ft depth.

transgressive shale can withstand a differential pressure of 4000 psi.

Thermal Maturation and Organic-Rich Shales.—

There is a significant change in the vitrinite reflectance/depth gradient at 2700 ± 300 m (9000 ± 1000 ft) present-day depth in the central Powder River Basin (Jiao and Surdam, 1993). Vitrinite reflectance increases slowly from 0.4 to 0.8% between depths of 1200 and 2700 m (4000 and 9000 ft), and more rapidly from 0.8 to 1.6% between 2700 and 4100 m (9000 and 13,500 ft). The classical liquid oil generation window is generally thought to lie between 0.5–0.7% and 1.0–1.3% reflectance (Tissot and Welte, 1984). At greater than 1.3% vitrinite reflectance, hydrocarbon generation is mainly wet gas.

There is a significant increase in the production index (PI, from anhydrous pyrolysis) at approximately 2700 m (9000 ft) present-day depth in the central Powder River Basin (Jiao, 1992). The trend in PI from shales in the Powder River Basin indicates a marked increase in hydrocarbons within the shales below a present-day depth of 2700 m (9000 ft). Above a depth of 2700 m (9000 ft), the PI is typically <0.1, but below that depth the PI is >0.1 and may rise to as high as >0.3 at 4300 m (14,000 ft).

Further insight into the mechanisms of overpressuring in the shale section may be gained from nuclear magnetic resonance

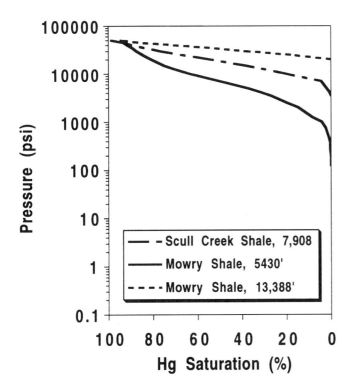

FIG. 21.—Plot of mercury saturation versus pressure for transgressive shales, Powder River Basin.

(NMR) studies of a sequence of Mowry Shale samples over the present-day depth interval 900 to 3400+ m (3000 to 11,000+ ft) reported by Jiao and Surdam (1993). Aliphatic functional groups have been cleaved off the kerogen in the Mowry Shale source rocks below a depth of 2700 m (9000 ft). The loss of aliphatic groups is associated with hydrocarbon production; thus, it appears that a significant percentage of the liquid hydrocarbon in these shales had been generated by the time of burial to a present-day depth of 2700 ± 300 m (9000 ± 1000 ft; i.e., a maximum burial depth of 3700 m (12,000 ft) or less).

These trends in vitrinite reflectance, production index, and nuclear magnetic resonance results have led us to suggest that above a depth of 2700 ± 300 m (9000 ± 1000 ft) in the shales there is a paucity of hydrocarbons; that below this depth, hydrocarbons have been generated in copious quantities; and that significant quantities of generated hydrocarbons have *not* been expelled from the shales.

Clay Diagenesis and Sealing Capacity of Pressure Seals.—

Clay diagenesis has played a role in decreasing porosity and permeability in the Cretaceous paleosols, clay-infiltrated sandstones, and transgressive shales in the Powder River Basin. The diagenesis of clay minerals in these rocks includes smectite altering to illite in mixed-layer illite/smectite (I/S) clays, smectite altering to chlorite in mixed-layer chlorite/smectite clays, and kaolinite reacting to chlorite.

The diagenesis of mixed-layer illite/smectite and chlorite/smectite clays during the progressive burial of a sedimentary sequence is widely recognized and is considered to be an important empirical diagenetic geothermometer; it may also be related to regional hydrocarbon generation (Burst, 1969; Hower et al., 1976; Boles and Franks, 1979; Hower, 1981; Pytte and Reynolds, 1989; Bruce, 1984; Hagen and Surdam, 1984). It has also been demonstrated that the diagenesis of I/S clay has a significant effect on porosity and permeability (Foster and Custard, 1980; Ahn and Peacor, 1986; Freed and Peacor, 1989). The cited studies have documented the main compositional and structural changes in the I/S burial diagenetic sequence to be: (1) increase in illite layers, (2) increase in interlayer potassium, (3) increase in aluminum substituted for silicon in the tetrahedral layer, and (4) release of Mg^{2+}, Fe^{2+}, Ca^{2+}, Si^{4+}, Na^+, and water. Two important points can be made concerning these changes. First, the structural changes and water release have a significant effect on the porosity and permeability of the shale. Second, the released cations may affect porosity and permeability by abetting the precipitation and deposition of quartz, chlorite, kaolinite and late carbonate.

We used X-ray diffractometry to determine the I/S diagenetic profile for the Mowry Shale in the Powder River Basin. The I/S composition changes with progressive burial, from approximately 10% illite at a present-day depth of 900 m (3000 ft) to 85% illite at 4100 m (13,500 ft; Fig. 23). This trend can also be shown for mixed-layer I/S in the paleosol and clay infiltrated sandstone samples from the Rozet unconformity zone (Fig. 20): with increasing burial depth, the percentage of illite in the mixed-layer I/S clays steadily increases from 30% at about 1200 m (4000 ft) present-day depth to 80% at 3000 m (10,000 ft), and the clay structure also becomes ordered.

The effect on the shale matrix of the smectite-to-illite transition can be explained in several ways. Foster and Custard

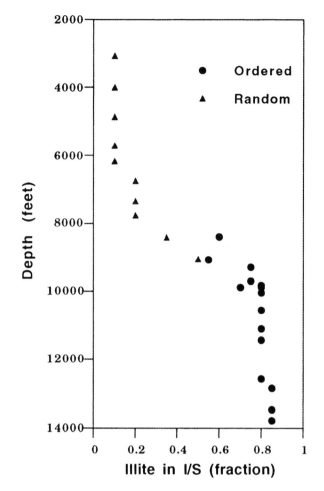

FIG. 23.—Plot of percent illite in mixed-layer smectite/illite clays versus depth in the Mowry Shale, Powder River Basin.

(1980) have suggested that finely-divided silica is produced during the transition, leading to loss of permeability. More recently, Freed and Peacor (1989) have proposed that ions and fluids are able to move through at least two different pathways other than fractures: (1) avenues provided by the abundant dislocations within smectite megacrystals and (2) irregular contacts between smectite megacrystals and adjacent mineral grains such as quartz and feldspar. Freed and Peacor (1989) have suggested that dislocations within smectite megacrystals constitute the major pathway for ion and fluid transport through shale.

On the basis of transmission electron microscopy observations, Ahn and Peacor (1986) reported that when smectite megacrystals dominate the clay matrix, abundant dislocations within the smectite provide pathways for ion transport, contributing to local permeability. As the transition of smectite to illite in the mixed-layer I/S proceeds, illite packets thicken, become dominant, and eventually coalesce. Pathways for ion transport become more and more limited. The result is a locally continuous matrix of subparallel illite packets; local permeability is reduced by this illite mat, and hydraulic continuity is severely restricted.

Freed and Peacor (1989) have proposed that the development of subparallel illite packets is sufficient to produce an effective hydraulic seal. At the onset of the smectite-to-illite transition, the hydraulic barrier begins to develop. The growth of coalescing illite packets leads to the gradual development of a more efficient barrier; and, as described above, in the Powder River Basin this barrier development hypothesis is consistent with the observation that the onset of overpressuring, hydrocarbon accumulation, and clay transition zones are spatially related. Also, the transformation of smectite to illite in the mixed-layer I/S may reduce the effective permeability by favoring carbonate and silica cementation: when smectite alters to illite, it takes up K and Al and releases Na, Ca, Fe, Mg, and Si.

In the Powder River Basin, progressive clay diagenesis and the consequent change in the structure of mixed-layer I/S from random to ordered coincides with increasing sealing capacity; the sealing capacity of Mowry shale increases from 300 m (1000 ft) to 1500 m (5000 ft) between present-day depths of 300 and 4000 m (1000 and 13,000 ft). Furthermore, the percentage of illite in the mixed-layer I/S in the shales increases from 20% at approximately 2400 m (8000 ft) deep to 75% at approximately 3000 m (10,000 ft) deep, and the clay structure becomes ordered in that interval. This depth range coincides with that of the transition between normal and overpressured rocks, and with that of single-phase to multiphase fluid-flow in the Cretaceous shales. This coincidence may indicate that the transition of smectite to illite in the mixed-layer I/S plays an important role in this increase in sealing capacity (or decrease in local permeability).

The coincidence of progressive clay diagenesis (and the consequent change in the clay structure from random to ordered) with markedly increased sealing capacity is also observed in the paleosol and clay-infiltrated sandstone. With increasing burial depth, the proportion of illite in the I/S clays increases from 30% to 80% and the chlorite/kaolinite ratio also increases; the sealing capacity of the paleosol and clay infiltrated sandstone increases by an order of magnitude, and the displacement pressure of the seals increases from 100 to 2000 psi. Coincident with clay diagenesis, Type-C seals were converted to Type-A seals during maximum burial (~10 Ma).

In the Mowry/Muddy system, and in the other Cretaceous shales in the Powder River Basin, there is concurrence of significant changes in fluid-flow regime, hydrocarbon generation, clay diagenesis, and sealing capacity. It is difficult to definitely assign roles of cause or effect to these processes. However, it is clear that as the fluid-flow regime evolves from single-phase to multiphase with the addition of hydrocarbons, the mixed-layer clay undergoes diagenesis from smectite-rich to illite-rich and from random to ordered, and the hydrocarbon sealing capacity and capillary displacement pressure of the shales increase by an order of magnitude.

As discussed above, in the Powder River Basin, the upper 2700 ± 300 m (9000 ± 1000 ft) of section is normally pressured and dominantly a water driven system (a single-phase fluid-flow regime), while below a depth of 2700 m (9000 ft) the shales are typically overpressured and characterized by a fluid-flow regime dominated by hydrocarbons (a multiphase fluid-flow regime). Therefore, it has been concluded that it is the relative permeability (permeability to a given fluid), not the absolute permeability, in flow barriers that plays a major role in the overpressuring and pressure compartmentalization of the shale-rich Cretaceous section in the Powder River Basin.

CONCLUSIONS

The large overpressured Cretaceous shale section in the Powder River Basin is a regional, gas-saturated, anomalously pressured compartment. The driving mechanism in its formation was the generation and storage of liquid hydrocarbon and subsequent reaction of the stored liquid hydrocarbon to gas. This mechanism ensured the conversion of the fluid-flow regime from a single-phase, water-dominated regime to a multiphase, hydrocarbon-dominated regime. As a consequence, the displacement pressure of the shale increased significantly as relatively low-permeability fluid-flow barriers were converted to capillary seals. In the Powder River Basin, the shale-rich Cretaceous section below a present-day depth of 2700 m (9000 ft) became a regional pressure compartment, probably at about the time of maximum burial (10–20 Ma). In contrast, above and below the regional pressure compartment within the shale-rich Cretaceous section, the fluid-flow regime is a water dominated single-phase regime characterized by strong water drive, at normal (hydrostatic) pressure.

The major difference between pressure compartmentalization in the Cretaceous sandstone and shale is one of scale. In both cases, the driving mechanism was the generation and storage of liquid hydrocarbons and the reaction of liquid hydrocarbons to gas. Consequently, fluid-flow barriers were converted to capillary seals. Where three-dimensional closure of the low-permeability fluid-flow barriers occurred, fluid compartmentalization may have occurred in the sandstone. When hydrocarbons saturated the compartments, the integrity of the three-dimensional bounding capillary seals was established, resulting in anomalously pressured gas accumulations.

Following the conversion of the fluid-flow regime from single-phase to multiphase, the sealing capacity of hydrocarbon/pressure seals increased significantly with increasing burial depth and progressive clay diagenesis, and finite displacement pressures developed in the Powder River Basin. Progressive clay diagenesis and the consequent change in the structure of mixed-layer I/S from random to ordered evidently coincide with markedly increasing sealing capacity in the shales, paleosols, and clay-infiltrated sandstones. The corresponding depth range, 2400 to 3000 m (8000 to 10,000 ft), coincides with the depth range of the conversion of the fluid-flow regime from single-phase to multiphase and the pressure regime from normal to overpressured in the Cretaceous section in the Powder River Basin.

The effects on sandstone diagenesis of the evolution of the fluid-flow regime have depended on position in the basin. In the basin center, the Muddy Sandstone has contained only marine connate waters. Its fluid-flow regime has evolved from single-phase to multiphase and its pressure regime from normally pressured to anomalously pressured in isolated compartments. The overpressure compartmentalization was accompanied by the arrival of hydrocarbons. The overpressuring reduced the effective rock stress, the rate of compaction, and mineral reaction rates, resulting in significant inhibition of porosity and permeability loss within the overpressured compartments between present-day depths of approximately 2400 and 4000 m (8000 and 13,000 ft). However, in the Muddy Sandstone close to the basin margin, early connate waters were later flushed by meteoric water, but always in a single-phase, normally pressured fluid-flow regime. Here, the formation fluid was in hydrologic communication with the surface meteoric water regime. Within such a fluid-flow regime, early carbonate cementation and some subsequent dissolution probably occurred during subsidence. After uplift and with meteoric water flushing, the early carbonate cement continuously dissolved, creating some enhanced porosity and permeability; some of the permeability and porosity was lost during the flushing, however, due to kaolinite precipitation.

We suggest that the evolution of the fluid-flow regime had a marked effect on the progressive diagenesis of both shale and sandstone, and that the arrival of hydrocarbons in the fluid-flow regime was responsible for the most significant changes that took place in reservoir quality and the sealing capacity of the shales during the progressive burial of the Cretaceous stratigraphic section in the Powder River Basin of Wyoming.

ACKNOWLEDGMENTS

The original manuscript was reviewed by David Copeland and Kathy Kirkaldie (Institute for Energy Research, University of Wyoming), who made numerous helpful suggestions. Suggestions from SEPM reviewers James Hickey, Julie Kupecz, and Robert Loucks further improved the manuscript. This study was funded by the Gas Research Institute under Contract No. 5089–260–1894 and the NSF-EPSCoR-ADP under Grant No. EHR-910–8774.

REFERENCES

AHN, J. H. AND PEACOR, D. R., 1986, Transmission and analytical electron microscopy of the smectite-to-illite transition: Clays and Clay Minerals, v. 34, p. 165–179.

BERG, R. R., 1975, Capillary pressure in stratigraphic traps: American Association of Petroleum Geologists Bulletin, v. 59, p. 939–956.

BOLES, J. R. AND FRANKS, S. G., 1979, Clay diagenesis in Wilcox sandstones of southwest Texas, implications of smectite diagenesis on sandstone cementation: Journal of Sedimentary Petrology, v. 49, p. 55–70.

BRUCE, C. H., 1984, Smectite dehydration—its relation to structural development and hydrocarbon accumulation in northern Gulf of Mexico Basin: American Association of Petroleum Geologists Bulletin, v. 68, p. 673–683.

BURST, J. R., Jr., 1969, Diagenesis of Gulf Coast clay sediments and its possible relationships to petroleum migration: American Association of Petroleum Geologists Bulletin, v. 53, p. 487–502.

DOLSON, J. C., MULLER, D. S., EVETTS, M. J., AND STEIN, J. A., 1991, Regional paleotopographic trends and production, Muddy Sandstone (Lower Cretaceous), Central and Northern Rocky Mountains: American Association of Petroleum Geologists Bulletin, v. 75, p. 409–435.

FOSTER, W. R. AND CUSTARD, H. C., 1980, Smectite-illite transformation, role in generating and maintaining geopressure: American Association of Petroleum Geologists Bulletin, v. 64, p. 708.

FREED, R. L. AND PEACOR, D. R., 1989, Geopressured shale and sealing effect of smectite to illite transition: American Association of Petroleum Geologists Bulletin, v. 73, p. 1223–1232.

HAGEN, E. S. AND SURDAM, R. C., 1984, Maturation history and thermal evolution of Cretaceous source rocks of the Big Horn Basin, Wyoming and Montana, in Woodward, J., Meissner, F., and Clayton, J., eds., Hydrocarbon Source Rocks of the Greater Rocky Mountain Region: Denver, Rocky Mountain Association of Geologists, p. 321–338.

HAM, H. H., 1987, A method of estimating formation pressures from Gulf Coast well logs, in Dutta, N., ed., Geopressure: Tulsa, Society of Exploration Geophysicists, Geophysics Reprint Series No. 7, p. 295–307.

HEASLER, H. P., GEORGE, J. P., AND SURDAM, R. C., 1995, Pressure compartments in the Powder River Basin, Wyoming and Montana as determined from drill-stem test data, in Ortoleva, P., ed., Basin Compartments and Seals: Tulsa, American Association of Petroleum Geologists Memoir 61, p. 235–262.

HOWER, J., 1981, Shale diagenesis, in Longstaff, F., ed., Clay Diagenesis and the Resource Geologists: Toronto, Mineralogical Association of Canada, p. 60–80.

HOWER, J., ESLINGER, E. V., HOWER, M. E., AND PERRY, E. A., 1976, Mechanism of burial and metamorphosis of argillaceous sediments: 1. Mineralogical and chemical evidence: Geological Society of America Bulletin, v. 87, p. 725–737.

IVERSON, W. P., MARTINSEN, R. S., AND SURDAM, R. C., 1995, Pressure seal permeability and two-phase flow, in Ortoleva, P., ed., Basin Compartments and Seals: Tulsa, American Association of Petroleum Geologists Memoir 61, p. 313–319.

JIAO, Z. S., 1992, Thermal maturation/diagenetic aspects of the abnormal pressure in the Cretaceous shales and sandstones in the Powder River Basin, Wyoming: Unpublished Ph.D. Dissertation, University of Wyoming, Laramie, 242 p.

JIAO, Z. S. AND SURDAM, R. C., 1993, Low-permeability rocks, capillary seals, and pressure compartment boundaries in the Cretaceous section of the Powder River Basin: Casper, Wyoming Geological Association Jubilee Anniversary Field Conference Guidebook, p. 297–310.

MACGOWAN, D. B., JIAO, Z. S., AND SURDAM, R. C., 1995, Formation water chemistry of the Muddy Sandstone, and organic geochemistry of the Mowry Shale, Powder River Basin, Wyoming: Evidence for mechanism of pressure compartment formation, in Ortoleva, P., ed., Basin Compartments and Seals: Tulsa, American Association of Petroleum Geologists Memoir 61, p. 321–331.

MAGOON, L. B., AND DOW, W. G., 1994, The petroleum system, in Magoon, L., and Dow, W., eds., The Petroleum System – From Source to Trap: Tulsa, American Association of Petroleum Geologists Memoir 60, p. 3–24.

MARTINSEN, R. S., 1995, Stratigraphic compartmentalization of reservoir sandstones: Examples from the Muddy Sandstone, Powder River Basin, Wyoming, in Ortoleva, P., ed., Basin Compartments and Seals: Tulsa, American Association of Petroleum Geologists Memoir 61, p. 273–296.

MAUCIONE, D. T., SEREBRYAKOV, V., VALASEK, P., WANG, Y., AND SMITHSON, S. B., 1995, A sonic log study of abnormally pressured zones in the Powder River Basin of Wyoming using sonic logs, in Ortoleva, P., ed., Basin Compartments and Seals: Tulsa, American Association of Petroleum Geologists Memoir 61, p. 333–348.

NORTH, F. K., 1985, Petroleum Geology: Boston, Allen and Unwin, 607 p.

ODLAND, S. K., PATTERSON, P. E., AND GUSTASON, E. R., 1988, Amos Draw Field: A diagenetic trap related to an intraformational unconformity in the Muddy Sandstone, Powder River Basin, Wyoming, in Diedrich, R., et al., eds., Eastern Powder River Basin—Black Hills: Casper, Wyoming Geological Association 39th Field Conference Guidebook, p. 147–160.

PURCELL, W. R., 1949, Capillary pressure—their measurement using mercury and the calculation of permeability therefrom: American Institute of Mining Engineers Transactions, v. 186, p. 39–46.

PYTTE, A. M. AND REYNOLDS, R. C., Jr., 1989, The thermal transformation of smectite to illite, in Naeser, N. and McCullok, T., eds, Thermal History of Sedimentary Basins: New York, Springer-Verlag, p. 133–140.

SCHOWALTER, T. T., 1979, Mechanics of secondary hydrocarbon migration and entrapment: American Association of Petroleum Geologists Bulletin, v. 63, p. 723–760.

SMITH, D. A., 1966, Theoretical considerations of sealing and non-sealing faults: American Association of Petroleum Geologists Bulletin, v. 50, p. 363–374.

SNEIDER, R. M., STOLPER, K., AND SNEIDER, J. S., 1991, Petrophysical properties of seals: American Association of Petroleum Geologists Bulletin, v. 75, p. 673–674.

SURDAM, R. C., JIAO, Z. S., AND MARTINSEN, R. S., 1995a, The regional pressure regime in Cretaceous sandstones and shales in the Powder River Basin, in Ortoleva, P., ed., Basin Compartments and Seals: Tulsa, American Association of Petroleum Geologists Memoir 61, p. 213–233.

SURDAM, R. C., JIAO, Z. S., AND HEASLER, H. P., 1995b, Pressure compartmentalization in the Cretaceous rocks of the Laramide basins of Wyoming: American Association of Petroleum Geologists Memoir, in press.

TISSOT, B. P. AND WELTE, D. H., 1984, Petroleum Formation and Occurrence, 2nd edition: Berlin, Springer-Verlag, 699 p.

WARING, J., 1976, Regional distribution of environments of the Lower Cretaceous Muddy Sandstone, southeastern Montana, in Laudon, R., ed., Geology and Energy Resources of the Powder River Basin: Casper, Wyoming Geological Association 28th Annual Field Conference Guidebook, p. 83–96.

WYOMING GEOLOGICAL ASSOCIATION, 1988, Stratigraphic Nomenclature Chart (compiled by R. Andrews, S. Dietrich, R. Galze, S. Ordonez, and W. Ruddiman), in Diedrich, R., et al., eds., Eastern Powder River Basin—Black Hills: Casper, 39th Annual Field Conference Guidebook, p. 16. [Annual WGA Guidebooks since 1986 have contained variations of this chart.]

RELATIONSHIP BETWEEN ORGANIC MATTER AND AUTHIGENIC ILLITE/SMECTITE IN DEVONIAN BLACK SHALES, MICHIGAN AND ILLINOIS BASINS, USA

VICTORIA C. HOVER, DONALD R. PEACOR, AND LYNN M. WALTER

Department of Geological Sciences, 2534 C.C. Little Building , University of Michigan, 425 E. University, Ann Arbor, MI 48109–1063

ABSTRACT: Organic-rich Late Devonian Antrim and New Albany Shales are important hydrocarbon source rocks in the Michigan and Illinois Basins. These shales have been investigated using STEM/AEM and SEM techniques to clarify textures and mutual genetic relations between the low to moderately mature organic matter (R_0 = 0.45–0.6 %) and authigenic illite-rich clays.

The Antrim and New Albany Shales contain up to 15 wt % TOC dominated by the marine algae *Tasmanites*. Organic matter forms an interconnected network within the clay-rich matrix with no detectable intergranular pore space even at the STEM scale. The clay-rich matrix is principally illite-rich mixed-layer I/S. Crystal sizes, compositions, defect states and presence of smectite interlayers are consistent with authigenesis of illite from precursor smectite with subsequent preservation of that immature, diagenetic illite after neoformation.

Textural relations verify that gas transport from the Antrim Shale source occurs via desorption from organic matter and diffusion through the interconnected organic and authigenic illite matrix to open fractures. In both the Antrim and New Albany Shales, plastic deformation of organic material entrains illite crystals, suggesting that illite formation preceded or was concurrent with thermal maturation of organic matter. Based on existing thermal maturity data, illite authigenesis occurred prior to Late Pennsylvanian/Early Permian time at depths less than ~2000 m and temperatures less than ~120°C. In contrast to the Gulf Coast Tertiary sequence, fluids derived from smectite dehydration were probably not a significant driving force for primary hydrocarbon migration from these intracratonic basin shales as a result of the presence of relatively thin shale sequences and normal pressures during burial compaction.

INTRODUCTION

The Devonian Antrim and New Albany Shales are important organic-rich source rocks for hydrocarbons in the intracratonic Michigan and Illinois Basins, respectively. Little is known, however, about the textural relationships between organic matter and authigenic clay minerals, particularly mixed-layer illite/smectite (I/S), or the extent to which early diagenetic textures and compositions of these clays have been modified or preserved since formation. The smectite-to-illite transformation is well characterized for relatively young sequences such as the Gulf Coast Tertiary section (Perry and Hower, 1970; Hower et al., 1976; Ahn and Peacor, 1986, 1989; Freed and Peacor, 1989a, b, 1992; Peacor, 1992; Eberl, 1993), the North Sea Mesozoic and Tertiary section (Pearson et al., 1982; Pearson and Small, 1988; Glasmann et al., 1989; Hansen and Lindgreen, 1989; Lindgreen et al., 1991) and for young sediments in other basins (e.g., Kisch, 1983, 1987).

The transformation of smectite-to-illite in Tertiary sediments of the Gulf Coast occurs via mixed layer I/S with increasing burial depth (e.g., Perry and Hower, 1970; Hower et al., 1976; Eberl, 1993). The proportion of illite in mixed layer I/S increases from <20% at pre-transition depths (less than ~2000 m) to ~80% at post transition depths (greater than ~3700 m) according to the generalized reaction: smectite + K^+ + Al^{3+} = illite + Si^{4+} (Hower et al., 1976). The inferred source of K and Al is dissolution of detrital K-feldspar and/or muscovite, the resulting Si produces authigenic quartz. Fe and Mg released during the transformation contribute to formation of authigenic chlorite (e.g., Hower et al., 1976; Boles and Franks, 1979). With increasing temperature and burial depth, the proportion of illite layers in I/S increases and the stacking sequence of illite and smectite evolves from random (R0) to fully ordered ISIS. . . . (R1); (Freed and Peacor, 1992).

Random ordering of I/S begins at temperatures of ~50°C (Perry and Hower, 1970) and fully R1-ordered I/S occurs at ~100–120°C (Hower et al., 1976). Mixed layering with ~80% illite layers persists to the maximum depths sampled (~5000 m); (Hower et al., 1976; Boles and Frank, 1979). The loss of interlayer water from smectite during the transformation has been tied to the development of overpressures in the Gulf Coast (Powers, 1967; Burst, 1969, Bruce, 1984; Bethke, 1986a) and may assist in primary hydrocarbon migration from shales (Powers, 1967; Burst, 1969; Perry and Hower, 1972; Bruce, 1984). Textural changes in these mudrocks during the neoformation of I/S from smectite contributes to the loss of porosity and permeability and may also be involved in the development of overpressure (Freed and Peacor, 1989a).

The smectite-to-illite transformation has been shown to occur at approximately the same time and over the same depth interval as the onset of hydrocarbon generation in source-rocks in many sedimentary basins (e.g., Kisch, 1987). Several studies have attempted to correlate the progressive thermal maturation of organic matter as indicated by vitrinite reflectance (or other thermal maturation parameters) with changes that occur in mixed layer I/S during burial. In particular, the transition from R0 to R1 I/S appears to coincide with vitrinite reflectance values of 0.45–0.65% (Powell et al., 1978; Foscolos and Powell, 1979; Pearson et al., 1982; Bruce, 1984; Burtner and Warner, 1986; Kisch, 1987; Pearson and Small, 1988; Hansen and Lindgreen, 1989; Pollastro, 1993). Those vitrinite reflectance values correspond to the onset ($R_0 \approx 0.45-0.5\%$) and early generation ($R_0 \leq 7\%$) of oil (Tissot and Welte, 1984; Waples, 1985).

The smectite-to-illite transformation in Tertiary sediments of the Gulf Coast appears to take place in relatively dilute (relative to seawater), organic acid-rich formation fluids. These fluids are derived in part from dehydration of smectite during the smectite-to-illite transformation (e.g., Morton and Land, 1987; Land and Macpherson, 1992) and from generation of organic acids during early stages of kerogen thermal maturation (e.g., Surdam et al., 1984, 1989; Surdam and Crossey, 1985; Crossey et al., 1986b; Lundegard and Kharaka, 1990; MacGowan and Surdam, 1990). The presence of organic acids may, in part, control the smectite-to-illite reaction by providing K and Al to pore fluids through enhanced dissolution of detrital smectite, K-feldspar, and muscovite and by promoting illite precipitation through pH changes driven by organic acid generation and decomposition (Eberl, 1993; Small and Manning, 1993; Small, 1993, 1994).

Siliciclastic Diagenesis and Fluid Flow: Concepts and Applications, SEPM Special Publication No. 55
Copyright © 1996, SEPM (Society for Sedimentary Geology), ISBN 1-56576-032-8

Whereas a general correlation among the timing of organic matter maturation, formation of organic acids, and the smectite-to-illite transition appears to exist, little is known about the microstructural and textural relationships between organic matter and authigenic clay components in known source rocks. This paper is part of a detailed investigation of core samples from the Devonian Antrim and New Albany Shales in the Michigan and Illinois Basins (Fig. 1) using X-ray diffraction (XRD), scanning electron microscopy (SEM) and scanning transmission/analytical electron microscopy (STEM/AEM) techniques. Detailed analyses and interpretations of mixed-layer I/S fabric (e.g., crystal size, defect state), polytypism and quantitative AEM-determined compositions are presented in Hover et al. (1996). This part of the study focuses on the textural interrelationships between organic matter and authigenic illite-rich matrix clay. The textural observations presented here have implications for the timing of illitization and porosity reduction relative to organic matter maturation, the relative timing of primary hydrocarbon migration and expulsion of aqueous fluids, and the mechanism of gas production from these black shales.

PETROPHYSICAL FRAMEWORK OF THE ANTRIM
AND NEW ALBANY SHALES

The Michigan and Illinois intracratonic basins are filled with mainly Paleozoic sediments up to 5 km thick in Michigan (Fig.

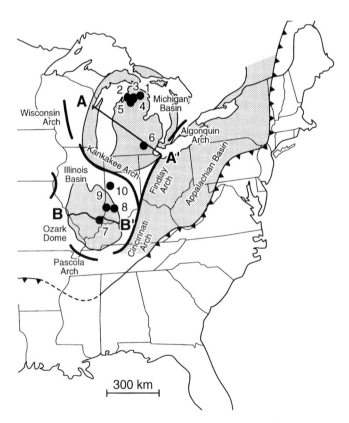

FIG. 1.—Well Location Map. 1. Hycrude #1, MI-1; 2. State Chester #18 (MI-2); 3. State Loud #D3–20; 4. State Loud #C2–21; 5. State South Branch #1–19; 6. Livingston-Dey #A1–15; 7. H. C. Ford #C-17; 8. Mellis #1; 9. Phegley #1; 10. J. Marks #1–4. Outlines of the Michigan, Illinois and Appalaction Basins are shaded. A-A' and B-B' cross sections shown in Figure 2.

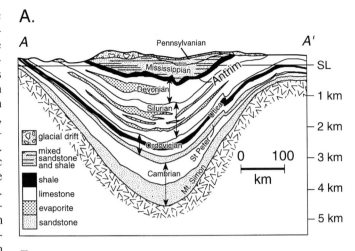

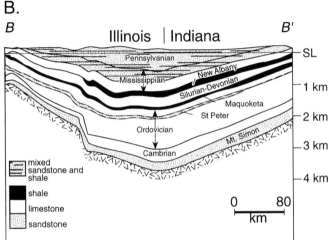

FIG. 2.—Schematic cross section through (A) the Michigan Basin (after Bally, 1989), and (B) the Illinois Basin (after Swann, 1968).

2A); (Bally, 1989; Catacosinos et al., 1990) and up to 6 km thick in Illinois (Fig. 2B; Swann, 1968). Burial history curves for basin strata (Cercone, 1984; Bethke et al., 1991); (Fig. 3) indicate that episodic subsidence occurred throughout Paleozoic time, related to tectonic events along the eastern and southern margins of the North American craton (Sleep et al., 1980, Howell and van der Pluijm, 1990; Kolata and Nelson, 1990a, b). Relatively slow average basin subsidence and sedimentation rates (0.01–0.03 mm/yr) and small shale volumes resulted in normal compaction during burial without significant development of overpressures (e.g., Bethke, 1985, 1986a, b; Bethke et al., 1988, 1991; Bethke and Marshak, 1990). Uplift and erosion began in Late Pennsylvanian to Early Permian time (Fig. 3) and continued through mid to late Mesozoic time (Cercone, 1984; Barrows and Cluff, 1984; Cluff and Byrnes, 1990; Bethke et al., 1991).

The Upper Devonian organic-rich Antrim and New Albany Shales are potential source-rocks for hydrocarbons in the Michigan (Rullkötter et al., 1986, 1992) and Illinois (Lineback, 1981; Barrows and Cluff, 1984; Hatch et al., 1990) Basins, respectively. The Antrim Shale, in particular, is a fractured gas-reservoir and is currently the target of extensive exploration and

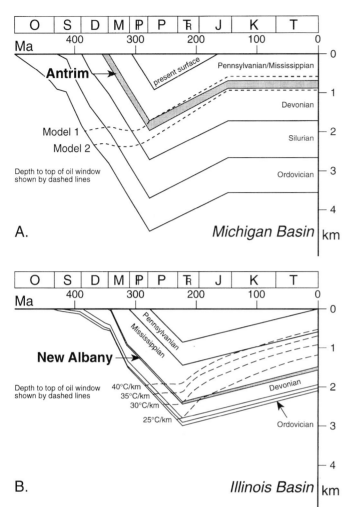

FIG. 3.—Burial history curves for (A) the central Michigan Basin (after Cerone, 1984) and (B) the central Illinois Basin (after Bethke et al., 1991). The Antrim and New Albany Shale horizons are shaded in Figures A and B, respectively. Depths to the top of the oil-window (Ro ≈ 0.5 %) are shown by dashed lines for various geothermal gradients. In the Michigan Basin, model 1 assumes an initial gradient of 45°C/km, and model 2 assumes an initial gradient of 35°C/km through Permian time. Both models assume a subsequent linear decrease to a present value of 25°C/km. Model 1 accurately predicts upper Paleozoic maturation but overestimates lower Paleozoic maturation. Conversely, model 2 accurately predicts lower Paleozoic maturation but underestimates upper Paleozoic maturation (Cercone, 1984).

kerogen source for derived hydrocarbons is hydrogen-rich, marine algal, Type II kerogen with minor Type III kerogen (Dellapenna, 1991; Barrows and Cluff, 1984; Hatch et al., 1990).

The Antrim Shale was sampled from the margin of the Michigan Basin and present burial depths range from about 18 to 640 m (Table 1). These sediments, however, may have been buried up to 1000 m deeper prior to post Pennsylvanian uplift (Cercone, 1984). Assuming geothermal gradients of 25–45°C/km (Cercone, 1984), the Antrim Shale in the study area reached maximum temperatures of ~50–100°C during deepest burial in Late Pennsylvanian time. Even if geothermal gradients were greater in the eroded strata (e.g., Cercone and Pollack, 1991), maximum temperatures were probably no greater than ~70–110°C.

The Antrim Shale in the study area reached thermal maturation levels corresponding to vitrinite reflectance values of $R_0 \approx 0.45-0.5\%$ based on comparison with nearby samples (Cercone and Pollack, 1991). These low reflectance values mark the onset of generation of liquid hydrocarbons for marine Type II kerogen (Tissot and Welte, 1984). Our samples from the basin margins have thermal maturities lower than those observed in the deepest part of the basin where the Antrim Shale reached maturation levels within the peak oil generation window (Tissot and Welte, 1984; Waples, 1985; $R_0 \approx 1.0\%$, Cercone and Pollack, 1991). In the central Michigan Basin, oil generation from the Antrim Shale began in latest Mississippian to earliest Pennsylvanian time (Fig. 3; Cercone, 1984) and would have occurred later at the basin margin.

Present burial depths of the New Albany Shale samples in the Illinois Basin examined in this study are 480 to 1440 m (Table 1). However, these samples may have been buried up to 1500 m deeper in the past (Barrows and Cluff, 1984; Cluff and Byrnes, 1990; Bethke et al., 1991). Assuming similar geothermal gradients (Cluff and Byrnes, 1990; Bethke et al., 1991), estimated maximum burial temperatures for the New Albany Shale samples of this study were ~75–90°C (M1575.56) to ~100–150°C (W4720). These basin margin samples reached thermal maturation levels corresponding to vitrinite reflectance values of $R_0 \approx 0.5-0.6\%$ (Barrows and Cluff, 1984; Cluff and Byrnes, 1990) and are, thus, slightly more mature than the Antrim Shale samples. Oil generation from the Antrim Shale in the deepest portion of the basin began in Pennsylvanian to Permian time (Fig. 3; Bethke et al., 1991). Significant migration from the source area has been inferred (Bethke and Marshak, 1990; Bethke et al., 1991). Present thermal maturation levels in the deep basin New Albany Shale ($R_0 \approx 1.2\%$) correspond to the onset of gas generation (Tissot and Welte, 1984; Waples, 1985).

ANALYTICAL METHODS

Core samples of approximately 66 Antrim and 4 New Albany black shale facies were examined by XRD methods to determine bulk mineral content and to serve as a basis for selection of a small number of representative samples for further SEM and STEM analyses. Analytical methods and representative XRD patterns are described in Hover et al. (1996). Table 1 lists identification numbers, full well names, sample depths and geographic locations of samples used specifically for SEM and STEM analyses.

production in northern Michigan (Decker et al., 1992). Similar lithologies in the New Albany Shale also produce gas (Lineback, 1981; Barrows and Cluff, 1984). Laminated black shale lithologies of the Norwood and Lachine Members of the Antrim Shale (Gutschick and Sandberg, 1991a, b) and Blocher Member of the New Albany Shale (Cluff, 1980; Lineback, 1981; Cluff et al., 1981) were investigated in this study (Table 1). The Antrim Shale contains up to 15 wt % total organic carbon (TOC; Dellapenna, 1991, Decker et al., 1992) and the New Albany Shale up to 9 wt % TOC (Barrows and Cluff, 1984; Cluff and Byrnes, 1990), composed dominantly of large fragments and spores of the algae *Tasmanites* (Boneham, 1967). ROCK-EVAL and traditional pyrolysis analyses of these shales indicate the

TABLE 1.—SAMPLE IDENTIFICATION AND LOCATION

Well Number/Name	Sample Number	Depth Meters (feet)	Formation Member	Location
MICHIGAN BASIN				
1. Hycrude #1, MI-1	HC 57.9	17.6 m (57.9')	Antrim/Lachine	Alpena Co., Michigan (Paxton Quarry)
1. Hycrude #1, MI-1	HC 107.6	32.8 m (107.6')	Antrim/Norwood	Alpena Co., Michigan (Paxton Quarry)
2. State Chester #18, MI-2	SC 1490.75	454.4 m (1490.75')	Antrim/Lachine	Otsego Co., Michigan
4. State Loud #C2–21	SLC2 1449.7	442 m (1449.7')	Antrim/Lachine	Montmorency, Co., Michigan
5. State South Branch #1-19	SB 2104a	641 m (2104')	Antrim/Lachine	Crawford Co., Michigan
6. Livingston-Dey #A1-15	A1 1679.75	512 m (1679.75')	Antrim/Norwood	Livingston Co., Michigan
ILLINOIS BASIN				
7. H. C. Ford #C-17	W4720	1439 m (4720')	New Albany/Blocher	White Co., Illinois
8. Mellis #1	M1575.56	480.2 m (1575.56')	New Albany/Blocher	Green Co., Indiana

Carbon-coated thin-sections and ion-milled samples of core material were examined using SEM techniques to characterize textural and mineralogical relationships between detrital and authigenic minerals and organic matter, and to locate areas of interest for further STEM studies. A Hitatchi Model S-570 SEM operating at 15 kV and ~ 0.2 µA specimen current in the backscatter and secondary electron modes was used. Qualitative analyses of minerals were obtained simultaneously using a Kevex Quantum energy dispersive X-ray analytical system. SEM images in Figure 4 are backscattered electron images (BSE) obtained on ion-milled samples.

Ion-thinned core samples were examined by STEM methods to determine microtextural relationships between authigenic clay minerals and associated organic matter, and to determine clay microstructure and chemical composition. Samples were prepared as described in Hover et al. (1996). Briefly, samples were oriented with bedding direction perpendicular to the electron beam. Lattice fringe images of (00l) planes of phyllosilicates perpendicular to c* can be imaged in this orientation. Samples were examined using a Philips CM-12 STEM operated at 120 kV. The TEM images and selected area electron diffraction (SAED) patterns shown in Figures 5–11 are typical of hundreds of TEM images obtained.

RESULTS

The Antrim and New Albany Shales consist mainly of quartz, mixed-layer I/S and/or illite, minor K-feldspar, plagioclase feldspar, chlorite, muscovite, minor to trace pyrite and dolomite, and rarely, calcite based on XRD results (Hover et al., 1996). In all but the finest (<0.2 µm) fraction, muscovite cannot be distinguished from illite or mixed-layer clays because of overlap of their ~10 − Å (00l) diffraction peaks. However, the finest fraction of these shales consists of a slightly expandable illite-rich mixed-layer I/S containing approximately 85–90% illite layers. This fine fraction is representative of the authigenic clay component as confirmed by TEM observations described below and will be referred to as authigenic I/S or illite in the following discussions.

SEM images (Fig. 4) show that the black shales contain abundant flat plates and spore-like bodies of algal *Tasmanites* up to 10 x 500 µm in size and other unidentifiable organic matter (Fig. 4A). The organic matter comprises an interconnected matrix within which authigenic and detrital grains occur (Figs. 4A-C). Pyrite framboids and equant grains are associated with the organic matter. A pre-compaction, early diagenetic origin for pyrite is indicated by the deformation of algal material and detrital matrix phyllosilicates around pyrite framboids and aggregates (Figs. 4B, C). Original bedding is defined by the subparallel orientation of silt-sized grains of chlorite, muscovite and *Tasmanites*. Subrounded grains of silt-sized quartz, K-feldspar and albite are also present (Figs. 4A-C). Because of their large sizes and irregular shapes, these silt-sized phases are assumed to be detrital. The inorganic matrix groundmass is composed principally of fine-grained, illite-rich clay as shown by TEM analyses described below.

Low magnification TEM images illustrate that organic matter forms a nearly continuous, interconnected matrix (Figs. 5–7). Pyrite framboids are completely surrounded by, and imbedded in, organic matter (Fig. 5). Isolated fine-grained, wispy, elongate single crystals of illite and clusters of subparallel illite crystals appear to "float" in the organic matter matrix (Figs. 5–7). All pore space is filled with organic matter and no intergranular porosity has been observed even in the highest resolution TEM images (see below). Fine-grained, euhedral quartz is commonly associated with illite (Fig. 6) and may have formed from Si released during the smectite-to-illite reaction. Although much illite is oriented subparallel to bedding, some illite crystals appear to have a random orientation in the vicinity of detrital grains suggesting accumulation in a pressure "shadow", free from compaction stress during burial (Fig. 6).

Higher magnification images of these black shales show the typical microstructure of illite crystals. These features are described in detail in Hover et al. (1996) and are briefly summarized here. Illite intergrowths, oriented subparallel to bedding, are composed of crystals (packets) as thin as three individual 10-Å layers (30 Å) and seldom greater than 20 layers thick (200 Å); (Figs. 8, 9). Individual crystals are generally less than 5000 Å long. Illite packets within intergrowths have low-angle grain boundaries and many layer terminations with a high proportion of defects. Packets with relatively straight 10-Å lattice fringes

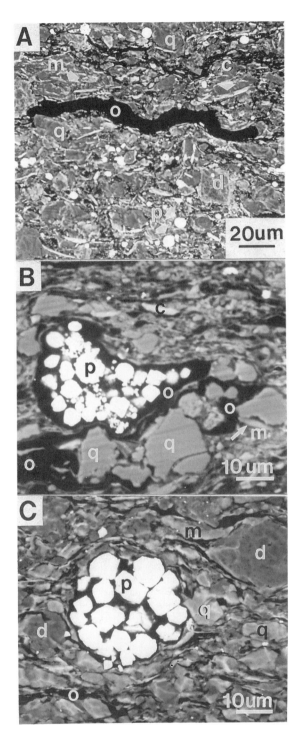

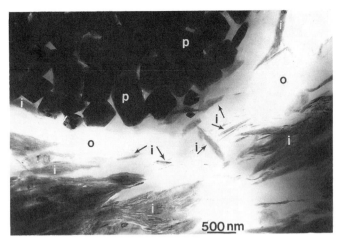

FIG. 5.—TEM image showing textural relationship of framboidal pyrite (p), organic matter (o), and wispy authigenic illite-rich I/S intergrowths (i) "floating" in the organic matrix (Antrim Shale, SB 2104a).

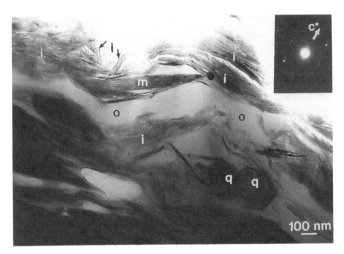

FIG. 6.—TEM image showing typical texture of organic matrix and authigenic illite intergrowths. Note random orientation of some illite with respect to the bedding caused by entrainment in organic matter during compaction. Fine-grained euhedral quartz (q) (1000–2000 Å) is present and interpreted to be authigenic. Silt-sized detrital mica (m) is also present (Antrim Shale, SB 2104a).

FIG. 4.—BSE images of laminated black shales. (A) Detrital *Tasmanites* (o), chlorite (c), and muscovite (m) grains are oriented subparallel to bedding. Quartz (q) occurs as subrounded silt-sized grains. Pyrite (p) is present as framboids and equant grains. Dolomite (d) occurs as euhedral rhombs. Fine-grained matrix is made up largely of illite-rich I/S (Antrim Shale, SLC2 1449.7). (B) An algal spore-like body of *Tasmanites* (o) encloses pyrite (p). Detrital quartz (q), muscovite (m) and chlorite (c) are present (Antrim Shale, HC 107.6). (C) Detrital muscovite (m) grains are oriented subparallel to bedding. Organic-rich matrix "wraps" around a cluster of equant pyrite (p) grains and a euhedral dolomite rhomb (d) indicating a pre-compaction origin for those phases. Subrounded quartz silt-sized grains are present in the matrix with fine-grained illite-rich I/S (New Albany Shale, M 1575.56).

and packets with somewhat wavy lattice fringes of approximately 10 Å are characteristic of (00l) illite or mixed-layer I/S interplaner spacings with smectite layers fully collapsed due to the high vacuum conditions of the STEM (Ahn and Peacor, 1986, 1989; Fig. 8B). Contrast in some lattice fringe images indicates that domains of ordered R1 and R2 I/S are also present (Fig. 8B).

Selected area electron diffraction (SAED) patterns of illite intergrowths (insets Figs. 8, 9) show the characteristic 10-Å spacing of the (00l) reflections in the c* direction. The fanning of (00l) reflections about c* indicates subparallel orientation of individual crystals in the illite intergrowths and diffuseness along c* reflects lattice spacings slightly different from 10 Å. The second row of spots in the SAED images corresponds to

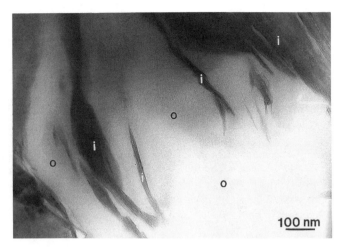

FIG. 7.—TEM image showing individual wispy illite crystals entrained in organic matrix. Note absence of pore space between organic matter and illite (New Albany Shale, M 1575).

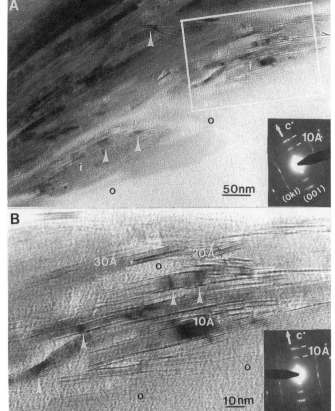

FIG. 8.—TEM image showing (A) interface between authigenic illite (i) and organic matter (o). Note absence of pore space. (B) Enlargement of A showing individual illite-rich crystals with 10-Å lattice fringe spacings. Contrast in some areas indicates presence of R1 and R2 ordered I/S consistent with interlayering of illite and collapsed smectite layers. Illite grain boundaries have many layer terminations and defects. Small arrows show locations where individual illite crystals have curved lattice fringes and mottled texture. Mottling results from beam damage at areas of increased lattice strain. These textures indicate deformation of illite subsequent to formation as a result of compaction and/or deformation of organic matter during continued maturation. The SAED pattern (insets) shows fanning of (00*l*) reflections resulting from subparallel orientation of illite crystals and diffuseness parallel to c* consistent with variable (00*l*) lattice spacing and mixed-layering. The (*hk*0) reflections are diffuse and streaked consistent with stacking disorder typical of 1M$_d$ polytypism (Antrim Shale, SB 2104a).

(0*kl*) reflections. These reflections show streaking and diffuseness in the c* direction with only a few distinct diffraction spots reflecting 10-Å periodicity (inset, Fig. 8A). This pattern is characteristic of the disordered 1M$_d$ polytype and is typical of authigenic mixed-layer I/S which has not undergone modification subsequent to formation (Ahn and Peacor, 1986 1989; Peacor, 1992).

Quantitative AEM analyses of matrix illite in the Antrim and New Albany black shales (Hover et al., 1996) indicate that compositions are consistent with dioctahedral illite-rich mixed-layer I/S and are distinct from compositions of detrital muscovite in these same samples. No silt-sized illite was observed, and TEM images, SAED patterns and AEM-determined compositions all confirm an authigenic origin for matrix illite.

Individual illite packets are often contorted and bent and appear to wrap around other authigenic and detrital grains as if entrained in the organic matter during deformation subsequent to illite formation (Figs. 6, 7). At higher magnification, some illite crystals show evidence of bending and curving of lattice fringes and mottling in contrast along fringes resulting from beam damage in areas of increased strain (small arrows, Figs. 8–11). At the interface between organic matter and illite, some illite crystals have been cleaved, and organic matter "intruded" between individual separated layers (see large arrow, Fig. 10). Virtually all intercrystalline microporosity between individual illite crystals is completely filled with organic matter. These observations indicate that subsequent to illite formation, the matrix organic matter and entrained authigenic illite continued to deform. This deformation occurred during subsequent compaction or was associated with continued thermal maturation of organic matter (plastic deformation) after illite formation. It is possible that these textures resulted from illite formation simultaneous with organic matter maturation and deformation, however, we would expect less evidence of illite deformation if this were the case.

There is no significant difference in matrix illite microtexture, polytypism or composition between the Antrim and New Albany Shale samples (compare Figs. 8, 9) even though esti-

mated paleoburial depths vary from ~1000–3000 m and temperatures from ~50–150°C.

In summary, textural and crystallographic features illustrated by the TEM images indicate: 1) matrix illite microstructures and compositions are essentially identical in the Antrim and New Albany Shales, 2) matrix illite is authigenic, having formed from an original smectite precursor, 3) original diagenetic textures typical of post-transition illite-rich I/S have been preserved since formation, and 4) illite formed prior to or concurrent with deformation of organic matter associated with maximum levels of thermal maturation.

IMPLICATIONS OF ILLITE-ORGANIC MATTER RELATIONSHIPS

Both the smectite-to-illite transformation and hydrocarbon maturation are kinetically controlled and, therefore, governed

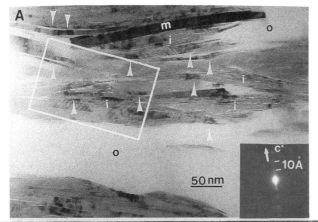

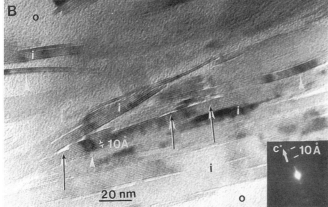

FIG. 9.—TEM image (A) showing the interface between authigenic illite (i) with curved lattice fringes and mottled texture (small arrows) and organic matter (o). Large grain at top is detrital muscovite (m). (B) Enlargement of A showing authigenic illite with many layer terminations and defects at grain boundaries. Some areas of possible microporosity between illite packets are indicated by very light contrast (black and white arrows). Other textures and SAED patterns as described in Figure 8 (New Albany Shale, M 1575).

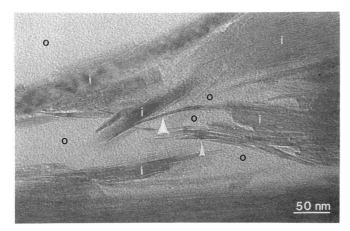

FIG. 10.—TEM image of the interface between organic matter (o) and authigenic illite (i) showing cleaving of illite layers and intrusion of organic matter (large arrow). Other grains are bent (small arrows). These textures indicate deformation of illite subsequent to formation. Some microporosity may be present at layer terminations (Antrim Shale, A1 1679.10).

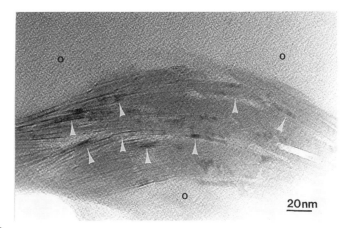

FIG. 11.—TEM image of an illite intergrowth within organic matter matrix showing the bending of lattice fringes in most grains (small arrows) (Antrim Shale, A1 1679.10).

not only by temperature, time and composition, but also by additional factors including water composition (e.g., K, Al, organic acid content, pH), availability of reactants (e.g., K-feldspar, muscovite), water-rock ratios and porosity/permeability (e.g., Hower et al., 1976; Srodon and Eberl, 1984; Whitney, 1990; Huang et al., 1993; Small, 1993, 1994). The degree of hydrocarbon maturation also depends on the type of organic matter present, with lipid-rich Type II organic matter maturing earlier than woody Type III organic matter (e.g., Tissot and Welte, 1984; Waples, 1985). Several studies have shown that there is not a unique relationship between the extent of the smectite-to-illite reaction and organic maturation levels (e.g., Bruce, 1984; Kisch, 1987; Scotchman, 1987; Hillier and Clayton, 1989) nor is there a consistent relationship between the extent of the I/S reaction and temperature or burial depth (e.g., McCubbin and Patton, 1981; Srodon and Eberl, 1984; Kisch, 1987; Freed and Peacor, 1989b; Velde and Espitalié, 1989; Pollastro, 1993; Velde and Lanson, 1993; Essene and Peacor, 1995). These studies have pointed out the importance of considering differences in diagenetic pathways between basins.

It is important to also consider the microtextural relationships between organic matter and authigenic illite as these have important implications for the extent of interaction between organic and inorganic phases and the relative timing of diagenetic events. The microtextural relationships revealed by this TEM study of these low to moderately mature Antrim and New Albany Shales can be used to constrain the timing of illitization relative to organic maturation and porosity loss, the relationship between illitization and organic acid formation in these shales and the mechanism of gas production in the Antrim Shale. Each of these relationships is discussed below.

Timing of Illite Authigenesis Relative to Organic Matter Maturation and Migration

The microtextural relationships between authigenic illite and organic matter in the Antrim and New Albany Shales indicate that illite formed prior to or concurrently with latest compaction and plastic deformation of organic matter associated with maximum thermal maturation. Evidence is found in the presence of deformed illite crystals and polycrystalline domains entrained within the organic matter (Figs. 8, 9), the "cleaving" of one or two illite layers from host grains by intruding organic matter (Figs. 10, 11), and the absence of porosity between authigenic and detrital phases caused by "intrusion" of organic matter into available pore space. The relatively low levels of thermal maturation for the Antrim and New Albany Shales ($R_0 = 0.45$–0.5% and $R_0 = 0.5$–0.6%, respectively) indicate only ~10–15% of possible hydrocarbons have been generated by these shales (Waples, 1981). The smectite-to-illite transformation in these samples preceded or was coincident with early onset of liquid hydrocarbon generation ($R_0 \approx 0.5\%$), but occurred well before peak oil generation which would have required higher maturation levels ($R_0 \approx 0.7$–1.0%). Similarly, in the deeper portion of the basin, it is probable that the smectite-to-illite reaction was complete at similar thermal maturation levels and must have, therefore, occurred prior to or at most concurrent with the onset of hydrocarbon generation during Late Pennsylvanian to Early Permian time (Fig. 3; Cercone, 1984; Cluff and Byrnes, 1990). This implies that the smectite-to-illite transformation must have occurred prior to maximum burial in Pennsylvanian time at depths less than ~2000 m and temperatures less than ~120°C. Subsequent to authigenic illite formation under these conditions, no further change in illite microstructure, polytypism or composition occurred except for the deformation (bending and cleaving) of illite grains during plastic deformation of organic matter during subsequent thermal maturation.

In the Gulf Coast, the smectite-to-illite transformation produces significant loss of porosity and coincides with the formation of overpressured zones (Freed and Peacor, 1989b). Post-transition illite-rich shales may act as permeability barriers and contribute to the overpressures. Hover et al. (1996) shows that the illite-rich I/S in the Antrim and New Albany Shales is essentially identical in microstructure, defect state, polytype and composition to post-transition illite-rich I/S in deeper Gulf Coast mudrocks. This suggests that the Antrim and New Albany Shales have experienced similar processes of illite formation and porosity loss and may have become effective permeability barriers prior to the onset of peak oil generation. It is unlikely, however, that significant overpressures occurred in the Michigan and Illinois Basins due to the relatively thin nature of these shales and slow depositional and subsidence rates (e.g., Bethke, 1985, 1986 a, b; Bethke and Marshak, 1990; Bethke et al., 1991).

Bruce (1984) has shown that for three delta systems of Miocene age, the onset of peak hydrocarbon (oil) generation may occur at depths above, within or below the depth interval of the smectite-to-illite transition depending on the burial history and time-temperature path of the sediments and in part, on the initial clay composition. He suggested that smectite dehydration water produced during the smectite-to-illite transition during or after peak liquid hydrocarbon generation, is likely to occur when basins become overpressured as in the Louisiana Gulf Coast. Smectite dehydration water is trapped due to rapid sedimentation and subsidence rates and its expulsion from overpressured sequences occurs after peak hydrocarbon generation and might flush hydrocarbons from source rocks (e.g., Powers, 1967; Burst, 1969; Perry and Hower, 1972; Bruce, 1984).

Other studies have shown that the transition from random R0 smectite-rich I/S to R1 ordered illite-rich I/S occurs within the

oil generating stage corresponding to vitrinite reflectance values of $R_0 \approx 0.6$–0.7% (e.g., Burtner and Warner, 1986; Pearson et al., 1982; Pearson and Small, 1988; Hansen and Lindgreen, 1989). Smectite dehydration water may be important in hydrocarbon expulsion in these situations (Bruce, 1984).

In basins undergoing normal compaction or with relatively thin shale sequences such as the Michigan and Illinois Basins, however, the smectite-to-illite transition and expulsion of hydration water may precede peak liquid hydrocarbon generation and may not in assist in primary hydrocarbon migration (Bruce, 1984). Powell et al. (1978) and Foscolos and Powell (1979) showed that in the Sverdrup Basin, Canada, the smectite-to-illite transformation occurred in source rocks with vitrinite reflectance values of $R_0 \approx 0.5\%$. and preceded peak oil generation. In that basin, interlayer water lost from smectite during the transition may not have played a significant role in primary liquid hydrocarbon migration. This situation is similar to that in the Michigan and Illinois Basins where the smectite-to-illite transition is also complete at low levels of maturation and preceded peak hydrocarbon generation during the Pennsylvanian even in the deepest, most mature part of the basins. This implies that smectite dehydration and shale dewatering cannot be invoked as principle driving forces for the expulsion of maturing oils from the Antrim and New Albany Shales. However, smectite dehydration and water expulsion during the smectite-to-illite transition may indirectly contribute to later hydrocarbon migration in these settings by causing development of microfractures in such shales (Bruce, 1984; Scotchman, 1987).

Timing of Illite Authigenesis and Organic Acid Formation

In the Gulf Coast, the timing of the smectite-to-illite transition has been linked to the formation of dilute, organic acid-rich formation fluids associated with mudrock-rich sequences (Eberl, 1993). Organic acid anions, principally acetate, make up a significant portion of the negative anions (up to 8000 mg/l) in Gulf Coast formations (e.g., Carothers and Kharaka, 1978; Hanor and Workman, 1986; Fisher, 1987; Lundegard and Kharaka, 1990; MacGowan and Surdam, 1990; Land and Macpherson, 1992). Experimental maturation of mudrocks and kerogens confirms the dominance of acetate anion formation (Lundegard and Senftle, 1987; Barth and Bjorlykke, 1993; Rae et al., 1993; Manning et al., 1994) and shows that oxalate and citrate anions also form (Manning et al., 1994). The ability of organic acids to form complexes with Al can enhance dissolution rates and increases solubility of aluminosilicates (e.g., Surdam and Crossey, 1985; Surdam et al., 1984, 1989; Fein, 1991a, b). Oxalate and citrate acid anions are more efficient complexers of Al than acetate anion (e.g., Surdam et al., 1984; Fein, 1991b) and may be important in promoting aluminosilicate dissolution despite their low concentrations.

Surdam et al. (1984, 1989) showed that peak organic acid formation occurs during the early stages of organic maturation corresponding to relatively low vitrinite reflectance values ($R_0 \leq 0.5\%$). The presence of these acids is confined to a narrow temperature range limited by bacterial degradation (<60°C) and thermal decarboxylation (>140°C); (Carothers and Kharaka, 1978; Surdam et al., 1989). Organic acids are not present in Antrim or New Albany Shale formation fluids today (Walter and Martini, 1995, pers. commun.), but were likely present in the shale pore fluids during early thermal maturation as a result of high concentrations of organic matter (e.g., Surdam et al., 1984, 1989; Barth and Bjorlykke, 1993; Small, 1994). As shown by Small (1993, 1994), organic acids could have enhanced the dissolution of detrital smectite, K-feldspar and muscovite by promoting complexation of Al and controlled the timing of precipitation through buffering of pH (a_K/a_H ratios) during generation and breakdown with increasing temperature. As implied in part by the necessary loss of pore fluids and water from original smectite, sufficient porosity must have existed in the Antrim and New Albany Shales to permit fluid flow during the smectite-to-illite reaction. Flow rates probably did not exceed those of compaction driven flow as these basins were not overpressured (e.g., Bethke, 1985, 1986a, b; Bethke et al., 1991). Even these moderate flow conditions can produce rapid and extensive illitization in sandstones (Small, 1993) and may have operated in the Antrim and New Albany Shales.

The thermal conditions and maturation levels ($R_0 \sim 0.45–0.6\%$) inferred for the Antrim and New Albany Shales are thus consistent with formation of organic acids during the early stages of organic matter maturation prior to or approximately concurrent with deepest burial of these samples. Subsequent thermal breakdown of these acids may have promoted illitization of smectite and compaction driven fluid flow was sufficient to transfer components within these shales. Textural relationships between matrix illite-rich I/S and organic matter, along with inferences that only a small fraction of potential hydrocarbons could have been produced from these samples, imply that illitization occurred prior to or concurrently with maximum burial in Pennsylvanian at depths less than ~2000 m and temperatures less than ~120°C. Organic matter continued to mature and deform in the Antrim and New Albany Shales subsequent to illite formation, but no further change in authigenic I/S occurred (Hover et al., 1996).

Implications for Gas Production from the Antrim Shale

TEM observations indicate there is no visible intercrystalline or intergranular matrix porosity in the Antrim and New Albany Shales, even at highest resolution. Organic matter completely fills intergranular pore spaces between all detrital and authigenic grains. Such textural observations are compatible with the low matrix permeability of the Antrim Shale and the inferred mode of gas production in northern Michigan. In this unique self-sourced gas reservoir, the bulk of generated gas is not stored in matrix porosity; rather it is stored adsorbed onto kerogen and/or dissolved in bitumen present in the black shales (Dellapenna, 1991; Decker et al., 1992). Up to 6–7 scf/ft³ (standard cubic foot gas/cubic foot rock) can be stored at reservoir pressures (Decker et al., 1992). Gas is produced when pressure is reduced through removal of formation water (residing mostly in open natural fractures) via pumping. The normally adsorbed gas desorbs under the reduced pressure conditions and migrates through the matrix to open natural microfractures and larger fracture conduits, ultimately to the well bore (Decker, et al., 1992). Clay minerals may also adsorb a significant amount of hydrocarbons as shown by pyrolysis experiments with natural kerogens (e.g., Crossey et al., 1986a; Tannenbaum et al., 1986) and may also be a source of stored gas.

The presence of authigenic illite intergrowths containing many layer terminations and defects within the nearly contin-

uous organic matter matrix of the Antrim Shale provides a microporous network for the migration of desorbed natural gas to microfracture conduits. A similar mechanism may also be invoked for gas produced from relatively shallow New Albany reservoirs in the southeastern area of the Illinois Basin.

CONCLUSIONS

The smectite-to-illite transformation in the Antrim and New Albany Shales is related temporally with the earliest stages of organic matter maturation. TEM images indicate the illite-rich I/S formed prior to or concurrent with deformation of organic matter associated with maximum levels of thermal maturation. Based on the low vitrinite reflectance values of $R_0 \leq 0.6\%$ and geothermal gradient data, illite authigenesis occurred prior to Late Pennsylvanian/Early Permian time at depths less than ~2000 m and temperatures less than ~120°C. The relatively thin shale sequences, the normal pressures predicted during burial, and the textural evidence for illitization of smectite prior to the bulk of hydrocarbon generation, suggests that smectite dehydration water was probably not a significant driving force for expulsion of liquid hydrocarbons from these organic-rich source rocks. The presence of organic acids generated during early organic matter maturation may have promoted smectite illitization by increasing the rate of dissolution or increasing the solubility of Al- and K-bearing mineral reactants. Subsequently, illite precipitation may have been promoted by the thermal breakdown of organic acids that controlled the a_K/a_H activity ratio during continued maturation. The microtextures observed at the STEM scale, e.g., the nearly continuous organic matter matrix, lack of intercrystalline porosity, and presence of illite crystallites containing many lattice defects suggest that the disseminated illite could provide the microporous network for diffusion of desorbed gas to open microfractures and larger fracture conduits in these organic-rich shales.

ACKNOWLEDGMENTS

Samples were obtained from Shell Western Exploration and Production Incorporated, the University of Michigan Department of Geological Sciences Subsurface Lab, Western Michigan Department of Geology Core Laboratory and the Indiana State Survey. W-T. Jiang, G. Li and C. Henderson provided training and advice during SEM and STEM/AEM analyses. The STEM and SEM used in the study were acquired under Grants #EAR-87–08276 and #BSR-83–14092, from the National Science Foundation, respectively. This work was partly funded by the National Science Foundation Grant #EAR-9104565 to D. R. P. Support was also provided by grants from Shell Development Company and Chevron Oil Field Research Company to L. M. W. Electron Microbeam Analyses were partly funded by the Robert B. Mitchell Fund, Department of Geological Sciences, The University of Michigan. Reviews by J. R. Boles, W. B. Simmons and M. W. Totten greatly improved the manuscript.

REFERENCES

AHN, J. H. AND PEACOR, D. R., 1986, Transmission and analytical electron microscopy of the smectite-to-illite transition: Clays and Clay Minerals, v. 34, p. 165–179.
AHN, J. H. AND PEACOR, D. R., 1989, Illite/smectite from Gulf Coast shales: A reappraisal of transmission electron microscope images: Clays and Clay Minerals, v. 37, p. 542–546.

BALLY, A. W., 1989, Phanerozoic basins of North America, *in* Bally, A. W. and Palmer A. R., eds., The Geology of North America—An Overview: Boulder, Colorado, Geological Society of America, The Geology of North America, v. A., p. 397–446.

BARTH, T. AND BJORLYKKE, K., 1993, Organic acids from source rock maturation: Generation potentials, transport mechanisms and relevance for mineral diagenesis: Applied Geochemistry, v. 8, p. 325–337.

BARROWS, M. H. AND CLUFF, R.M., 1984, New Albany Shale Group (Devonian-Mississippian) source rocks and hydrocarbon generation in the Illinois Basin, *in* Demaison, G. and Murris, R J. J., eds., Petroleum Geochemistry and Basin Evaluation: Tulsa, American Association of Petroleum Geologists Memoir 35, p. 111–138.

BETHKE, C. M., 1985, A numerical model of compaction-driven groundwater flow and heat transfer and its application to the paleohydrology of intracratonic sedimentary basins: Journal of Geophysical Research, v. 90, p. 6817–6828.

BETHKE, C. M., 1986a, Inverse hydrologic analysis of the distribution and origin of Gulf Coast-type geopressured zones: Journal of Geophysical Research, v. 91, p. 6535–6545.

BETHKE, C. M., 1986b, Hydrologic constraints on the genesis of the Upper Mississippi Valley mineral district from Illinois Basin brines: Economic Geology, v. 81, p. 233–249.

BETHKE, C. M., HARRISON, W. J., UPSON, C., AND ALTANER, S. P., 1988, Supercomputer analysis of sedimentary basins: Science, v. 239, p. 261–267.

BETHKE, C. M. AND MARSHAK, S., 1990, Brine migrations across North America—The plate tectonics of groundwater: Annual Reviews of Earth and Planetary Science, v. 18, p. 287–315.

BETHKE, C. M., REED, J. D., AND OLTZ, D. F., 1991, Long-range petroleum migration in the Illinois Basin: The American Association of Petroleum Geologists Bulletin, v. 75, p. 925–945.

BOLES, J. R. AND FRANKS, S. G., 1979, Clay diagenesis in Wilcox sandstones of southwest Texas: Implications of smectite diagenesis on sandstone cementation: Journal of Sedimentary Petrology, v. 49, p. 55–70.

BONEHAM, R. F., 1967, Devonian *Tasmanites* from Michigan, Ontario, and northern Ohio: Michigan Academy of Science, Arts and Letters, v. 52, p. 163–172.

BRUCE, C. H., 1984, Smectite dehydration—Its relation to structural development and hydrocarbon accumulation in northern Gulf of Mexico Basin: The American Association of Petroleum Geologists Bulletin, v. 68, p. 673–683.

BURST, J. F., 1969, Diagenesis of Gulf Coast clayey sediments and its possible relation to petroleum migration: The American Association of Petroleum Geologists Bulletin, v. 53, p. 73–93.

BURTNER, R. L. AND WARNER, M. A., 1986, Relationship between illite/smectite diagenesis and hydrocarbon generation in Lower Cretaceous Mowry and Skull Creek Shales of the northern Rocky Mountain area: Clays and Clay Minerals, v. 34, p. 390–402.

CAROTHERS, W. W. AND KHARAKA, Y. K., 1978, Aliphatic acid anions in oilfield waters—Implications for origin of natural gas: The American Association of Petroleum Geologists Bulletin, v. 62, p. 2441–2453.

CATACOSINOS, P. A., DANIELS, P. A., Jr., AND HARRISON W. B., III, 1990, Structure, stratigraphy, and petroleum geology of the Michigan Basin, *in* Leighton, M. W., Kolata, D. R., Oltz, D. F., and Eidel, J. J., eds., Interior Cratonic Basins: Tulsa, American Association of Petroleum Geologists Memoir 51, p. 561–601.

CERCONE, K. R., 1984, Thermal history of Michigan Basin: The American Association of Petroleum Geologists Bulletin, v. 68, p. 130–136.

CERCONE, K. R., AND POLLACK, H. N., 1991, Thermal maturity of the Michigan Basin, *in* Catacosinos, P. A. and Daniels, P. A., Jr., eds., Early Sedimentary Evolution of the Michigan Basin: Boulder, Geological Society of America Special Paper 256, p. 1–11.

CLUFF, R. M. 1980, Paleoenvironment of the New Albany Shale Group (Devonian-Mississippian) of Illinois: Journal of Sedimentary Petrology, v. 50, p. 767–780.

CLUFF, R. M., AND BYRNES, A. P., 1990, Lopatin analysis of maturation and petroleum generation *in* the Illinois Basin, in Leighton, M. W., Kolata, D. R., Oltz, D. F., and Eidel, J. J., eds., Interior Cratonic Basins: Tulsa, American Association of Petroleum Geologists Memoir 51, p. 425–454.

CLUFF, R. M., REINBOLD, M. L., AND LINEBACK, J. A., 1981, The New Albany Shale Group of Illinois: Champaign, Illinois State Geological Survey Circular 518, 83 p.

CROSSEY, L. J., HAGEN, E. S., AND SURDAM, R. C., AND LaPOINT, T. W., 1986a, Correlation of organic parameters derived from elemental analysis and pro-

grammed pyrolysis of kerogen, *in* Gautier, D. L., ed., Roles of Organic Matter in Sediment Diagenesis: Tulsa, Society of Economic Paleontologists and Mineralogists Special Publication 38, p. 35–45.

CROSSEY, L. J., SURDAM, R., C., AND LAHANN, R., 1986b, Application of organic/inorganic diagenesis to porosity prediction in Gautier, D. L., ed., Roles of Organic Matter in Sediment Diagenesis: Tulsa, Society of Economic Paleontologists and Mineralogists Special Publication 38, p. 147–155.

DECKER, D., COATES, J-M. P., AND WICKS, D., 1992, Stratigraphy, gas occurrence, formation evaluation and fracture characterization of the Antrim Shale, Michigan Basin, Topical Report, January 1990–March 1992: Lakewood, Advanced Resources International, Inc., prepared for the Gas Research Institute, Contract No. 5091–213–2305, 101 p.

DELLAPENNA, T. M., 1991, Sedimentological, structural, and organic geochemical controls on natural gas occurrence in the Antrim Formation in Otsego County, Michigan: Unpublished M. S. Thesis, Western Michigan University, Kalamazoo, 147 p.

EBERL, D. D., 1993, Three zones for illite formation during burial diagenesis and metamorphism: Clays and Clay Minerals, v. 41, p. 26–37.

ESSENE, E. J. AND PEACOR, D. R., 1995, Clay mineral thermometry—A critical perspective: Clays and Clay Minerals, v. 43, p. 540–553.

FEIN, J. B., 1991a, Experimental study of Al-, Ca-, and Mg-acetate complexing at 80°C: Geochimica et Cosmochimica Acta, v. 55, p. 955–964.

FEIN, J. B., 1991b, Experimental study of aluminum-oxalate complexing at 80°C: Implications for the formation of secondary porosity within sedimentary reservoirs: Geology, v. 19, p. 1037–1040.

FISHER, J. B., 1987, Distribution and occurrence of aliphatic acid anions in deep subsurface waters: Geochimica et Cosmochimica Acta, v. 51, p. 2459–2468.

FOSCOLOS, A. E. AND POWELL, T. G., 1979, Mineralogical and geochemical transformation of clays during burial-diagenesis (catagenesis): Relation to oil generation, *in* Mortland, M. M. and Farmer, V. C., eds., Proceedings of the VI International Clay Conference, 1978—Oxford: Amsterdam, Elsevier Scientific Publishing Company, p. 261–270.

FREED, R. L. AND PEACOR, D. R., 1989a, Geopressured shale and sealing effect of smectite to illite transition: The American Association of Petroleum Geologists Bulletin, v. 73, p. 1223–1232.

FREED, R. L. AND PEACOR, D. R., 1989b, Variability in temperature of the smectite/illite reaction in Gulf Coast sediments: Clay Minerals, v. 24, p. 171–180.

FREED, R. L. AND PEACOR, D. R., 1992, Diagenesis and the formation of authigenic illite-rich I/S crystals in Gulf Coast shales: TEM study of clay separates: Journal of Sedimentary Petrology, v. 62, p. 220–234.

GLASMANN, J. R., LARTER, S., BRIEDIS, N. A., AND LUNDEGARD, P. D., 1989, Shale diagenesis in the Bergen High area, North Sea: Clays and Clay Minerals, v. 37, p. 97–112.

GUTSCHICK, R. C. AND SANDBERG, C. A., 1991a, Upper Devonian biostratigraphy of Michigan Basin, *in* Catacosinos, P. A. and Daniels, P. A., Jr., eds., Early Sedimentary Evolution of the Michigan Basin: Boulder, Geological Society of America Special Paper 256, p. 155–179.

GUTSCHICK, R. C. AND SANDBERG, C. A., 1991b, Late Devonian history of Michigan Basin, *in* Catacosinos, P. A. and Daniels, P. A., Jr., eds., Early Sedimentary Evolution of the Michigan Basin: Boulder, Geological Society of America Special Paper 256, p. 181–202.

HANOR, J. S. AND WORKMAN, A. L., 1986, Distribution of dissolved volatile fatty acids in some Louisiana oil field brines: Applied Geochemistry, v. 1, p. 37–46.

HANSEN, P. L. AND LINDGREEN, H., 1989, Mixed-layer illite/smectite diagenesis in Upper Jurassic claystones from the North Sea and onshore Denmark: Clay Minerals, v. 24, p. 197–213.

HATCH, J. R., RISATTI, J. B., AND KING, J. D., 1990, Geochemistry of Illinois Basin oils and hydrocarbon source rocks, *in* Leighton, M. W., Kolata, D. R., Oltz, D. F., and Eidel, J. J., eds., Interior Cratonic Basins: Tulsa, American Association of Petroleum Geologists Memoir 51, p. 403–423.

HILLIER, S. AND CLAYTON, T., 1989, Illite/smectite diagenesis in Devonian lacustrine mudrocks from northern Scotland and its relationship to organic matter maturity indicators: Clay Minerals, v. 24, p. 181–196.

HOVER, V. C., PEACOR, D. R., AND WALTER, L. M., 1996, STEM/AEM evidence for preservation of burial diagenetic fabrics in Devonian shales: Implications for fluid/rock interaction in cratonic basins (USA): Journal of Sedimentary Research, p. 519–530.

HOWELL, P. D. AND van der PLUIJM, B. A., 1990, Early history of the Michigan basin: Subsidence and Appalachian tectonics: Geology, v. 18, p. 1195–1198.

HOWER, J., ESLINGER, E. V., HOWER, M. E., AND PERRY, E. A., 1976, Mechanism of burial metamorphism of argillaceous sediment: 1. Mineralogical and chemical evidence: Geological Society of America Bulletin, v. 87, p. 725–737.

HUANG, W-L., LONGO, J. M., AND PEVEAR, D. R., 1993, An experimentally derived kinetic model for smectite-to-illite conversion and it use as a geothermometer: Clays and Clay Minerals, v. 41, p. 162–177.

KISCH, H. J., 1983, Mineralogy and petrology of burial diagenesis (burial metamorphism) and incipient metamorphism in clastic rocks, in Larsen G. and Chilingar, G. V., eds., Diagenesis in Sediments and Sedimentary Rocks, 2: Amsterdam, Elsevier Scientific Publishing Company, Developments in Sedimentology 25B p. 289–493.

KISCH, H. J., 1987, Correlation between indicators of very low-grade metamorphism, in Frey, M., ed., Low Temperature Metamorphism: Glasgow, Blackie and Son Limited, p. 227–300.

KOLATA, D. R. AND NELSON, W. J., 1990a, Tectonic history of the Illinois Basin, in Leighton, M. W., Kolata, D. R., Oltz, D. F., and Eidel, J. J., eds., Interior Cratonic Basins: Tulsa, American Association of Petroleum Geologists Memoir 51, p. 263–285.

KOLATA, D. R. AND NELSON, W. J., 1990b, Basin-forming Mechanisms of the Illinois Basin, in Leighton, M. W., Kolata, D. R., Oltz, D. F., and Eidel, J. J., eds., Interior Cratonic Basins: Tulsa, American Association of Petroleum Geologists Memoir 51, p. 287–292.

LAND, L. S. AND MACPHERSON, G. L., 1992, Origin of saline formation waters, Cenozoic section, Gulf of Mexico sedimentary basin: The American Association of Petroleum Geologists Bulletin, v. 76, p. 1344–1362.

LINDGREEN, H., JACOBSEN, H., AND JAKOBSEN, H. J., 1991, Diagenetic structural transformations in North Sea Jurassic illite/smectite: Clays and Clay Minerals, v. 39, p. 54–69.

LINEBACK, J. A., 1981, Coordinated study of the Devonian black shale in the Illinois Basin: Illinois, Indiana, and western Kentucky: Champaign, Illinois State Geological Survey, Contract/Grant Report: 1981–1, U. S. Department of Energy Report DE-AS21–78MC08214, 36 p.

LUNDEGARD, P. D. AND KHARAKA, Y. K., 1990 Geochemistry of Organic acids in subsurface waters—field data, experimental data and models, in Melchoir, D. C. and Bassett, R. L., eds., Chemical Modeling of Aqueous Systems II: Washington, American Chemical Society, ACS Symposium Series 416, p. 169–189.

LUNDEGARD, P. D. AND SENFTLE, J. T., 1987, Hydrous pyrolysis: a tool for the study of organic acid synthesis: Applied Geochemistry v. 2, p. 605–612.

MACGOWAN, D. B. AND SURDAM R. C., 1990, Carboxylic acid anions in formation waters, San Joaquin Basin and Louisiana Gulf Coast, U.S.A.—Implications for clastic diagenesis: Applied Geochemistry, v. 5, p. 667–701.

MANNING, D. A. C., RAE, E. I. C., AND GESTSDÓTTIR, K., 1994, Appraisal of the use of experimental and analogue studies in the assessment of the role of organic acid anions in diagenesis: Marine and Petroleum Geology, v. 11, p. 10–19.

MCCUBBIN, D. G. AND PATTON, J. W., 1981, Burial diagenesis of illite/smectite, a kinetic model, The American Association of Petroleum Geologists Bulletin, v. 65, p. 956.

MORTON, R. A. AND LAND, L. S., 1987, Regional variations in formation water chemistry, Frio Formation (Oligocene), Texas Gulf Coast: The American Association of Petroleum Geologists Bulletin, v. 71, p. 191–106.

PEACOR, D. R., 1992, Diagenesis and low-grade metamorphism of shales and slates, in Buseck, P. R., ed., Minerals and Reactions at the Atomic Scale: Transmission Electron Microscopy: Washington, DC, Mineralogical Society of America Reviews in Mineralogy, v. 27, p. 335–380.

PEARSON, M. J. AND SMALL, J. S., 1988, Illite-smectite diagenesis and palaeotemperatures in northern North Sea Quaternary to Mesozoic shale sequences: Clay Minerals, v. 23, p. 109–132.

PEARSON, M. J., WATKINS, D., AND SMALL, J. S., 1982, Clay diagenesis and organic maturation in northern North Sea sediments, in Van Olphen H. and Veniale, F., eds., Proceedings of the VII International Clay Conference, 1981—Bologna and Pavia: Amsterdam, Elsevier Scientific Publishing Company, p. 665–675.

PERRY, E. AND HOWER, J., 1970, Burial diagenesis in Gulf Coast pelitic sediments: Clays and Clay Minerals, v. 18, p. 165–177.

PERRY, E. AND HOWER, J., 1972, Late stage dehydration in deeply buried pelitic sediments: The American Association of Petroleum Geologists Bulletin, v. 56, p. 2013–2021.

POLLASTRO, R. M., 1993, Considerations and applications of the illite/smectite geothermometer in hydrocarbon-bearing rocks of Miocene to Mississippian age: Clays and Clay Minerals, v. 41, p. 119–133.

POWELL, T. G., FOSCOLOS, A. E., GUNTHER, P. R., AND SNOWDON, L. R., 1978, Diagenesis of organic matter and fine clay minerals: A comparative study: Geochimica et Cosmochimica Acta, v. 42, p. 1181–1197.

POWERS, M. C., 1967, Fluid-release mechanisms in compacting marine mudrocks and their importance in oil exploration: The American Association of Petroleum Geologists Bulletin, v. 51, p. 1240–1254.

RAE, E. I. C., MANNING, D. A. C., AND HUGHES, C. R., 1993, Experimental diagenesis of mudrocks, in Manning, D. A. C., Hall, P. L., and Hughes, C. R., eds., Geochemistry of Clay-Pore Fluid Interactions: London, Mineralogical Society/Chapman and Hall, p. 213–242.

RULLKÖTTER, J., MARZI, R., AND MEYERS, P. A., 1992, Biological markers in Paleozoic sedimentary rocks and crude oils from the Michigan Basin: Reassessment of sources and thermal history of organic matter, in Schidlowski, M., et al., eds., Early Organic Evolution: Implications for Mineral and Energy Resources: Berlin, Heidelberg, Springer-Verlag, p. 324–335.

RULLKÖTTER, J., MEYERS, P. A., SCHAEFER, R. G., AND DUNHAM, K. W., 1986, Oil generation in the Michigan Basin: A biological marker and carbon isotope approach: Advances in Organic Geochemistry—1985, Organic Geochemistry, v. 10, p. 359–375.

SCOTCHMAN, I. C., 1987, Clay diagenesis in the Kimmeridge Clay Formation, onshore UK, and its relation to organic maturation: Mineralogical Magazine, v. 51, p. 535–551.

SLEEP, H. N., NUNN, J. A., AND CHOU, L., 1980, Platform Basins: Annual Review of Earth and Planetary Science, v. 8, p. 17–34.

SMALL, J. S., 1993, Experimental determination of the rates of precipitation of authigenic illite and kaolinite in the presence of aqueous oxalate and comparison to the K/Ar ages of authigenic illite in reservoir sandstones: Clays and Clay Minerals, v. 41, p, 191–208.

SMALL, J. S., 1994, Fluid composition, mineralogy and morphological changes associated with the smectite-to-illite reaction: An experimental investigation of the effect of organic acid anions: Clay Minerals, v. 29, p. 539–554.

SMALL, J. S. AND MANNING, D. A. C., 1993, Laboratory reproduction of morphological variation in petroleum reservoir clays: Monitoring of fluid composition during illite precipitation, in Manning, D. A. C., Hall, P. L., and Hughes, C. R., eds., Geochemistry of Clay-Pore Fluid Interactions: London, Mineralogical Society/Chapman and Hall, p. 181–212.

SRODON, J. AND EBERL, D. D., 1984, Illite, in Bailey, S. W., ed., Micas: Washington, DC, Mineralogical Society of America Reviews in Mineralogy v. 13, p. 495–544.

SURDAM, R. C., BOESE, S. W., AND CROSSEY, L. J., 1984, The Chemistry of Secondary Porosity, in McDonald, D. A. and Surdam. R. C., eds., Clastic Diagenesis: Tulsa, American Association of Petroleum Geologists Memoir 37, p. 127–149.

SURDAM R. C. AND CROSSEY, L. J., 1985, Organic-inorganic reactions during progressive burial: key to porosity and permeability enhancement and preservation: Philosophical Transactions of the Royal Society of London, v. A 315, p. 135–156.

SURDAM, R. C., CROSSEY, L. J., HAGEN, E. S., AND HEASLER, H. P., 1989, Organic-inorganic reactions and sandstone diagenesis: The American Association of Petroleum Geologists Bulletin, v. 76, 1–23.

SWANN, D. H., 1968, A summary geologic history of the Illinois Basin, in Geology and Petroleum Production of the Illinois Basin: A Symposium: Cooperative Publication of the Illinois, Indiana and Kentucky Geological Societies, p. 3–22.

TANNENBAUM, E., HUIZINGA, B. J., AND KAPLAN, I. R., 1986, Role of minerals in the thermal alteration of organic matter—II. A material balance: The American Association of Petroleum Geologists Bulletin, v. 70, p. 1156–1165.

TISSOT, B. P. AND WELTE, D. H., 1984, Petroleum Formation and Occurrence, second edition: Berlin, Springer-Verlag, 699 p.

VELDE, B. AND ESPITALIÉ, J., 1989, Comparison of kerogen maturation and illite/smectite composition in diagenesis: Journal of Petroleum Geology, v. 12, p. 103–110.

VELDE, B. AND LANSON, 1993, Comparison of I/S transformation and maturity of organic matter at elevated temperatures: Clays and Clay Minerals, v. 41, p. 178–183.

WAPLES, D. W., 1981, Organic geochemistry for exploration geologists: Minneapolis, Burgess, 151 p.

WAPLES, D. W., 1985, Geochemistry in Petroleum Exploration: Boston, International Human Resources Development Corporation, 232 p.

WHITNEY, G., 1990, The role of water in the smectite-to-illite reaction: Clays and Clay Minerals, v. 38, p. 343–350.

SOURCES OF SILICA FROM THE ILLITE TO MUSCOVITE TRANSFORMATION DURING LATE-STAGE DIAGENESIS OF SHALES

MATTHEW W. TOTTEN

Department of Geology and Geophysics, University of New Orleans, New Orleans, Louisiana 70148

AND

HARVEY BLATT

School of Geology and Geophysics, University of Oklahoma, Norman, Oklahoma 73019

ABSTRACT: The nature of clay-mineral diagenesis in shales has received much attention because of its possible role in providing material to interbedded sandstones. In particular, the smectite to illite transformation has been cited by many authors as a source of silica during mudrock diagenesis. We suggest that illite is not an end-member itself but a transition between smectite and muscovite. This is supported by chemical and physical differences between the two minerals from the literature.

Mudrocks from the Stanley Formation (Mississippian) of the Ouachita Mountains of Oklahoma and Arkansas were investigated both chemically and petrographically to test the hypothesis that illite undergoes a continuous transformation to muscovite. Results were related to thermal maturity as determined from vitrinite reflectance data of Houseknecht and Matthews (1985) on the same samples. Within the phyllosilicate fraction, the concentration of silica was found to decrease with increasing thermal maturity. Loss of water from interlayer positions also showed a linear decrease across the same interval. This is consistent with the change of illite into muscovite mica.

Coincident with the observed changes in the phyllosilicate fraction, a corresponding increase in the amount, grain-size and percentage of composite grains was found in the non-phyllosilicate fraction. This is interpreted as the result of authigenic growth of solid silica released during clay-mineral transformations. This is consistent with reported increases in the amount of polycrystalline quartz in metapelites compared to their precursors. Because of the increase in the percentage of quartz in metapelites, they are a likely source of abundant silt-size quartz. The approximate mass-balance of silica between the phyllosilicate and non-phyllosilicate fractions also suggests that the Stanley mudrocks behaved as closed systems during diagenesis.

INTRODUCTION

The smectite to illite clay-mineral transformation is well known in the literature of diagenesis. The transition appears to be a continuous one, varying from 100% smectite layers through a mixed-layer sequence until 100% illite layers are found within the clay mineral. After the loss of expandable clay layers (smectite layers), illite becomes more ordered and therefore more crystalline with increasing burial. Consequently, illite is often considered an end-member of this transition (Srodon and others, 1986).

A precise definition for illite has not been agreed upon. It is generally agreed that illite is similar in structure to muscovite, sharing similar polytypes based on the type of layer stacking. Most illite tends to be of the 1Md polytype (Eslinger and Pevear, 1988), while most muscovite is of the $2M_1$ polytype (Deer et al., 1967). Because of their similar structure, illite and muscovite share a basal spacing of 10 Å and are not discriminated by conventional XRD methods. Consequently, the terms mica and illite have been used interchangeably in many instances for fine-grained, potassium-rich phyllosilicates.

There are several distinctions between illite and muscovite. Both minerals have a distinct chemistry. Illite has more silicon and therefore a smaller negative layer charge than muscovite. Illite therefore requires less potassium to balance this charge. Chemical data compiled from the literature indicate that there is almost as much difference between the chemistry of illite and muscovite as exists between illite and smectite (Table 1). Muscovite is generally coarser-grained than illite, which is itself coarser-grained than smectite. There are also very slight differences in the cell edges, which are extremely difficult to detect except with pure phases. Another important distinction between the two minerals is their thermal stability. Illite generally begins dehydroxylation near 550°C, while muscovite is thermally stable to near 700°C (Todor, 1976). The variation in the temperature of dehydroxylation correlates with illite crystallinity, suggesting a gradational recrystallization of illite into muscovite

TABLE 1.—Si^{+4} AND K^+ CONCENTRATIONS OF END-MEMBER PHYSILS (IONS/10 OXYGENS)

Smectite: Si^{+4}	K^+		Source
4.0	0.0	(ideal montmorillinite)	Ross and Hendricks (1945)
3.86	.06	(avg. of 6 mont.)	Deer et al. (1967)
3.96	—		Grim and Kulbicki (1961)
3.85	.08		Ramseyer and Boles (1986)
3.87	0.0	(avg. of 8 mont.)	Weaver (1989)
Illite:			
3.50	.75	(ideal illite)	Bailey (1984)
3.32	.57		Frey (1970)
3.39	.74	(avg of 4 $2M_1$ illites)	Weaver (1989)
3.40	.62		Weaver and Pollard (1973)
3.35	.65	(extrap. 100% illite)	Weaver (1989)
3.4	.6		Velde (1985)
3.42	.75		Warren and Curtis (1989)
Muscovite:			
3.0	1.0	(ideal muscovite)	Bailey (1984)
3.17	.78	(avg. 15 muscovites)	Deer et al. (1967)
3.09	.85		Rothbauer (1971)
3.02	.86		Güven (1971)
3.06	.93		Bailey (1984)

(Hunziker et al., 1986). These gradational variations between illite and muscovite strongly imply that illite is not itself an end-member but is only a phase halfway between smectite and muscovite. It has only rarely been recognized as such because of the similar positions of peaks on X-ray diffractograms.

The consequences of the continued recrystallization of illite into mica with increasing thermal maturity are important to sedimentary petrologists. The same elements released during illitization of smectite would continue to be released. This includes silicon, iron, magnesium and water. The reaction would also consume aluminum and potassium.

The fate of these elements is not fully agreed upon even during the smectite-illite transformation. As early as 1962 it was suggested that clay-mineral transformations could provide silica to sandstones (Towe, 1962). Many authors feel this is also

Siliciclastic Diagenesis and Fluid Flow: Concepts and Applications, SEPM Special Publication No. 55

the case for other released cations as well as silica (Land and Macpherson, 1992; Curtis, 1978). Other work suggests that shales behave as closed systems (Totten and Blatt, 1993; Sullivan and McBride, 1991; Bloch and Hutcheon, 1992). In either case, the continued release of silica during recrystallization of illite almost doubles the amount available for authigenic quartz growth. If the silica indeed remains within the shale, it could be a major contributor to the silt-size quartz population.

It is the aim of this research to assess the possibility that phyllosilicate recrystallization during diagenesis continues after the smectite to illite transition is complete, with muscovite mica as the final product. The fate of the released silica is also investigated, and the contribution to the silt-sized quartz population estimated. A method to determine the stage of illite recrystallization is proposed.

GEOLOGIC SETTING

The pelites studied are from the Stanley Formation (Mississippian), which crops out in the Ouachita Mountains of southwestern Arkansas and southeastern Oklahoma. The Stanley Formation has a maximum thickness of 3,350 m (Morris, 1989) and is part of a much thicker sequence of deep-water flysch deposited off the southern margin of the North American continent during late Paleozoic time. Many previous studies of burial diagenesis have focused on the Tertiary sequence of the U.S. Gulf Coast and rely upon the assumption of relatively constant provenance. The sedimentation rate during deposition of the Stanley Formation was very similar to the sedimentation rates of the Cenozoic Gulf Coast. The much shorter interval of Stanley deposition (10 Ma) minimizes the effect of variations in source area.

The Stanley Formation is composed of over 70% shale, with interbedded sandstones and a few tuff beds (Johnson, 1968). The shale varies in color from olive green to black, and is generally blocky, although in some localities it is finely fissile. Stanley shales are very fine-grained, massive, and were deposited as distal turbidites in an abyssal setting (Houseknecht and Matthews, 1985). The deep-marine depositional environment of the Stanley shales limits the effect of sea-level changes on the detrital mineralogy.

In the core area of the Ouachita Mountains in Arkansas, the Stanley shales have been metamorphosed to phyllite and perhaps lower greenschist facies. Slate and unmetamorphosed shale occur outward from the core area. Houseknecht and Matthews (1985) reported that vitrinite reflectance values range from 0.24 in unmetamorphosed shales to a maximum of 4.28 in phyllites from the core area. Using the relation of Barker (1988), this represents a maximum temperature variation approaching 300°C. Guthrie et al. (1986) reported a relation between illite crystallinity (Kubler Index) and vitrinite reflectance with a correlation coefficient of .82. Based upon XRD patterns of glycolated samples, mixed layer clays are still present in shales of low thermal maturity ($R_0 < 1.5\%$), although illite is the dominant clay mineral throughout. A large proportion of this illite may be detrital because a major source of the sediment supplied to the Ouachita basin was the stable craton. The detrital illite would not be involved in the smectite to illite transition during diagenesis of the Stanley shales but could be involved in thermally driven recrystallization toward micas. The

excellent correlation between illite crystallinity and thermal maturity found by Guthrie et al. (1986) supports this concept.

METHODS

Sample Collection

Samples used for this study are the same as those used in Houseknecht and Matthews (1985), Guthrie et al. (1986), and Totten and Blatt (1993). Grab bag samples were collected by Matthews at least one foot below the surface of the outcrop to avoid weathering influences. Sample locations are shown in Figure 1.

Analytical Procedures

Whole-rock major-element composition was determined by XRF analysis of fused discs. The discs were prepared using a La_2O_3-doped lithium tetraborate/lithium carbonate flux. Analyses were performed at the University of Oklahoma's Texaco X-ray Laboratory on a Rigaku SMAX XRF using a rhodium anode X-ray tube operated at 40 kV and 65 Ma.

Quartz and feldspar were separated from the bulk rock using a modified sodium pyrosulfate fusion method of Blatt et al. (1982). Total quartz and feldspar were calculated by weight remaining after the fusion. Two separate fusions were performed for each sample. The fused residue from one run was fashioned into a fused disc and analyzed for major element concentrations by XRF. Quartz percentage and modal feldspar abundance were calculated from the resulting XRF data. The chemical composition of the clay-mineral fraction was calculated by subtracting the non-clay composition from the whole rock data. The fused residue of the second run was separated into size fractions. An aliquot of each sized separate was reserved for petrographic investigation.

Oriented, sedimented mounts were made of the less than 2-μm fraction of disaggregated whole-rock samples and analyzed by XRD. The minerals present confirmed the results of Guthrie et al. (1986) on the same samples. The samples were then placed in an oven at 500°C for two hours and reanalyzed by XRD. Peak areas for the 10-Å peak were calculated from both runs and the difference in peak area calculated.

Petrography

The size-separated quartz and feldspar were examined using a polarizing petrographic microscope. Grain mounts were made using Petropoxy 154 (index of refraction = 1.540 ± 0.001) and examined with a Leitz microscope under both plane and polarized light. For each size fraction, 200 grains were point counted to determine the percentage of polycrystalline grains. Chert is commonly defined as polycrystalline quartz with each individual crystal being less than 20–30 μm in diameter. For the purpose of this study the term chert includes all polycrystalline quartz. Because of the small size of the grains in our samples, very few quartz aggregates had individual crystals coarser than 30 μm.

Because sodic plagioclase has an index of refraction similar to quartz and because twinning and cleavage are usually not visible in silt-sized feldspar grains, petrographic identification of feldspar was not considered reliable. The quantity of plagioclase in each size fraction was determined by XRD, by com-

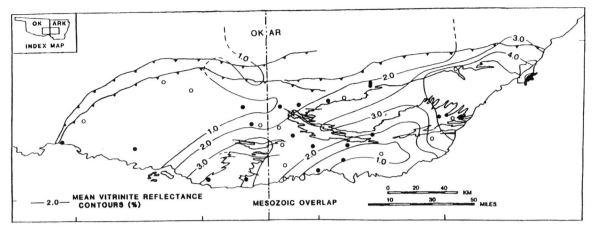

FIG. 1.—Sample locations in relation to vitrinite reflectance contours (after Houseknecht and Matthews, 1985).

parison to diffractograms of prepared standards with known proportions of quartz and albite. The most sensitive peak (highest intensity) for plagioclase determinations is the reflection of the 002 plane, with a d-spacing of 3.197 Å. The technique is sensitive to approximately 1% albite, below which the peak is not statistically detectable above background. Additional information on analytical procedures is found in Totten (1992).

RESULTS

The major element chemical analyses are presented in Table 2. Major-element oxides are reported on a 100% anhydrous basis and are in weight percent. Several analyzed samples were omitted from this study based on their anomalous trace-element characteristics (Totten and Blatt, 1993). The anomalous signature is interpreted as indicating a different provenance, hence obscuring diagenetic alteration. Grain-size parameters and major-element concentrations of the non-phyllosilicate fraction are given in Table 3.

The results of this study were evaluated with respect to maximum temperature as calculated from the vitrinite reflectance data of Houseknecht and Matthews (1985). These values were obtained using the relation reported by Barker (1988) assuming a minimum time period at maximum burial temperatures.

DISCUSSION

Phyllosilicate Fraction

A decrease in the amount of silica present in the clay-mineral fraction with increasing thermal maturity is apparent in Figure 2. This decrease appears linear, and continues to maximum temperatures near 300°C, a temperature beyond any reported for complete illitization of smectite. The linear decrease of silica in this fraction suggests that recrystallization of illite into mica with lower silica concentrations continues after illitization. Ideal illite contains 55% SiO_2, while muscovite contains 45% SiO_2. The lower SiO_2 values from the clay-mineral fraction of

TABLE 2.—WHOLE ROCK CONCENTRATION DATA

Spl	SiO_2	TiO_2	Al_2O_3	Fe_2O_3	MnO	MgO	CaO	Na_2O	K_2O	P_2O_5	H_2O^+	H_2O^-
15	63.46	0.88	20.27	7.92	0.03	2.26	0.22	1.20	3.56	0.20	0.66	4.61
16	65.30	0.84	20.13	6.24	0.03	1.99	0.32	1.46	3.50	0.19	0.66	4.25
24	61.27	0.82	21.36	8.30	0.05	2.39	0.36	1.29	3.93	0.23	0.59	5.18
25	64.82	0.86	20.04	7.11	0.04	1.76	0.26	1.04	3.93	0.14	1.34	5.59
27	65.83	0.81	17.99	7.13	0.07	2.39	0.31	1.39	4.10	0.18	0.31	3.89
31	67.35	0.90	21.33	3.76	0.01	1.28	0.03	0.21	5.06	0.07	1.43	4.91
32	65.79	0.86	18.80	6.56	0.07	2.17	0.47	1.14	3.99	0.15	1.68	4.85
35	60.49	0.94	21.93	7.72	0.05	2.55	0.30	1.41	4.42	0.19	0.88	5.11
36	61.52	1.04	23.41	7.32	0.03	1.85	0.08	0.27	4.33	0.15	2.25	7.04
40	67.26	0.78	17.56	6.59	0.05	1.74	0.19	2.32	3.38	0.13	0.92	4.39
41	59.03	0.84	20.10	6.95	0.04	1.86	2.80	1.14	4.77	2.47	1.55	5.79
42	63.34	0.87	22.01	4.77	0.02	1.57	0.11	1.72	5.46	0.13	1.28	5.04
51	65.54	0.86	19.81	6.55	0.06	1.89	0.24	0.82	4.09	0.14	1.45	5.11
52	59.77	0.90	22.09	8.02	0.09	1.80	0.15	2.28	4.74	0.16	3.88	5.80
56	62.60	0.87	20.06	7.33	0.07	2.05	0.26	2.70	3.89	0.17	0.67	4.72
57	63.22	0.88	20.09	7.94	0.04	2.22	0.23	1.56	3.62	0.20	1.75	5.46
65	70.52	0.72	17.54	3.14	0.07	2.17	0.59	2.08	3.01	0.16	0.55	3.61
113	63.65	0.85	21.19	6.15	0.02	2.13	0.50	1.13	4.22	0.16	2.50	5.47
120	64.41	0.89	19.36	6.44	0.02	2.52	0.71	0.73	4.69	0.23	3.51	4.67
127	65.71	0.79	18.52	6.48	0.08	2.02	0.57	2.27	3.40	0.16	1.73	4.88
128	61.33	0.88	21.47	7.11	0.03	2.61	0.76	1.34	4.28	0.19	4.03	5.57

Major element oxides reported on a 100% anydrous basis

TABLE 3.—MAJOR-ELEMENT CONCENTRATIONS AND GRAIN-SIZE PARAMETERS OF THE NON-PHYLLOSILICATE FRACTION.

Sample	Fused Residue Concentration Data								Calculated Mineralogy				Statistical Values							Mineralogy of Size Fractions					
	SiO_2	Na_2	K_2O	CaO	P_2O_2	Fe_2O_3	Al_2O_3	TiO_2	Quartz	Albite	Orthoclase	Anorthite	ϕ(5%)	ϕ(16%)	ϕ(25%)	ϕ(50%)	ϕ(75%)	ϕ(84%)	ϕ(95%)	20–30 μm QTZ	CHT	SPAR	10–20 μm QTZ	CHT	SPAR
15 Ms	91.88	2.34	<.01	0.07	0.02	<.01	4.95	0.67	80.74	18.32	0.00	0.21	5.00	5.60	5.95	7.00	8.35	8.75	9.80	55	45	?	37	46	17
16 Ms	91.67	1.76	0.22	0.10	0.02	0.60	5.00	0.42	82.97	13.87	1.74	0.35	4.60	5.30	5.60	6.60	8.75	9.40	10.95				42	48	10
24 Ms	97.42	0.66	<.01	0.03	<.01	0.11	1.63	0.04	94.59	5.12	0.00	0.14	3.55	5.00	5.35	6.30	7.95	8.60	9.80	33	56	11	48	52	?
25 Ms	97.27	0.58	0.03	0.07	<.01	0.27	1.61	0.06	94.61	4.50	0.23	0.33	3.50	4.45	5.20	6.95	8.10	8.50	9.15	47	53	?	39	52	9
27 Ms	92.74	2.17	<.01	0.03	<.01	0.25	4.59	0.09	82.53	16.99	0.00	0.14	4.60	5.60	5.90	7.00	8.40	8.90	10.35	38	21	9	65	35	<1
31 Ms	99.57	<.10	<.01	0.04	<.01	<.01	0.34	0.07	99.78	0.00	0.00	0.13	3.85	4.75	5.65	7.00	8.40	8.80	10.10	79	28	<1	57	37	6
32 Ms	90.21	1.47	0.19	0.19	0.03	0.64	6.72	0.34	84.83	11.83	1.53	0.73								72		?			
35 Ms	98.13	0.51	0.02	0.04	<.01	0.23	0.83	0.14	95.36	3.93	0.15	0.19	4.60	6.10	6.75	7.80	8.85	9.40	10.30				53	47	<1
36 Ms	98.21	<.10	<.01	0.05	<.01	0.22	0.54	0.93	98.65	0.00	0.00	0.17	4.15	4.80	5.40	7.40	8.70	9.20	10.60	41	59	<1	40	52	8
39 Ms	95.95	0.84	0.02	0.05	<.01	<.01	2.80	0.12	92.90	6.58	0.16	0.24	5.40	6.50	6.90	7.85	8.80	9.20	10.10	46	54	?	44	48	8
40 Ms	94.25	1.76	<.01	0.10	<.01	<.01	3.70	0.06	85.75	13.72	0.00	0.47	4.50	5.60	5.50	6.20	7.40	8.10	9.25	53	37	10	49	51	<1
41 Ms	96.36	0.18	0.31	0.31	0.07	0.18	2.20	0.16	94.64	1.40	2.42	1.03	4.50	4.85	6.00	7.40	8.60	9.10	10.50				55	42	3
42 Ms	95.11	1.00	0.35	0.08	<.01	0.15	3.07	0.09	88.87	7.78	2.73	0.38	4.50	4.80	5.20	6.00	7.40	8.30	10.00	46	47	7	57	40	3
51 Ms	97.20	0.70	<.01	0.07	<.01	0.15	1.57	0.23	93.87	5.42	0.00	0.33	3.60	5.00	4.45	6.95	8.35	8.85	10.05	67	29	4	45	55	<1
52 Ms	96.07	0.89	0.11	0.07	<.01	0.31	2.38	0.05	91.54	6.92	0.86	0.33	4.25	5.75	5.85	7.50	9.20	9.70	11.10	55	45	<1	17	82	1
56 Ms	98.44	0.38	<.01	0.05	<.01	0.10	0.93	0.04	96.69	2.94	0.00	0.23	5.10	6.00	6.10	7.15	8.25	8.60	9.40	21	79	<1	45	51	4
57 Ms	96.92	0.64	0.03	<.01	<.01	0.18	1.98	0.11	94.50	4.99	0.23	0.00	5.30	5.75	6.40	7.60	8.80	9.35	10.60	43	57	?	34	57	9
65 Ms	91.60	2.42	0.03	0.12	0.03	0.19	5.23	0.29	79.86	18.97	0.24	0.38	3.30	3.75	4.10	5.80	7.80	8.80	10.30	60	40	?	74	23	3
113 Ms	97.83	0.70	0.02	0.06	0.02	<.01	1.09	0.17	94.07	5.40	0.15	0.16	4.60	6.15	6.15	7.85	9.65	10.30	11.50	79	18	3	70	27	3
120 Ms	97.17	0.73	0.02	0.19	<.01	0.14	1.46	0.07	92.36	5.62	0.93	0.89	5.20	6.15	6.70	7.95	9.25	9.80	11.25	79	21	?	69	24	7
127 Ms	94.69	1.00	0.41	0.11	<.01	0.33	3.32	0.04	88.13	7.79	3.20	0.52	4.60	5.40	5.82	7.00	8.25	8.70	10.10	64	27	9	61	32	7
128 Ms	97.89	0.48	0.12	0.19	<.01	<.01	1.04	0.13	94.36	3.70	0.93	0.89	4.60	6.10	6.75	7.95	9.00	9.65	11.15						

? = not enough sample to mount on XRD

blanks = not enough sample or grains too charged to make reliable grain mount

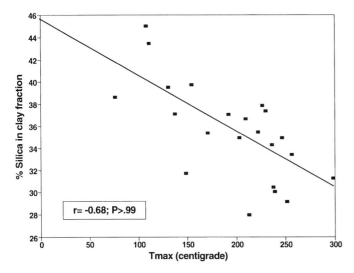

FIG. 2.—Percentage of silica in the clay-mineral fraction versus temperature. The decrease in silica is consistent with the transformation of illite to muscovite in this fraction.

the Stanley shales are due to dilution by lower SiO_2-containing phases such as chlorite.

Because conventional X-ray techniques do not discriminate between illite and muscovite, a means to detect this transition was needed. As mentioned previously, illite contains more water in the interlayer than does mica. To drive off this interlayer water, the samples were heated to 500°C and X-rayed. The patterns were then compared to diffractograms of unheated samples. Loss of interlayer water will result in a decrease in the area of the 10-Å peak representing the basal spacing of the clay. The larger the difference in peak area, the more water that was lost during heating.

Figure 3 illustrates the good correlation between the loss of peak area and increasing thermal maturity. The relationship is linear, suggesting that interlayer water is continually expelled

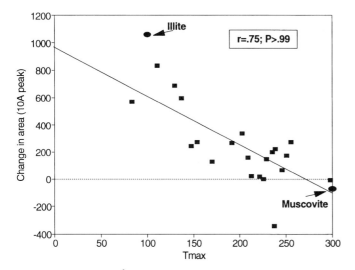

FIG. 3.—Loss of 10-Å peak area with heating versus maximum temperature. Loss in area was calculated by subtracting peak areas from samples X-rayed before and after heating to 500°C for 2 hours.

across this entire range of thermal maturity. To validate this interpretation, samples of muscovite from the University of New Orleans mineralogy collection and illite from the clay minerals repository were treated in a similar manner. The results are also shown on Figure 3 and are consistent with the results from the Stanley shale samples. To insure that adsorbed water was not causing this phenomenon, weight loss after heating to 100°C overnight was compared to thermal maturity, and no correlation was found. Because a major difference between illite and muscovite is the amount of interlayer water present, the expulsion of this water during late diagenesis reflects recrystallization of illite into muscovite.

Differences in the character of the phyllosilicate fraction with increasing temperature are also seen in SEM as shown in Figures 4A and 4B. Samples from areas of lower thermal maturity are of much finer grain size than those of higher maturity. This result is similar to the findings of Guthrie (1985).

Non-phyllosilicate Mineral Fraction

Corresponding changes in the non-phyllosilicate fraction relative to the phyllosilicate fraction were observed. A possible increase in the percentage of the non-phyllosilicate fraction with increasing temperature is shown on Figure 5. This has been reported to be a consequence of quartz growth during clay-

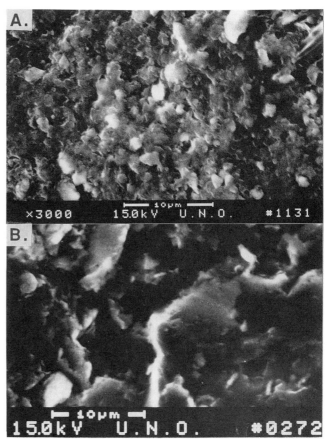

FIG. 4.—SEM images of samples from low thermal maturity (A) and (B) high thermal maturity. Grain size of individual clay particles are significantly coarser in shales of higher maturity.

TABLE 4.—CORRELATION COEFFICIENTS FOR GRAIN-SIZE
PARAMETERS VERSUS TMAX.

Parameter	corr. coef. & sign. level
	n = 21
Mean	**−.49** (>95%)
phi(95%)	**−.59** (>99%)
phi(84%)	**−.52** (>95%)
phi(75%)	**−.49** (>95%)
phi(50%)	**−.43** (>95%)
phi(25%)	−.35
phi(16%)	−.35
phi(5%)	−.27

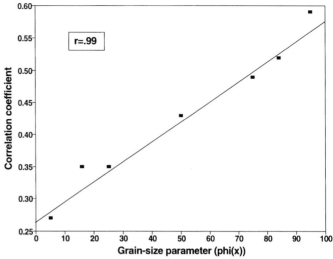

FIG. 6.—Grain-size parameters versus correlation coefficients with maximum temperature. The finest sizes correlate the strongest, suggesting authigenic growth in the finest size fractions with increasing temperature.

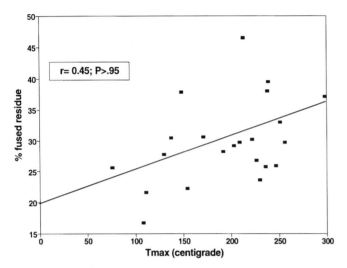

FIG. 5.—Percent of quartz and feldspar fraction versus maximum temperature. Thermal maturity only explains a portion of the percentage of this fraction ($r^2 = .20$). Other factors controlling this percentage include distance of transport and position in turbidite flow.

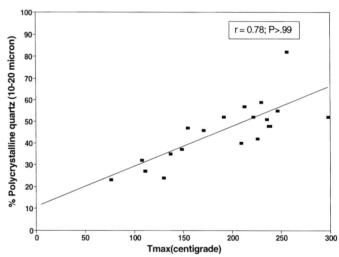

FIG. 7.—Percent polycrystalline quartz in the 10- to 20-μm fraction versus temperature. There is a continuous increase in the percentage of polycrystalline grains from about 10% to 60% across the range in maximum temperature. The coefficient of causation ($r^2 = .61$) suggests that over 60% of the percentage of polycrystalline grains can be explained by thermal maturity. The remaining 40% is due to other factors, such as the initial percentage when deposited.

mineral diagenesis (Totten and Blatt, 1993). Strong evidence for this conclusion is the increased percentage of fine grained quartz in the samples of high thermal maturity. Table 4 lists the correlation coefficients for different grain-size parameters with maximum temperature. Each grain-size parameter reflects a specific position on a cumulative grain-size frequency curve. For example, the phi (95%) corresponds to 95% on the cumulative curve and represents the finest 5% of the sediment. The parameters representing the finest grain-sizes correlate with maximum temperature much better than those representative of coarser size. These correlations decrease in a systematic fashion with increasing grain-size; in fact, the correlation coefficients themselves correlate extremely well with the grain-size parameters (Fig. 6).

In addition to the quartz fraction increasing in both size and amount during shale diagenesis, this fraction also exhibits an increase in polycrystallinity. This is best seen in the 10- to 20-μm grain-size fraction (Fig. 7). Extension of the regression line to 300°C results in a value of 60% polycrystalline quartz grains. This is consistent with the character of quartz released from schists (Blatt, 1967). The composite nature of quartz grains from samples of high thermal maturity as opposed to the monocrystalline quartz of low thermal maturity is seen in Figure 8.

Mass Balance Between Grain-size Fractions

From the regression line of Figure 2 the amount of silica released from the clay-mineral fraction across the level of diagenesis encountered in the Stanley is approximately 10%. This amount coincides with the quantity of silica expected to be released during recrystallization of illite to muscovite (Table 1). This compares very well with the total amount of material gained in the non-phyllosilicate fraction (Fig. 9). Because over 90% of this fraction is composed of quartz and chert, it seems apparent that the gain in silica in this fraction is directly related to the loss of silica in the phyllosilicate fraction.

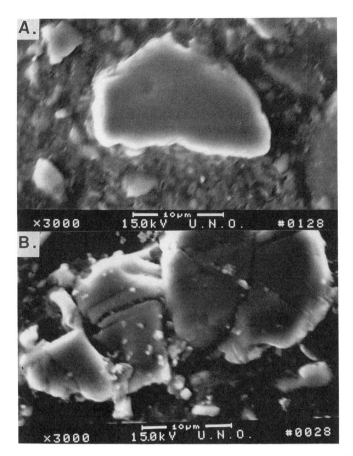

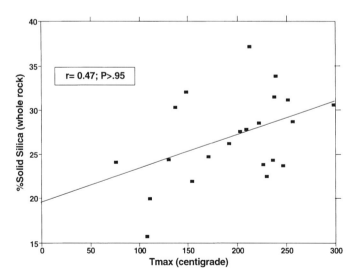

FIG. 8.—SEM images of quartz from samples of (A) low thermal maturity and (B) high thermal maturity. Quartz grains from shales of low thermal maturity are mostly monocrystalline, in contrast with quartz grains from shales of high thermal maturity which are mostly polycrystalline.

FIG. 9.—Whole-rock percentage of solid silica versus maximum temperature. This fraction increases 10% over the thermal maturity range of this study. The coefficient of causation ($r^2 = .22$) suggests that about 20% of the percentage of this fraction is due to thermal maturity. Other factors controlling this percentage include distance of transport and position in turbidite flow.

The implication of the mass balance within these rocks suggests that silica is not exported from shales during diagenesis. This is in agreement with whole-rock chemical data from the same rocks (Totten and Blatt, 1993). Even though the continued recrystallization of illite to muscovite releases more silica than previously expected during shale diagenesis, it appears that the silica remains within the shale system and is not exported. While this does not solve the problem of sources of silica for sandstone cementation, it does contribute to our understanding of the origin of silt-sized quartz. A considerable amount of silica precipitates within mudrocks during diagenesis, all within the silt-sized range. This supports the suggestion by Blatt (1987) that metapelites are the source of much of the silt-sized quartz population.

CONCLUSIONS

Results suggest that illite is not the end-member mineral that it is often considered to be, but is a transition phase as clay recrystallizes into mica. Because of the similarity between illite and muscovite diffraction patterns, this is often overlooked. To erase some of the confusion regarding a precise definition for illite, a chemical definition should be applied, and illite could be defined as a high silica, low potassium, more hydrous precursor to muscovite.

In the absence of chemical analyses of the phyllosilicate fraction, a method to determine the relative proportions of illite and muscovite is to measure the difference in peak area of basal plane reflections both before and after heating the sample to 500°C. The varying effect of this treatment probably reflects dehydroxylation of interlayer water. Because illite holds more water than muscovite, the amount of peak area change is greater after heating an illite. Our results show a continual decrease in the area of this peak with increasing thermal maturity, which we interpret as varying degrees of recrystallization of illite to muscovite. One can imagine a mixed-layer illite/muscovite analogous to mixed-layer illite/smectite.

Coincident with the continual expulsion of water after expandable clays are gone, the phyllosilicate fraction exhibits a linear decrease in silica. This is also consistent with recrystallization of illite to muscovite.

During clay diagenesis, silica released precipitates as authigenic solid silica. Both the percentage of quartz and the grain-size of this fraction increase with increasing thermal maturity. An approximate mass-balance for silica exists between the two fractions, supporting previous work that interpreted these shales as closed systems during diagenesis. In addition to the solid-silica fraction increasing in size and amount, the degree of polycrystallinity dramatically increases during shale diagenesis. The percentage of composite grains approaches the percentage reported from schists. The growth of solid-silica in this fraction is consistent with metapelites being a major source of silt-size quartz.

REFERENCES

BAILEY, S. W., 1984, Crystal chemistry of the true micas, *in* Bailey, S. W., ed., Micas: Washington, D. C., Reviews in Mineralogy, v. 13, p. 13–60.
BARKER, C. E., 1988, Geothermics of Petroleum Systems: Implications of the stabilization of kerogen thermal maturation after a geologically brief heating duration at peak temperature, *in* Magoon, L. B., ed., Petroleum Systems of

the United States: Washington, D.C., United States Geological Survey Bulletin 1870, p. 26–29.

BLATT, H., 1967, Original characteristics of clastic quartz grains: Journal of Sedimentary Petrology, v. 37, p. 181–193.

BLATT, H., 1987, Oxygen isotopes and the origin of quartz: Journal of Sedimentary Petrology, v. 57, p. 373–377.

BLATT, H., JONES, R. L., AND CHARLES, R. G., 1982, Separation of quartz and feldspars from mudrocks: Journal of Sedimentary Petrology, v. 52, p. 660–662.

BLOCH, J. AND HUTCHEON, I. E., 1992, Shale diagenesis: A case study from the Albian Harmon Member (Peace River Formation), western Canada: Clays and Clay Minerals, v. 40, p. 682–699.

CURTIS, C. D., 1978, Possible links between sandstone diagenesis and depth-related geochemical reactions occurring in enclosing mudrocks: Journal of the Geological Society of London, v. 135, p. 107–117.

DEER, W. A., HOWIE, R. A., AND ZUSSMAN, J., 1967, Rock-forming Minerals; Sheet Silicates: London, Longmans, 270 p.

ESLINGER, E. AND PEVEAR, D., 1988, Clay Minerals for Petroleum Geologists and Engineers: Tulsa, Society of Economic Paleontologists and Mineralogists Short Course Notes 22, 413 p.

FREY, M., 1970, The step from diagenesis to metamorphism in pelitic rocks during Alpine orogenesis: Sedimentology, v. 15, p. 261–279.

GRIM, R. E., AND KULBICKI, G., 1961, Montmorillinite: High temperature reactions and classification: American Mineralogist, v. 46, p. 1329–1369.

GUTHRIE, J. M., 1985, Clay mineralogy as an indicator of thermal maturity of Carboniferous strata, Ouachita Mountains: Unpublished M.S. Thesis, University of Missouri-Columbia, Columbia, 74 p.

GUTHRIE, J. M., HOUSEKNECHT, D. W., AND JOHNS, W. D., 1986, Relationships among vitrinite reflectance, illite crystallinity, and organic geochemistry in Carboniferous strata, Ouachita Mountains, Oklahoma and Arkansas: American Association of Petroleum Geologists Bulletin, v. 70, p. 26–33.

GÜVEN, N., 1971, The crystal structure of $2M_1$ phengite and $2M_1$ muscovite: Zeitschrift fuer Kristallographie, Kristallgeometrie, Kristallphysik, Kristallchemie, v. 134, p. 196–212.

HOUSEKNECHT, D. W. AND MATTHEWS, S. M., 1985, Thermal maturity of Carboniferous strata, Ouachita Mountains: American Association of Petroleum Geologists Bulletin, v. 69, p. 335–345.

HUNZIKER, J. C., FREY, M., CLAUER, N., DALLMEYER, R. D., FRIEDRICHSEN, H., FLEHMIG, W., HOCHSTRASSER, K., ROGGWILER, P., AND SCHWANDER, H., 1986, The evolution of illite to muscovite: mineralogical and isotopic data from the Glarus Alps, Switzerland: Contributions to Mineralogy and Petrology, v. 92, p. 157–180.

JOHNSON, K. E., 1968, Sedimentary environment of the Stanley Group of the Ouachita Mountains of Oklahoma: Journal of Sedimentary Petrology, v. 38, p. 723–733.

LAND, L. S. AND MACPHERSON, G. L., 1992, Origin of saline formation waters, Cenozoic Section, Gulf of Mexico sedimentary basin: American Association of Petroleum Geologists Bulletin, v. 76, p. 1344–1362.

MORRIS, R. C., 1989, Stratigraphy and sedimentary history of post-Arkansas Novaculite Carboniferous rocks of the Ouachita Mountains, in Hatcher, R. D., Jr., Thomas, W. A., and Viele, G. W., eds., The Geology of North America, The Appalachian-Ouachita Orogen in the United States: Boulder, Geological Society of America, v. F-2, p. 591–602.

RAMSEYER, K. AND BOLES, J. R., 1986, Mixed layer illite/smectite minerals in Tertiary sandstones and shales, San Joaquin Basin, California: Clays and Clay Minerals, v. 34, p. 115–124.

ROSS, C. S. AND HENDRICKS, S. B., 1945, Minerals of the montmorillinite group: Washington, D. C., United States Geological Survey Professional Paper 205-B, 79 p.

ROTHBAUER, R., 1971, Untersuchung eines $2M_1$-muscovits mit neutronenstrahlen: Neues Jahrbuch für Mineralogie Monatshefte, p. 143–154.

SRODON, J., MORGAN, D. J., ESLINGER, E. V., EBERL, D. D., AND KARLINGER, M. R., 1986, Chemistry of illite/smectite and end-member illite: Clays and Clay Minerals, v. 34, p. 368–378.

SULLIVAN, K. B. AND MCBRIDE, E. F., 1991, Diagenesis of sandstones at shale contacts and diagenetic heterogeneity, Frio Formation, Texas: American Association of Petroleum Geologists Bulletin, v. 75, p. 121–138.

TODOR, D. N., 1976, Thermal Analysis of Minerals: Kent, Abacus Press, 256 p.

TOTTEN, M. W., 1992, Diagenetic investigation of pelitic rocks across the shale-slate-phyllite transformation, Stanley Shale, Ouachita Mountains, Oklahoma and Arkansas: Unpublished Ph.D. Dissertation, University of Oklahoma, Norman, 182 p.

TOTTEN, M. W. AND BLATT, H., 1993, Alterations in the non-clay-mineral fraction of pelitic rocks across the diagenetic to low-grade metamorphic transition, Ouachita Mountains, Oklahoma and Arkansas: Journal of Sedimentary Petrology, v. 63, p. 899–908.

TOWE, K. M., 1962, Clay mineral diagenesis as a possible source of silica cement in sedimentary rocks: Journal of Sedimentary Petrology, v. 32, p. 26–28.

VELDE, B., 1985, Clay Minerals: A Physico-chemical Explanation of Their Occurrence: Amsterdam, Elsevier, 427 p.

WARREN, E. A. AND CURTIS, C. D., 1989, The chemical composition of authigenic illite within two sandstone reservoirs as analyzed by ATEM: Clay Minerals, v. 24, p. 137–156.

WEAVER, C. E., 1989, Clays, Muds, and Shales: Amsterdam, Elsevier, 819 p.

WEAVER, C. E. AND POLLARD, L. D., 1973, The Chemistry of Clay Minerals: Amsterdam, Elsevier, 213 p.

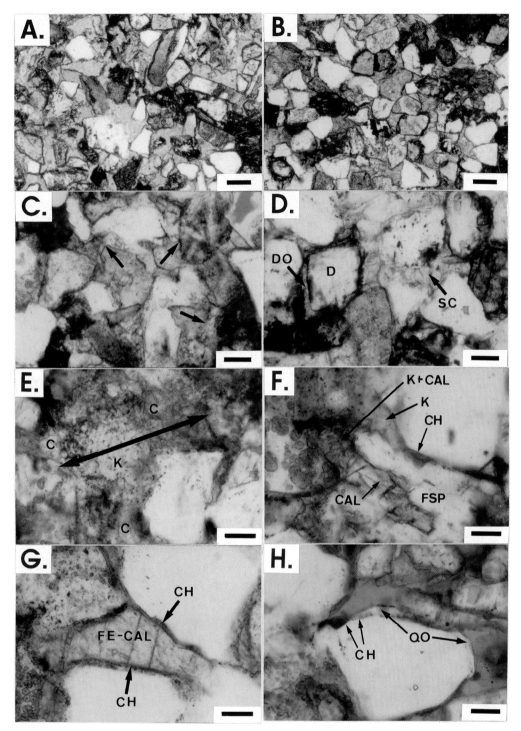

FIG. 7.—Petrographic photomicrographs of Point Lookout sandstones. (A) shows grain packing and cementation inside a concretion (scale bar = 100 μm). The blue-colored cement is poikilotopic Fe-calcite stained with the potassium ferricyanide. (B) shows grain packing and cementation outside of a concretion (scale bar = 100 μm); this is the host sandstone from the R4 unit at Sec. 10. Again, blue-colored cement is poikilotopic Fe-calcite. (C) is also a sandstone from R4. Here, a patch of poikilotopic Fe-calcite is shown in the center of the photo (stained blue with arrows), with poikilotopic calcite filling pores around the outer edge of the Fe-calcite patch (scale bar = 50 μm). The poikilotopic calcite is in optical continuity with the poikilotopic Fe-calcite. (D) has sparry calcite (SC) cementing the sandstone (scale bar = 50 μm). Also shown are intrabasinal dolomite grains (D) with Fe-dolomite overgrowths (DO). While there is unidentifiable oxidized iron material in places throughout the sandstone, it is almost always seen associated with the dolomite, as shown in this photo. The arrow in (E) spans the diameter of a kaolinite-filled pore (scale bar = 25 μm). Poikilotopic calcite (C) has been caught in the act of trying to fill the kaolinite-filled pore. The clean area of kaolinite (K) is in the center of the pore. (F) is another photo showing kaolinite (K) in poikilotopic calcite (CAL) (scale bar = 25 μm). Here, the poikilotopic calcite is filling in a partially dissolved feldspar (FSP). Additionally, authigenic chlorite (CH) is rimming the kaolinite-filled pore. Authigenic rims of chlorite (CH) preceded the poikilotopic Fe-calcite in (G) (scale bar = 25 μm). (H) shows how authigenic chlorite rims (CH) have affected the precipitiation of quartz overgrowths (QO) (scale bar = 25 μm). Where the chlorite rims are less developed, the quartz overgrowths were able to grow. Growth of the quartz was inhibited over thicker patches of chlorite. (See Loomis and Crossey, p. 27–30)

Pressure/Gas Profile
Powder River Basin
From Sonic Logs

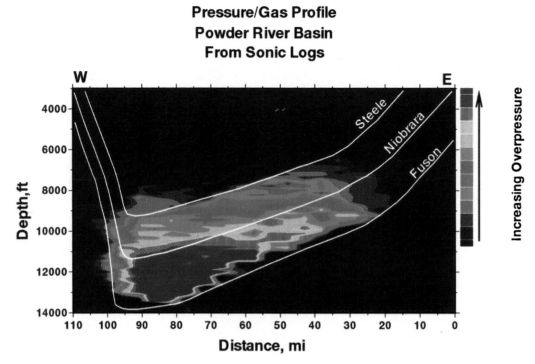

FIG. 4.—West-east pressure (gas) profile across the southern Powder River Basin, line of section A-A′ in Figure 1. The profile shows computed pressure in the Cretaceous shales derived from a panel of sonic logs. The derived pressure values are based the "equivalent depth" method as used by Ham (1987). From Surdam et al. (1995b). (See Surdam, Jiao, and Yin, p. 61)

Powder River Basin

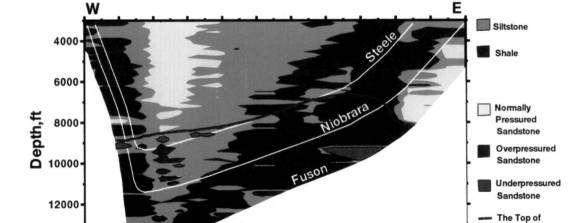

FIG. 8.—West-east hydrocarbon/pressure panel through the southern Powder River Basin (line of section A-A′ in Fig. 1). Compare with Figure 4. The lithology designations are based on gamma-ray responses, and pressure and gas saturation characteristics are based on sonic transit time edited for either sandstone or shale and DST data. The top of the Steele, Niobrara, and Fuson Formations are shown. The top of overpressuring in the shales (top of wet gas expulsion zone or GASEX) is shown as a solid red line. Normally pressured and underpressured sandstones are shown in yellow and green, respectively. Note that sandstones near but below the top of overpressuring in the shales are underpressured. These sandstones are interpreted as gas conduits or migration pathways out of the GASEX zone. In contrast, the sandstones shown in red are overpressured, compartmentalized, and gas-saturated. These sandstones represent gas accumulations in capillary traps; the position of these gas accumulations is not dependent on structural closure. From Surdam et al. (1995b). (See Surdam, Jiao, and Yin, p. 62)

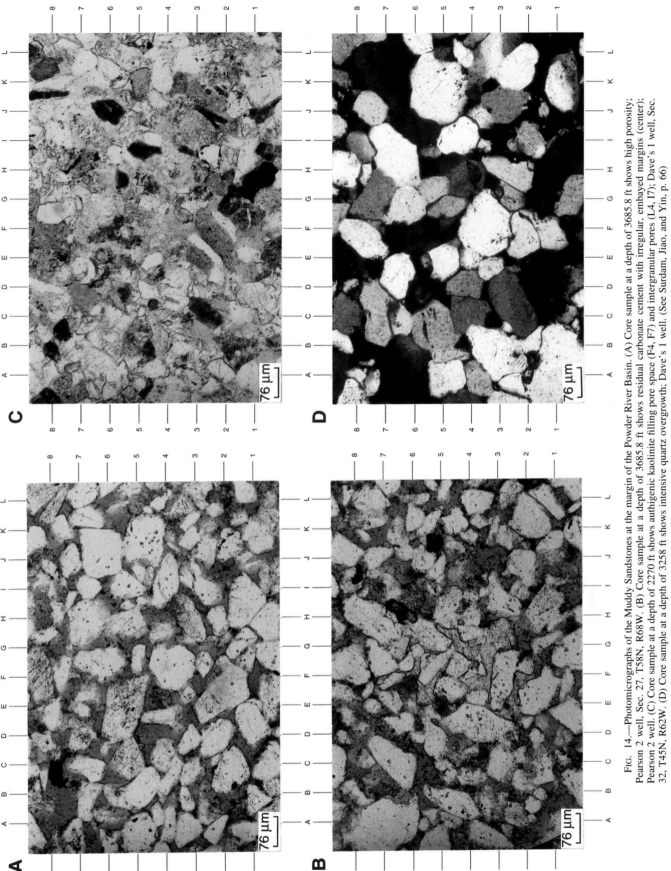

FIG. 14.—Photomicrographs of the Muddy Sandstones at the margin of the Powder River Basin. (A) Core sample at a depth of 3685.8 ft shows high porosity; Pearson 2 well, Sec. 27, T58N, R68W. (B) Core sample at a depth of 3685.8 ft shows residual carbonate cement with irregular, embayed margins (center); Pearson 2 well. (C) Core sample at a depth of 2270 ft shows authigenic kaolinite filling pore space (F4, F7) and intergranular pores (L4, I7); Dave's 1 well, Sec. 32, T45N, R62W. (D) Core sample at a depth of 3258 ft shows intensive quartz overgrowth; Dave's 1 well. (See Surdam, Jiao, and Yin, p. 66)

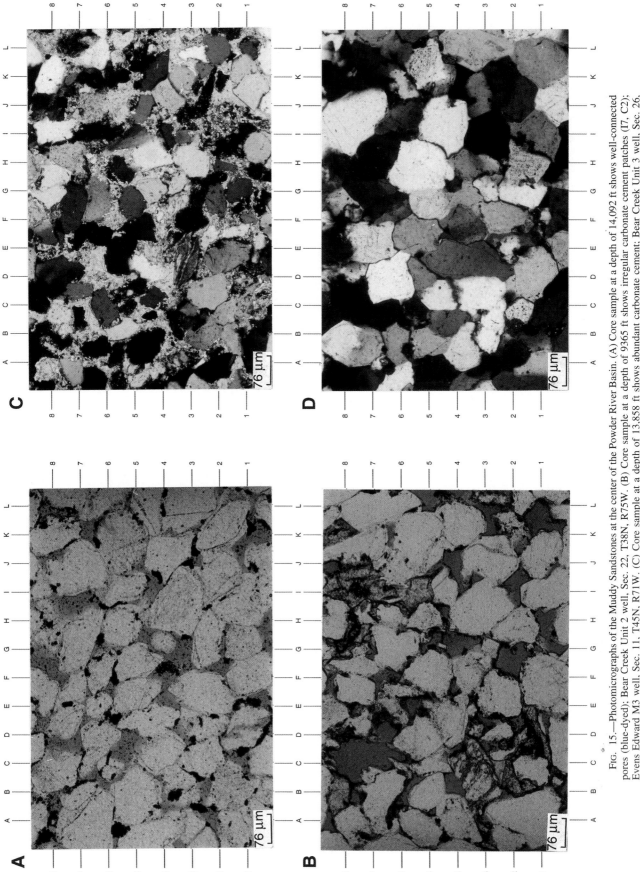

FIG. 15.—Photomicrographs of the Muddy Sandstones at the center of the Powder River Basin. (A) Core sample at a depth of 14,092 ft shows well-connected pores (blue-dyed); Bear Creek Unit 2 well, Sec. 22, T38N, R75W. (B) Core sample at a depth of 9365 ft shows irregular carbonate cement patches (I7, C2); Evens Edward M3 well, Sec. 11, T45N, R71W. (C) Core sample at a depth of 13,858 ft shows abundant carbonate cement; Bear Creek Unit 3 well, Sec. 26, T38N, R75W. (D) Core sample at a depth of 9212 ft shows intensive quartz overgrowth; Simpson 1 well, Sec. 6, T50N, R73W. (See Surdam, Jiao, and Yin, p. 66)

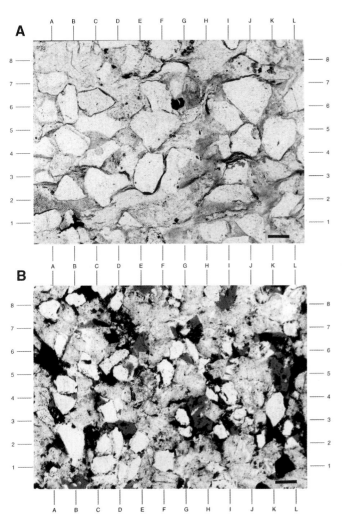

FIG. 7.—CL photo illustrating grain fractures subsequently healed by quartz cement. Bright-orange-red-luminescent cement is dolomite. Scale bar = 100 microns. (See Stone and Siever, p. 133)

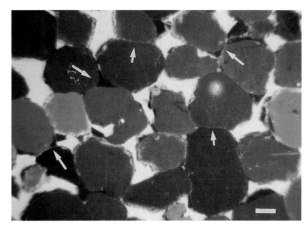

FIG. 16.—(A) Photomicrograph showing abundant clay matrix and absence of intergranular porosity in the Rozet sandstone in the unconformity zone. (B) Photomicrograph showing much less clay matrix and higher porosity (blue dyed) in the underlying Lazy B sandstone. Scale bars equal 76 μm on both photos. (See Surdam, Jiao, and Yin, p. 67)

FIG. 11.—CL photo illustrating grain-packing stabilization from grain-contact enlargement, which is achieved by pressure solution (short arrows) and quartz cementation (long arrows). Brightly-luminescent intergranular material is pore-filling epoxy. This sample is presently at 4270 m (14,010 ft) and has 15% intergranular porosity, yet is not overpressured, indicating that grain-packing arrangements are stable at high lithostatic pressures when small amounts of pressure solution and quartz cementation have occurred. Scale bar = 100 microns. (See Stone and Siever, p. 135)

FIG. 4.—CL photographs. (A) Quartz-cemented sandstone with very few pressure solution contacts. Quartz cement is non-luminescent and very dull-red-luminescent. Brightly-luminescent intergranular material is pore-filling epoxy. Pores and quartz cement on the right side of the photo contain bright-blue-luminescent kaolinite cement. (B) Intensely pressure-dissolved, quartz-cemented sandstone. Brightly-luminescent grains are K-feldspar, one of which is heavily-altered. Scale bars = 100 microns. (See Stone and Siever, p. 132)

FIG. 10.—Photomicrograph of clay-rich siltstone showing silica dissolution between fine-grained quartz grains, 16/7a-A3, 13072′5″, 3987.1 m (TVDSS). Arrows highlight interpenetrating sutured grain boundaries which have reduced pore connectedness. (See McLaughlin, Haszeldine, and Fallick, p. 109)

FIG. 11.—Photomicrograph of quartz overgrowths in a coarse grained sandstone where there is some slight dissolution between the overgrowths but none between the detrital grains, 16/7a-A11, 12712′6″, 3877.3 m (TVDSS). (See McLaughlin, Haszeldine, and Fallick, p. 109)

FIG. 4.—Thin-section photomicrographs of sandstone samples taken in plane light; scale bars represent 50 mm. (A) shows the nature of occurrence of the important detrital constituents in the sandstones, Q = quartz, F = feldspar, L = lithic fragments. (B) illustrates the diverse character of the suite of authigenic minerals present, D = dolomite, C = chlorite, Q = quartz overgrowths. (C) shows a typical example of an organic-rich siltstone sample. (D) shows a sandstone sample in which 100% of intergranular volume is occluded with calcite cement. (See Hays, Walling, and Tieh, p. 168–169)

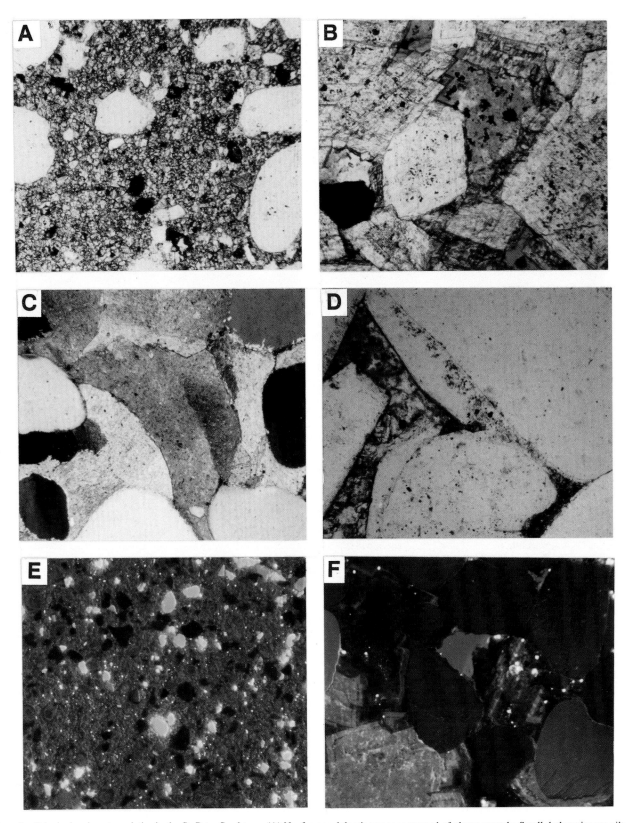

FIG. 5.—Principal carbonate varieties in the St. Peter Sandstone. (A) Nonferroan dolomicrospar composed of planar crystals. Small dark grains are silt-size, detrital K-feldspar crystals. (B) Planar dolospar distributed as euhedral crystals with ferroan rims (stained blue). (C) Coarse-crystalline, baroque dolospar with slightly- to moderately-curved crystal faces and sweeping extinction. (D) Poikilotopic, slightly ferroan, calcite distributed as a pore-fill cement. The mineral more commonly occurs as a nonferroan, pore-fill cement. Note that calcite formed after secondary quartz. (E) Dolomicrospar showing unzoned, orange-red CL. Tiny bright blue luminescent grains are detrital K-feldspar. (F) Baroque dolospar displaying an orange-red luminescent core (equivalent to planar dolospar) and a ferroan dull-brown luminescent rim. (See Pitman and Spötl, p. 190)

PART II
APPLICATIONS

QUARTZ DIAGENESIS IN LAYERED FLUIDS IN THE SOUTH BRAE OILFIELD, NORTH SEA

ÓRLA M. McLAUGHLIN, R. STUART HASZELDINE
Department of Geology and Applied Geology, University of Glasgow, Lilybank Gardens, Glasgow, G12 8QQ, Scotland
AND
ANTHONY E. FALLICK
Scottish Universities Research and Reactor Centre, Isotope Geosciences Unit, East Kilbride, G75 0QU, Scotland

ABSTRACT: The South Brae reservoir sandstones were cemented by quartz late in their diagenetic history. The volume of quartz cement in the South Brae sandstones ranges from 0 (in early calcite cemented concretions) to approximately 11%, with a mean of 3.5%. In four studied wells, quartz overgrowth $\delta^{18}O$ decreases with increasing depth and demonstrates a control by the sedimentological reservoir layering of the field. The data indicate that all overgrowths in any one well cannot have precipitated from a single pore fluid. Overgrowths in shallow reservoir sandstones have precipitated from a higher $\delta^{18}O$ fluid (basinal, $\delta^{18}O \approx +5‰$) than overgrowths in deeper reservoir sandstones (evolved meteoric, $\delta^{18}O \approx -7$ to $+5‰$). This long-term layering of diagenetic fluids also implies that quartz cement can be formed from local sources: advective transport by warm fluids is not required. The greatest volume of quartz is interpreted to have precipitated at temperatures between 70°C and 110°C (2.3 to 3.7 km). Quartz cementation probably occurred most rapidly during periods of overpressure release. Volumes of quartz cement are not significantly different in shallower and deeper reservoir sandstones.

INTRODUCTION

Cementation of reservoir sandstones by quartz is an important and common diagenetic process that is not yet fully understood. Many different and opposing interpretations have been presented concerning the origins of quartz cement, timing of cementation, controls on cement distribution and the hydrologic conditions which transport silica to the site of cementation. These unresolved problems were recently reviewed by McBride (1989) who concluded that more quantitative case studies were necessary in order to establish criteria that govern quartz cementation.

In the South Brae reservoir sandstones, it is apparent from thin-section examination that well-formed syntaxial authigenic quartz overgrowths reduce primary porosity but also help preserve porosity by mechanically strengthening the reservoir and preventing further compaction, thus possibly contributing to maintaining an open network of pores for fluid migration. This phenomenon is particularly common in coarse-grained sandstones and has also been recognised in the Middle Jurassic Brent Group sandstones by Harris (1989). Extensive formation of quartz overgrowths which have become large enough to have destroyed reservoir quality is limited to the thinner sandstones, which occur in predominantly shaly sequences (McLaughlin, 1992).

The principle aims of this study were to test the hypothesis of fault related diagenesis due to the proximity of the South Brae reservoir to the graben margin. Initially it was presumed that the diagenetic history would be similar to the Tartan oil field, where the faults acted as conduits for migrating hot fluids (Burley et al., 1989). A systematic examination of the variation in conditions of quartz overgrowth precipitation between wells, reservoir zones and depth was carried out. The method adopted was to measure $\delta^{18}O$ of quartz overgrowths for each of four wells (16/7a-A3, 16/7a-A11, 16/7a-A19 and 16/7a-A27) which form a dip cross section through the South Brae oil field (Figs. 1A and 1B). Formation conditions were constrained by integrating the isotope results with cement stratigraphy and temperature estimates based on fluid inclusion homogenization measurements. The influence on quartz cementation of the depositional environment and facies, grain size, sorting and depth was evaluated. Using point count data (McLaughlin, 1992) the

amount of internally derived silica was estimated and compared with the volume of silica imported into the sandstones.

GEOLOGICAL AND STRUCTURAL SETTING

In the South Brae oil field (Fig. 1B), the reservoir sequence is the Brae Formation. It is of Late Jurassic age and comprises thick units of sand-matrix conglomerate and sandstone, alternating with thick units of mudstone and sandstone (Fig. 1B). These sediments are commonly combined into large-scale fining upward sequences (Turner et al., 1987). The South Brae reservoir sequence is interpreted to have been deposited as the proximal part of a complex submarine fan system (Turner et al., 1987). In three dimensions, the coarse-grained packages often occur as channel-like bodies which radiate basin ward (becoming less conglomeratic and more sandy), and are separated by fine-grained interchannel areas (C. Turner, Marathon Oil UK, pers. commun., 1991); (Figs. 2A and 2B). These submarine fan sediments were deposited as syntectonic graben fill (Roberts, 1991); they are overlain by, and are laterally equivalent to, the organic-rich hemipelagic mudstones of the Kimmeridge Clay Formation (KCF).

A combination of lithostratigraphic, structural, biostratigraphic and reservoir pressure-correlation techniques has been utilized by Marathon to delineate 4 major zones in South Brae, designated A, B, C and D. Zone A is made up of the Upper Brae Formation, Zone B the Middle Brae Formation and Zones C/D make up the Lower Brae Formation. Each of these major zones have been subdivided into layers (e.g., Aa, Ab, Ac and Ad) and sublayers (e.g., Aa1, Aa$_{2-4}$). This refined lithology designation facilitates more accurate modelling of fluid flow in the reservoir as the zones are thought to follow the depositional stratification (Roberts, 1991).

METHODS

Petrographic Methods

Forty thin sections from the four studied South Brae wells were examined using a standard polarizing microscope. Textural relationships were examined in detail using a Cambridge Stereo Scan 360 Scanning Electron Microscope (SEM) equipped with a LINK energy dispersive analyzer. Cathodoluminescence (CL) analyses were performed on the SEM with a

Siliciclastic Diagenesis and Fluid Flow: Concepts and Applications, SEPM Special Publication No. 55

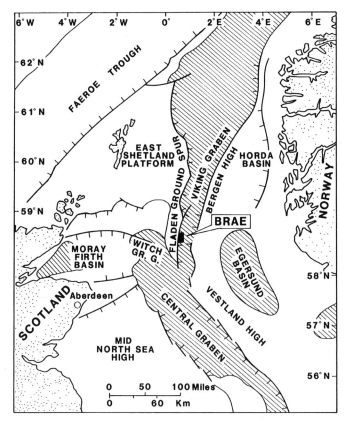

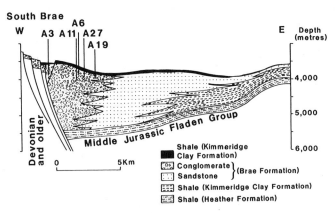

FIG. 1.—(A) Location of Brae relative to major tectonic elements. The shaded area contains thick Jurassic deposits (after Ziegler, 1978). (B) Schematic cross section of the South Brae submarine fan, showing approximate position of the wells sampled, 16/7a-A3, A6, A11, A19 and A27 (after Stoker and Brown, 1986).

Monochrome Oxford Instruments CL. Analyses of minerals in bulk sandstone and separates were accomplished using a Philips X-Ray Diffractometer.

Isotopic Analyses of Diagenetic Quartz

Approximately 10 mg of quartz sample was loaded into nickel reaction vessels, then outgassed under high vacuum at 250°C for one hour. Oxygen was liberated from the samples by overnight reaction with ClF_3 (Borthwick and Harmon, 1982) at

650°C, before being purified and reduced to CO_2 using a vacuum extraction line similar to that described by Clayton and Mayeda (1963). The CO_2 gas was analyzed on a VG-SIRA 10 mass spectrometer and $\delta^{18}O$ compositions of the quartz are quoted relative to SMOW (Standard Mean Ocean Water). Replicate analyses of NBS 28 produced an average $\delta^{18}O$ value of $9.6‰ \pm 0.3‰$, n = 9. The oxygen isotope compositions of quartz overgrowths were calculated using the mass balance technique of Milliken et al. (1981).

Fluid Inclusion Analyses and Results

Fluid inclusion microthermometric studies on quartz overgrowths can provide information on the temperature of the diagenetic fluid from which authigenic quartz precipitated. Only primary two phase aqueous liquid-vapor fluid inclusions were observed in the South Brae reservoir samples; no hydrocarbon fluid inclusions were found. Homogenization temperatures were made on inclusions lying along the boundaries between overgrowths and detrital grains. Most of the inclusions for which temperatures were measured had diameters >5 μm but <10 μm. All the inclusions were isolated and primary, with no signs of necking or decrepitation. Fluid inclusion temperatures were measured from different reservoir zones (see Table 1) and incorporated into the model for quartz cementation. Homogenization temperatures ranged from 82° to 114°C.

DETRITAL MINERALOGY AND DIAGENESIS

The sandstones examined from the South Brae reservoir formation are dominantly quartz arenites according to the scheme of McBride (1963) (Fig. 3). Quartz is the most abundant component comprising 27 to 85% of the detrital mineral assemblage. Monocrystalline quartz is dominant, with polycrystalline quartz grains constituting up to 5% of the samples examined. As expected, polycrystalline varieties become more common as grain size increases. In some coarser sandstones, small quartz granules and pebbles are common. Feldspar comprises 0 to 5% of the bulk composition. K-feldspar is the most common type in these sandstones, although also small amounts of plagioclase exist. Muscovite and a few heavy minerals (zircon, apatite, rutile and sphene) occur in trace amounts. Comminuted bivalve, gastropod and echinoderm shells, together with carbonaceous woody debris are present locally. Rock fragments are mainly clasts of Devonian sandstones and quartzites and some fragments of schists and mudstones (Stow et al., 1982). The sequence of diagenetic events in the South Brae reservoir was determined using cement stratigraphy from polarizing microscope, CL and SEM studies as detailed below. The generalized sequence of diagenetic events is similar above and below the oil-water contact (Fig. 4).

Early Cements

Calcite cements were the first to precipitate following deposition of the sediments. Calcite concretions formed in coarse-grained sandstones and conglomerates and range from 0.8 to 4.0 m in diameter. This cement is poikilotopic; enclosed detrital grains have an open texture and appear to float within the concretions. The volume of cement decreases from the centers to the margins of the concretions, ranging from 45 to 19%. This

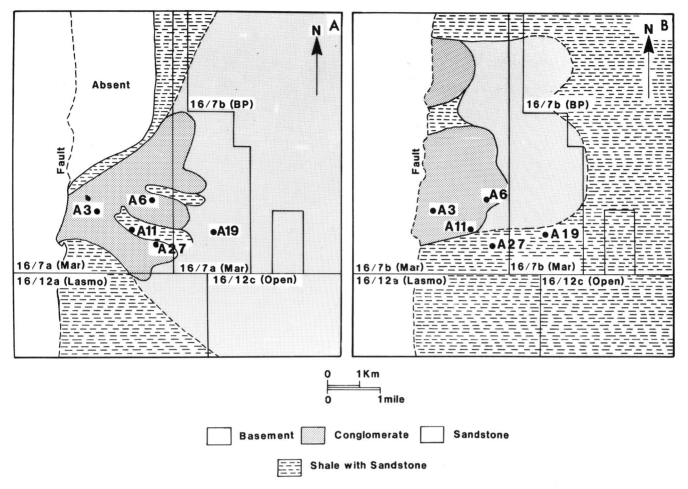

FIG. 2.—(A) Generalized facies map of Upper Brae Formation (Zone A) reservoir, comprized of channelized conglomerates which are separated by shaly islands. (B) Generalized facies map of Lower Brae Formation (Zone C) reservoir defined by two apron-shaped conglomeratic complexes.

TABLE 1.—FLUID INCLUSION HOMOGENIZATION TEMPERATURE
MEASUREMENTS FROM DIFFERENT RESERVOIR ZONES
IN WELLS 16/7a-A11 AND A19.*

16/7a-A11		16/7a-A19	
Zone	$T_hC°$ Average	Zone	$T_hC°$ Average
A_{a1}	106.1	A_{a2-4}	95.8
A_{a2-4}	112	B_{a1-4}	99.3
C_d	96.9		

*Measurements were made on inclusions lying along the boundary between the overgrowth and the detrital grain.

decrease in the cement volume towards concretion margins is possibly due to continuing compaction over time. Within the concretions, some authigenic pyrite occurs as framboids, but authigenic quartz and authigenic clays were not found. The fluid which precipitated the calcite cement appears to have only slightly etched the detrital quartz grains, while feldspar grains and shell fragments (high Mg) were partially to totally etched and subsequently replaced by low-Mg calcite (McLaughlin, 1992). Illite and chlorite precipitated prior to quartz cementation and occur locally as coatings on detrital quartz grains in a number of quartz cemented sandstones.

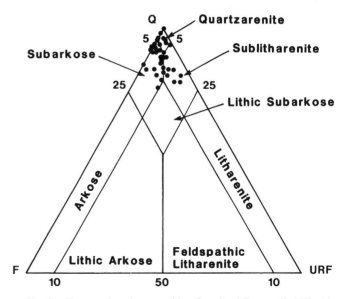

FIG. 3.—Framework-grain compositions from South Brae, wells 16/7a-A3, A6, A11, A19 and A27. Classification after McBride (1963).

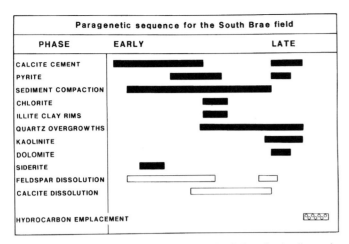

FIG. 4.—Paragenetic sequence showing relative timing of major diagenetic events in the South Brae oil field. Diagenesis above and below the O.W.C. appears to be identical.

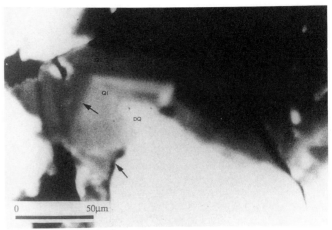

FIG. 5.—Scanning Electron Microscope cathodoluminescence photomicrograph of polyphase quartz cementation, 16/7a-A11, 13791'4", 4204.5 m (TVDSS). Arrows indicate demarcation of possible overgrowth zones. DQ is detrital quartz, Q1 indicates the first overgrowth generation and Q2 the second generation. The growth appears to be continuous as there is no corroded growth terminations.

Quartz Cement

Quartz cementation was widespread throughout the reservoir. The volume of quartz cement in South Brae sandstones ranges from 0 to 11%. Fine-grained sandstones show more compaction effects (with abundant evidence of long grain contacts and suturing) than coarse-grained sandstones. The overgrowths range in size from individual tiny crystal euhedra of 5–30 μm to grain-rimming cements 60 μm thick. Large euhedral overgrowths have been observed protruding into secondary pores, and overgrowths on quartz granules and pebbles in the coarser sandstones and conglomerates are also quite common. In some thin sections, there are small porous areas with no quartz overgrowths but with minor patches of calcite. This suggests that calcite may have previously existed in larger quantities and inhibited overgrowth development but has been partially removed by later dissolution.

CL of polished thin sections showed both dark red/brown quartz grains (indicative of a metamorphic source) and blue colors (indicative of a plutonic source; Zinkernagel, 1978). The quartz overgrowths were nonluminescent. SEM cathodoluminescence studies of the same samples allowed distinction between authigenic and detrital quartz. The detrital grains luminesced brightly, whereas the overgrowths did not. Quartz cementation was seen occasionally to be polyphase, with at least two distinct CL zones noted (Fig. 5). These zones are not abraded, and it appears that there was no hiatus during their precipitation.

Late Dissolution and Cements

Following quartz overgrowth formation, late dissolution created variable quantities of secondary porosity. The margins of calcite-cemented concretions were dissolved, and carbonate clasts, shell fragments and detrital feldspar grains also experienced varying amounts of dissolution. Most feldspar dissolution occurred in coarse-grained sandstones. Subsequent to formation of the secondary porosity, small amounts of pore-filling kaolinite and illite were precipitated. Some late stage cubic pyrite can

also be found in secondary pores as well as minor amounts of quartz, calcite and rhombic dolomite.

OXYGEN ISOTOPE RESULTS

$\delta^{18}O$ values measured for 24 samples of authigenic quartz for the four wells in the South Brae reservoir vary from 17.2‰ to 25.3‰ (Table 2). The initial quartz samples analyzed from well 16/7a-A19 comprised the size fractions, 160–250 mm, 250–500 mm and 500–1000 mm. Slight variations in the calculated $\delta^{18}O$ for the overgrowths existed, the differences ranging from +0.7 to 0.9‰. For the remaining samples the 250 to 750-mm size fraction was analyzed as this was most representative. Using this analytical technique, estimates of quartz overgrowth $\delta^{18}O$ are considered accurate to ±2‰ (Brint et al., 1991). Duplicates of each sample were analyzed.

The wide range (17.2 to 25.3‰) in overgrowth $\delta^{18}O$ (Table 2) suggests that these values could have been produced by several combinations of pore fluid composition and temperature. These conditions can be constrained somewhat by previous quantitative data for reservoirs worldwide, which indicate that most quartz cement was introduced after sandstones had been buried 1 to 2 km at temperatures of 50°C and greater (McBride, 1989).

INTERPRETATION

Formation Waters

The pH of present day South Brae formation brine is estimated by Marathon to be between 4.5 and 4.8 (J. Hardy, Marathon, pers. commun., 1991). Formation water analysis revealed that water composition varies with depth (see Table 3). With increasing depth, there is an increase in all the dissolved solids; these include sodium, calcium, magnesium, barium and strontium (see Table 3). Only four water samples were available for radiogenic isotopic analyses: $^{87}Sr/^{86}Sr$ values suggest that the pore waters may have been layered, with the high ratios due to

TABLE 2.—SUMMARY OF PERCENTAGE OVERGROWTH IN THE LACHED AND UNLEACHED ALIQUOTS, THE RESPECTIVE δ^{18}O MINERAL VALUES AND THE EXTRAPOLATED CORE AND OVERGROWTH VALUES FOR QUARTZ CEMENTED SANDSTONES IN THE SOUTH BRAE OIL FIELD

Size Fraction	Depth TVDSS (ft)	δ^{18}O‰ Leached	δ^{18}O‰ Unleached	% Overgrowth Fraction	% Overgrowth Fraction	Extrapolated Detrital	Extrapolated Authigenic	Reservoir Zone
16/7a-A3								
250–750 mm	13444'1"	12.68	14.46	11.80	31.27	11.6	20.8	Cc_{1-3}
250–750 mm	13086'2"	13.29	13.99	10.66	29.45	10.2	20.7	Ca_{1-4}
250–750 mm	12433'2"	12.76	13.50	17.44	24.00	10.8	22.1	Ad
250–750 mm	12111'5"	13.42	15.50	7.20	25.34	11.8	24.1	Aa_{2-4}
250–750 mm	12080'	13.00	14.92	8.30	22.89	11.9	25.1	Aa_1
16/7a-A11								
250–750 mm	13791'4"	11.95	13.24	19.00	39.30	10.7	17.1	Cd
250–750 mm	13238'	13.58	14.70	7.50	39.16	13.2	18.4	Ba_{1-4}
250–750 mm	13190'3"	13.97	14.82	26.61	39.21	12.1	18.6	Ba_{1-4}
250–750 mm	13164'6"	12.16	13.65	25.45	41.00	9.7	19.3	Ba_{1-4}
250–750 mm	12813'	12.18	13.20	18.05	28.87	10.5	19.9	Aa_{2-4}
250–750 mm	12790'	12.26	13.37	24.49	34.72	9.8	20.2	Aa_{2-4}
250–750 mm	12750'	12.97	14.51	24.10	33.87	9.1	24.6	Aa_1
250–750 mm	12712'6"	11.36	12.18	8.13	19.70	10.1	25.3	Aa_1
16/7a-A19								
250–750 mm	13660'	12.39	12.92	20.28	28.0	11.0	17.9	Aa_{2-4}
160–250 mm	13660'	12.73	13.29	17.34	28.22	11.8	17.2	Aa_{2-4}
500–750 mm	13630'	11.79	12.36	21.01	28.37	10.2	18.0	Aa_{2-4}
250–500 mm	13630'	12.11	12.81	20.73	30.17	10.6	18.0	Aa_{2-4}
160–250 mm	13630'	12.21	12.90	16.71	25.25	10.9	18.8	Aa_{2-4}
250–750 mm	13460'3"	12.24	13.99	10.94	27.74	11.1	21.6	Aa_{2-4}
500–750 mm	13415'	13.30	13.50	27.52	31.11	9.7	22.8	Aa_{2-4}
250–500 mm	13415'	13.85	15.46	26.20	39.60	10.7	22.8	Aa_{2-4}
160–250 mm	13415'	14.08	14.98	24.54	30.70	11.0	23.7	Aa_{2-4}
250–750 mm	13355'	11.75	13.66	15.53	30.40	9.8	22.6	Aa_1
16/7a-A27								
250–750 mm	13554'4"	14.03	15.15	21.22	34.06	12.2	20.9	Ba_{1-4}
250–750 mm	13413'6"	13.05	15.12	19.28	37.77	10.8	22.6	Ad
250–750 mm	13240'	10.93	13.59	3.06	25.05	10.6	22.7	Ac_1
250–750 mm	13061'	11.24	13.26	6.50	21.33	10.4	24.0	Aa_1
250–750 mm	12987'	11.97	14.00	2.00	16.70	11.8	24.8	Kimm.
250–750 mm	12978'	12.70	14.63	14.78	28.37	10.6	24.8	Kimm.

The mass balance equation used for calculating the core and overgrowth values is $\delta^{18}O_T = \delta^{18}O_C \cdot x_C + \delta^{18}O_{OG} \cdot x_{OG}$ (from Fisher 1982). T = Total sample (quartz + overgrowths), OG = Overgrowth, C = Detrital core, x = Mole fraction.

TABLE 3.—COMPOSITION OF FORMATION WATER SAMPLES

Well Zone	16/7a-A5 Aa_{1-4}	16/7a-A5 Ac_{1-6}	16/7a-A8 Aa_{1-4}	16/7a-A8 Ba_{1-4}	16/7a-A27 Aa_{1-4}	16/7a-A27 Bc_{1-4}	16/7a-A31 Ba_{1-4}	16/7a-A31 Cc_{1-3}
Sodium	17300	26180	22650	29210	23700	35220	27800	28600
Potassium	675	860	900	1240	920	1490	930	990
Calcium	345	230	210	350	345	705	385	350
Magnesium	69	37	43	68	33	77	46	49
Barium	290	670	550	780	490	1610	1170	1330
Strontium	12	20	26	67	19	145	91	95
Chloride	29500	39720	34790	46800	36730	58680	48270	46830
Sulphate	12	11	30	10	17	14	110	300
Bicarbonate	2980	3360	3270	3090	3620	1620	2980	3120
δ D (SMOW)	−11	−24	—	—	—	—	−20	−22
$^{87}Sr/^{86}Sr$	0.72456 ± 4	0.72641 ± 4	0.72470 ± 3	—	—	—	—	0.71770 ± 5
Rubidium	3.37	5.34	5.0	—	—	—	—	5.49
Strontium	12.2	16.4	31	—	—	—	—	84.3
δ^{18}O (SMOW)	−0.22	+0.76	—	—	—	—	−0.58	+0.58

Total dissolved solid analyses (ppm) were carried out by Marathon Oil, U.K. Isotopic analyses were measured in this study.

water/rock interaction and the dissolution of radiogenic feldspar and mica. δD and δ^{18}O show no obvious trends but are very similar to values for Mesozoic seawater and present day formation water values measured in the Magnus oil field (Fallick et al., 1993). This suggests that the layered pore water which precipitated quartz has been replaced by later diagenetic fluids and therefore δ^{18}O layering has been lost.

Oxygen Isotope Data

δ^{18}O values of the quartz overgrowths vary systematically from top to bottom in each well. In every case, δ^{18}O decreases with increasing depth (Fig. 6). With the range in overgrowth δ^{18}O being from 3.9‰ (16/7a-A3) to 8.1‰ (16/7a-A11). These data are considered reliable, as the extrapolated detrital core quartz δ^{18}O group tightly (Fig. 6), with a mean of 10.7 ± 1.4‰, n = 24.

For these overgrowths to have precipitated from an isotopically homogenous fluid throughout the reservoir, a temperature difference from top to bottom of between 20°C and 45°C would be required. Since the largest depth sampling interval is only 400 m, the data are incompatible with a conductive thermal regime and a (typical North Sea) geothermal gradient of 30°C

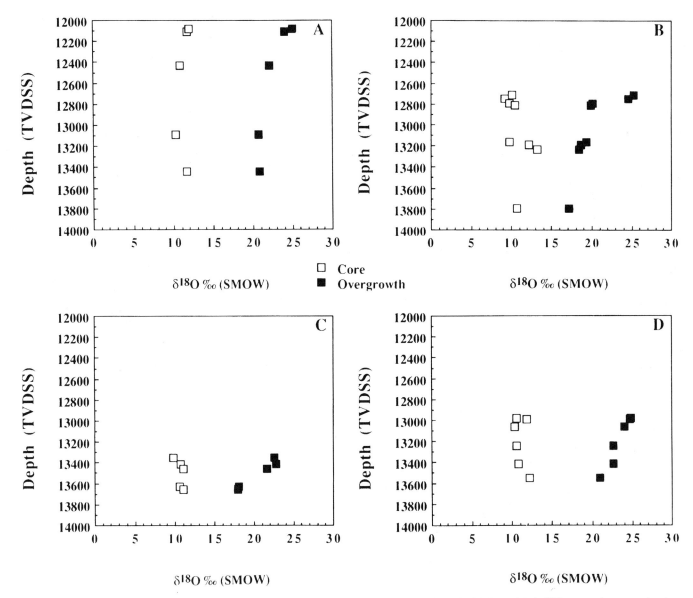

Fig. 6.—Plot showing the depths of quartz samples from wells 16/7a-A3 (A), A11 (B), A19 (C) and A27 (D) and their δ18O core and overgrowth values. While δ18O of the core samples remain constant, the overgrowth values decrease with increasing depth.

km^{-1}; the gradient would have to be at least a factor of two higher. South Brae fluid inclusion homogenization temperatures (from 82°C to 114°C) show no evidence for quartz cement precipitation at temperatures higher than those predicted for the ambient temperature gradient. These observations indicate that the overgrowths in any one well cannot all have precipitated from an isotopically homogenous pore fluid. Overgrowths in shallow reservoir sandstones are therefore inferred to have precipitated from an isotopically different fluid than overgrowths formed deeper.

In the South Brae reservoir, quartz overgrowths formed after calcite cementation. The depths at which cementation occurred were estimated using minus cement porosity volume point count values and calculated compaction curves (Baldwin and Butler, 1985). Our calculations suggest that the calcite cements precipitated at temperatures <70°C and to depths of 2.5 km

(McLaughlin et al., 1994). This would imply that the succeeding quartz overgrowths precipitated at temperatures greater than 70°C and at depths greater than 2.5 km. The present day bottom hole temperature collated from logging runs is approximately 120°C (J. Crane, Marathon Oil UK, pers. commun., 1991). The burial history for South Brae suggests that it is probably at the highest temperature, the deepest burial the reservoir has experienced and is actively expelling petroleum (Mackenzie et al., 1987).

From the diagenetic sequence (Fig. 4), it is inferred that dissolution and minor late calcite cementation post-date quartz overgrowth formation, which limits quartz precipitation to no hotter than 120°C (the approximate bottom hole temperature). By combining the quartz oxygen isotope data with the calcite isotope values (pre- and post-quartz) and homogenization temperatures for fluid inclusions in the quartz overgrowths, the type of pore fluid and temperature of quartz formation can be constrained (calculated porewater $\delta^{18}O$ given in Table 4).

In Figure 7, $\delta^{18}O$ values of the quartz overgrowths are plotted in their respective reservoir zones; in all wells there is a grouping of $\delta^{18}O$ for individual zones (i.e., a crude form of compartmentalization). The overgrowths sampled are from reservoir layers Aa_1, Aa_{2-4} and Ac_1 (Upper Brae) which are separated from Middle Brae sediments by a major fieldwide shale layer Ad. Layer Ba_{1-4} is a reservoir layer separated from the Lower Brae by another shaly interval. Ca_{1-4} and Cc_{1-3} are also reservoir layers.

An ad-hoc model (Fig. 8) was constructed to illustrate the temperature and $\delta^{18}O$ pore fluid principles which satisfy the data. To do this, both the pore fluid temperature and $\delta^{18}O$ are varied, the temperature is maintained above 70–80°C and below 120°C and the end member pore fluid boundaries of $-7‰$ (meteoric) to $+5‰$ (basinal) are observed. The most satisfactory model produced is one of warm basinal water overlying cooler meteoric-derived water (Fig. 9). This model agrees with the calcite isotope data which indicate calcite cementation from an evolved meteoric fluid (McLaughlin et al., 1994).

The range of overgrowth $\delta^{18}O$ suggests a possible model in which a hot basinal fluid entered the South Brae reservoir. The basinal fluid is inferred to have cooled en route to the western edge of the graben, due to mixing with interstitial fluids, while quartz cementation occurred. Because Zone A is the most permeable, the fluid is modeled to have passed through this layer first and the overgrowths have the highest $\delta^{18}O$, having precipitated from a hot, isotopically heavy fluid ($+5.0$ to $+2.0‰$ SMOW). As the fluid continued westward it cooled slightly and mixed with interstitial water thus becoming isotopically lighter. In the deeper, less permeable layers the water became isotopically lighter due to mixing, and consequently overgrowth $\delta^{18}O$ values are lower. It is unlikely that silica was carried into the reservoir by the hot basinal fluid as no additional silica cement has been observed along the flow paths; it is more likely that the silica was internally derived. Possible silica sources are discussed below.

SILICA SOURCES

Four possible silica sources for this cement have been identified:

(1) Dissolution of quartz grains by pressure solution. Pressure solution in the South Brae reservoir is apparent in very fine grained sandstones and at contacts between quartzose rock fragments (also noted by McBride et al., 1988).
(2) Silica released due to the compaction (and passive dissolution) of interdigitating shales and siltstones (Sharma, 1965). In the South Brae reservoir, the shale/sandstone ratio is high and the whole reservoir is overlain by the KCF. Gluyas and Coleman (1992) suggested that where these two lithologies interdigitate, ready access would be provided for fluids that moved into the sandstones. Shaw and Primmer (1991) noted that, in the KCF in the Brae area, quartz grains showed evidence of corrosion yet there was never any evidence of quartz overgrowths. They suggested that the silica had not been locally reprecipitated but had been mobilized and could be a potential source for quartz cementation in adjacent sandstones.

(3) Silica liberated during mineral reactions. Feldspar and detrital quartz dissolution occurred contemporaneously with the formation of calcite concretions but prior to quartz cementation. Later (post-calcite) dissolution of feldspar is particularly prevalent in the top reservoir zones, so it is possible that late feldspar dissolution may have provided a silica source.
(4) Silica which may be dissolved in circulating pore fluids due to flow over quartz grains in the sandstones (Leder and Park, 1986).

Discussion

Since the South Brae oil field lies on the westernmost edge of the Viking Graben (Fig. 1A), it is possible that any Graben-wide circulation of hot waters up major faults would be most noticeable here, by analogy with the situation proposed by Burley et al. (1989) for the Tartan oil field which lies on a major E-W fault system. To distinguish between the four silica sources above is not possible with the present data set. However, the hypothesized layering of the formation water indicated by $\delta^{18}O$, and present-day pore water salinity and $^{87}Sr/^{86}Sr$ values suggest that pore waters remained layered during the time span of quartz growth. Additionally, the growth temperatures deduced from measurements of $\delta^{18}O$ and from fluid inclusion analyses are compatible with normal burial temperatures. Thus, we can reasonably exclude large-scale advective transport of silica (Leder and Park, 1986; Burley et al., 1989).

All three remaining options have probably contributed silica to the South Brae quartz cement; (1) Quartz-quartz pressure solution is seen in thin sections (Fig. 10, see p. 98) and is estimated to have caused dissolution of 1–2% of the rock. Comparing Figures 10 and 11 (see p. 98), it can be seen that silica dissolution is more intense in finer grained sandstones, and silica must have therefore been transported to sites of precipitation in coarser sandstones. As the pore fluids are, and have been, layered this implies some type of diffusive or ionic drive. Although pressure solution appears capable of producing almost half of the observed quartz cement, the question of timing of pressure solution relative to quartz cementation remains. Because compaction of fine-grained beds would have proceeded faster than that of the coarse-grained beds, it is likely that the internally derived silica from clay rich silts was available about the time of quartz cementation in cleaner sandstones. (2) Silica released from compaction in mudrocks is demonstrated by the more intense cementation of the KCF sandstone compared to that in the main reservoir sequence. (3) Silica from early (pre- and syn-concretion) feldspar dissolution has been lost to the system, along with aluminum (McLaughlin et al., 1994). Late dissolution of up to 4% feldspar, could have provided 1–2% silica, which accords with the amount of quartz overgrowth in secondary pores.

INFLUENCE OF OVERPRESSURE

The current formation pressure in the South Brae oil field at 3884 m TVDSS (12740 ft) is 7128 psi (Roberts, 1991). The reservoir is overpressured by 1350 psi (Roberts, 1991) which is interpreted by Buhrig (1989) as a moderately geopressured reservoir. Overpressure development occurred predominantly during Late Cretaceous and Cenozoic times when the fault sys-

TABLE 4.—FLUID ORIGIN AND $\delta^{18}O$ (‰SMOW) FROM WHICH THE SOUTH BRAE QUARTZ OVERGROWTHS MAY HAVE BEEN PRECIPITATED.*

Well	Kimm.	Aa1	Aa2–4	Ac1	Ad	Ba1–4	Ca1–4	Cc1–3	Cd
A27	basinal +4	basinal +4	—	marine 0	marine 0	marine 1.5	—	—	—
A3	—	basinal +2	marine 1.5	—	mar/met −2.8	—	mar/met −3.2	mar/met −2.8	—
A11	—	basinal +3	marine 2.5	—	—	mar/met −4	—	—	mar/met −4.5
A19	—	basinal +3.5	basinal +2 Marine 2.5	—	—	—	—	—	—

*The oxygen isotope fractionation equation for quartz-water was taken from Friedman and O'Neil (1977).

(δ quartz − δ water) = 3.38 10⁶ (T⁻²) − 2.90

Where δ quartz = $\delta^{18}O$ of quartz (relative to SMOW), δ water = $\delta^{18}O$ of the water from which the quartz precipitated (relative to SMOW) and T = the temperature of formation in Kelvin scale. Temperatures used for the calculations were derived from the fluid inclusion study (see Table 1). End member possible fluids are (− 1.2‰) marine (Shackleton and Kennett, 1975), (− 7.0‰) meteoric (Hamilton et al., 1987, Hudson and Andrews, 1987) and (+ 2 to + 5.0‰) basinal (Egeberg and Aagaard, 1989, Wilkinson et al., 1992).

tems became inactive and rapid sediment loading began. Buhrig (1989) suggested that this excess pressure is a result of incomplete dissipation towards the adjacent platform along the graben. The overpressure today could be entirely due to lithostatic pressure buildup during the very rapid 280 m of subsidence in the past 2 Ma (Fig. 12). Episodes of partial release of overpressure prior to 2 Ma would have permitted fluids to drain out of the reservoir. Driven by sediment compaction, these fluids could have moved upward and outward from the KCF to the margins of the South Viking Graben. These inferred fluid movements could have produced the isotopic and temperature layering deduced from quartz overgrowths. Episodic quartz cementation related to overpressure release and pressure solution in the South Brae sandstone is likely to have resulted in the incremental growth zonation seen in the quartz overgrowths on CL (Fig. 5). For the past 2 Ma, the South Brae reservoir has apparently been effectively sealed, and its pore fluids and oil have been temporarily trapped.

Large overpressures minimize grain to grain overburden stress, and so minimize pressure solution. It is suggested that quartz cementation occurred most rapidly during periods of overpressure release. The CL stratigraphy seen on a sub micron scale probably records this overpressure release (Fig. 5). Luminescent zones in quartz are commonly attributed to variations in aluminum and trace element concentrations (Sprunt, 1981; Matter and Ramseyer, 1985) and thus could suggest episodic cementation from formation waters of variable chemistry (Land et al., 1987; McBride, 1989).

Overpressure release would also drain pressure and pore fluid from the surrounding KCF mudstones, via the permeable aquifers of the Brae Formation reservoirs which intimately finger into the less permeable mudrocks. During overpressure release, these basinal fluids must have flowed laterally out of the Graben via Brae sandstones, presumably up the western boundary fault of the South Brae oil field. Such fluid would be (1) warmer than South Brae formation waters (coming from deeper in the graben) and (2) less saline (having not interacted with so much mica and feldspar) and not been exposed to the saline fluids

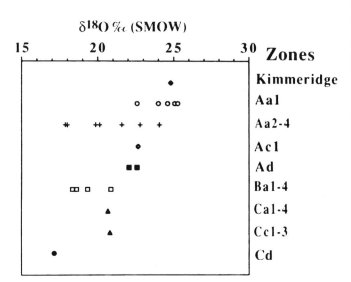

FIG. 7.—$\delta^{18}O$ values of authigenic quartz for wells 16/7a-A3, A11, A27 and A19 are plotted in their respective reservoir zones. There exists a broad grouping of $\delta^{18}O$ values for several zones.

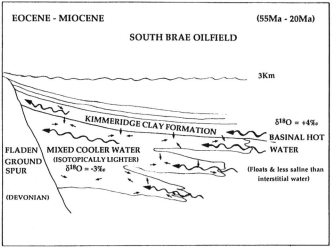

FIG. 8.—Cartoon illustrating the pore fluid regime existing in South Brae from Eocene to Miocene time when quartz cementation was taking place. Wavy arrows indicate large flows of basinal fluid entering and penetrating the aquifer. The basinal fluid floats as it is less saline and hotter than the surrounding interstitial pore fluids. The short arrows indicate no flow and they represent the diffusion of silica on a small scale.

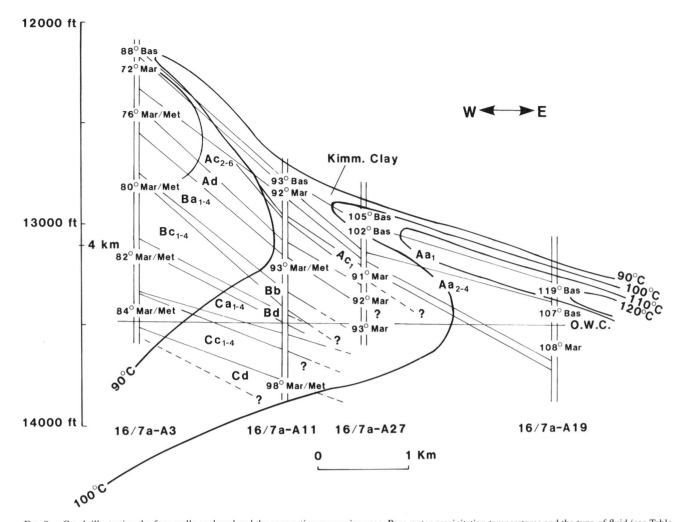

FIG. 9.—Graph illustrating the four wells analyzed and the connecting reservoir zones. Pore water precipitation temperatures and the type of fluid (see Table 4) have been calculated using the $\delta^{18}O$ mineral values from Table 2. The temperatures for cementation are contoured. To the east the fluid finger was at its hottest, as it travels west towards the graben margin it cools. The pore fluids were isotopically heavy in the upper reservoir zones and isotopically lighter in the deeper zones, particularly close to the western graben margin.

arriving upwards through the South Brae sands from Zechstein evaporite dissolution deeper at the graben edges. Thus the fluids expelled from the KCF would be expected to 'float' buoyantly over the cooler, denser South Brae pore waters. It is inferred that a record of these expelled fluids is seen in the $\delta^{18}O$ profiles deduced from the South Brae aquifers. Here the higher $\delta^{18}O$ basinal fluids occur preferentially towards the top of the South Brae reservoir package and also occur towards the top of each unit within that package (e.g., Aa_{2-4} in 16/7a-A19). A model is envisaged where basinal pore water layers extend up and out at the basin edge through the aquifers (Fig. 8). It is important to emphasize that no additional silica cement has been observed along these flow paths. This suggests that water moved but was neither hot enough, nor voluminous enough to transport large volumes of silica. Silica sources were decoupled from water sources, and silica was probably derived diffusively from the sources 1–3 listed in the preceding section.

MODEL FOR QUARTZ CEMENTATION

Petrographic, isotopic and fluid inclusion data all suggest that quartz cementation in the South Brae reservoir occurred under

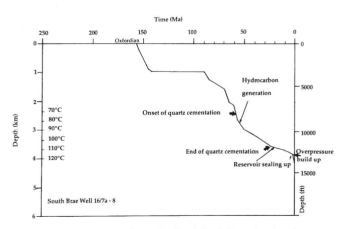

FIG. 12.—Burial curve for well 16/7a-8 South Brae showing time spans and relevant temperatures of quartz cementation, hydrocarbon generation and overpressure buildup. Sediments has been decompacted according to the porosity/depth relationships deduced by Sclater and Christie (1980), with the subsidence curve plotted against the timescale of Harland et al. (1989).

layered temperature and fluid conditions. A simple schematic model for quartz cementation (Fig. 8) has been proposed. The warm basinal water present during quartz precipitation initially fingered into the most porous zone Aa$_1$, consequently the temperatures for cementation are much higher in this zone than in the deeper zones. The basinal fluid rose buoyantly through the top of the reservoir as this fluid was hotter and less dense than the interstitial pore fluids. As the pore fluid moved towards the western margin of the Viking graben, it cooled and became isotopically lighter due to increased mixing with surrounding interstitial pore fluids. The overgrowths in the western section of the reservoir are isotopically lighter and precipitated from a cooler, lower δ^{18}O pore fluid. Quartz in the lower zones precipitated from a similar evolving basinal fluid which became isotopically light with depth and distance from the center of the basin. Internal silica sources, as discussed previously, appear to have been sufficient to supply the total volume of quartz present in South Brae. The silica did not move on any large scale, but was redistributed on a local scale.

CONCLUSIONS

(1) Silica cementation is preceded by calcite precipitation, the temperature of which is constrained to be 70–120°C by the burial history. This implies a porefluid δ^{18}O$_{H2O}$ which, taken together with δ^{18}O$_{OG}$, permits an estimate of the temperature of silica diagenesis. Such temperatures vary throughout the reservoir but are in accord with corresponding measurements of homogenization temperatures in fluid inclusions in the quartz overgrowths.
(2) Quartz overgrowth δ^{18}O in the four studied wells exhibit a similar trend (i.e., δ^{18}O decreases with increasing depth). It is unlikely these overgrowths precipitated from a single fluid, hence a model is invoked where the overgrowths precipitated from a compactional driven fluid that moved laterally upwards out of the basin. This fluid was layered in the oil field and followed the sedimentologically controlled reservoir layering. This gradually mixed with interstitial pore fluids to become isotopically lighter and cooler with depth and distance from the basin center.
(3) Internal silica sources appear to have been sufficient to supply the total volume of authigenic quartz present in the South Brae reservoir. No enhanced silica cement occurs at the reservoir top or base, suggesting silica transport was decoupled from fluid type and fluid flow.
(4) Quartz cementation occurred most rapidly during periods of overpressure release. Cementation may have been halted due to the reservoir becoming sealed and overpressured as this would have limited fluid movement through the oil field and permitted hydrocarbon accumulation to occur.

ACKNOWLEDGMENTS

Marathon Oil (U.K.) Ltd. and the Brae Group Partners provided funding for this project and allowed access to their data from the South Brae oil field. This research was carried out at the Department of Geology and Applied Geology, University of Glasgow and the Isotope Geosciences Unit, Scottish Universities Research and Reactor Centre (SURRC) in East Kilbride. The IGU is supported by NERC and the Universities of the Scottish Consortium. The authors thank Peter Ainsworth for his assistance with mineral separation and SEM analyses, and Douglas Maclean for his help with the production of photomicrographs. Special thanks to Calum 'Fish' Macaulay for trawling through an early version. The authors' interpretations of the data are not necessarily the views of Marathon Oil (U.K.) Ltd. and the Brae Group Partners.

REFERENCES

BALDWIN, B. AND BUTLER, C. O., 1985, Compaction curves: Bulletin of the American Association of Petroleum Geologists, v. 69, p. 622–626.
BORTHWICK, J. AND HARMON, R. S., 1982, A note regarding ClF$_3$ as an alternative to BrF$_5$ for oxygen isotope analysis: Geochimica et Cosmichima Acta, v. 46, p. 1665–1668.
BRINT, J. F., HAMILTON, P. J., HASZELDINE, R. S., FALLICK, A. E., AND BROWN, S., 1991, Oxygen isotopic analysis of diagenetic quartz overgrowths from the Brent sands: A comparison of two preparation methods: Journal of Sedimentary Petrology, v. 61, p. 527–533.
BUHRIG, C., 1989, Geopressured Jurassic reservoirs in the Viking Graben: modelling andgeological significance: Marine and Petroleum Geology, v. 6, p. 31–48.
BURLEY, S. D., MULLIS, J., AND MATTER, A., 1989, Timing diagenesis in the Tartan Reservoir (UK North Sea): constraints from combined cathodoluminescence microscopy and fluid inclusion studies: Marine and Petroleum Geology, v. 6, p. 98–120.
CLAYTON, R. N. AND MAYEDA, T. K., 1963, The use of bromine pentafluoride in the extraction of oxygen from oxides and silicates for isotope analysis: Geochimica et Cosmochimica Acta, v. 27, p. 43–52.
EGEBERG, P. K. AND AAGAARD, P., 1989, Origin and evolution of formation waters from oil fields on the Norwegian shelf: Applied Geochemistry, v. 4, p. 131–142.
FALLICK, A. E., MACAULAY, C. I., AND HASZELDINE, R. S., 1993, Implications of linearly correlated oxygen and hydrogen isotopic compositions for kaolinite and illite in the Magnus Sandstone, North Sea: Clays and Clay Minerals, v. 41, p. 184–190.
FISHER, R. S., 1982, Diagenetic history of Eocene Wilcox sandstones and associated formation waters, South-Central Texas: Unpublished Ph.D. Dissertation, University of Texas at Austin, 185 p.
FRIEDMAN, I. AND O'NEIL, J. R., 1977, Compilation of stable isotope fractionation factors of geochemical interest, in Fleischer, M., ed., Data of Geochemistry (sixth ed.): Washington DC, United States Geological Survey Professional Paper, 440-kk, 12 p.
GLUYAS, G. AND COLEMAN, M., 1992, Material flux and porosity changes during sediment diagenesis: Nature, v. 356, p. 52–54.
HAMILTON, P. J., FALLICK, A. E., MACINTYRE, R. M., AND ELLIOTT, S., 1987, Isotopic tracing of the provenance and diagenesis of Lower Brent Group Sands, North Sea, in Brooks, J. and Glennie, K., eds., Petroleum Geology of North West Europe: London, Graham and Trotman, p. 939–949.
HARLAND, W. B., ARMSTRONG, R. L., COX, A. V., CRAIG, L. E., SMITH, A. G., AND SMITH, D. G., 1989, A Geological Time Scale (2nd ed.): Cambridge, Cambridge University Press, 263 p.
HARRIS, N. B., 1989, Diagenetic quartz arenite and destruction of secondary porosity: an example from the Middle Jurassic Brent sandstone of Northwest Europe: Geology, v. 17, p. 361–364.
HUDSON, J. D. AND ANDREWS, J. E., 1987, The diagenesis of the Great Estuarine Group, Middle Jurassic, Inner Hebrides, Scotland, in Marshall, J. D., ed., Diagenesis of Sedimentary Sequences: Oxford, Blackwell, p. 259–276.
LAND, L. S., MILLIKEN, K. L., AND McBRIDE, E. F., 1987, Diagenetic evolution of Cenozoic sandstones, Gulf Coast of Mexico sedimentary basin: Sedimentary Geology, v. 50, p. 195–225.
LEDER, F. AND PARK, W. C., 1986, Porosity reduction in sandstone by quartz overgrowth: Bulletin of the American Association of Petroleum Geologists, v. 70, p. 1713–1728.
MACKENZIE, A. S., PRICE, I., LEYTHAEUSER, D., MÜLLER, P., RADKE, M., AND SCHAEFER, R. G., 1987, The expulsion of petroleum from Kimmeridge clay source-rocks in the area of the Brae Oilfield, UK continental shelf, in Brooks, J. and Glennie, K., eds., Petroleum Geology of North West Europe: London, Graham and Trotman, p. 865–877.
MATTER, A. AND RAMSEYER, K., 1985, Cathodoluminescence microscopy as a tool for provenance studies of sandstones, in Zuffa, G. G., ed., Provenance of Arenites: Dordrecht, Reidel, p. 191–211.

McBRIDE, E. F., 1963, A classification of common sandstones: Journal of Sedimentary Petrology, v. 33, p. 664–669.

McBRIDE, E. F., 1989, Quartz cement in sandstone: A Review: Earth Science Reviews, v. 26, p. 69–112.

McBRIDE, E. F., DIGGS, T. N., AND WILSON, J. C., 1988, Compaction of Wilcox and Carrizo sandstones (Paleocene-Eocene) to 4420 m, Texas Gulf Coast: Journal of Sedimentary Petrology, v. 61, p. 73–85.

McLAUGHLIN, Ó. M., 1992, Isotopic and textural evidence for diagenetic fluid mixing in the South Brae oilfield, North Sea: Unpublished Ph.D. Dissertation, University of Glasgow, Glasgow, 298 p.

McLAUGHLIN, Ó. M., HASZELDINE, R. S., FALLICK, A. E., AND ROGERS, G., 1994, The case of the missing clay, aluminium loss and secondary porosity, South Brae Oilfield, North Sea: Clay Minerals, v. 29, p. 651–663.

MILLIKEN, K. L., LAND, L. S., AND LOUCKS, R. G., 1981, History of burial diagenesis determined from isotopic geochemistry, Frio Formation, Brazoria County, Texas: Bulletin of the American Association of Petroleum Geologists, v. 65, p. 1397–1413.

ROBERTS, M. J., 1991, The South Brae Field, Block 16/7a, UK North Sea, *in* Abbotts, I. L., ed., United Kingdom Oil and Gas Fields: London, Geology Society of London Press, Geological Society Memoirs No. 14. p. 55–62.

SCLATER, J. G. AND CHRISTIE, P. A. F., 1980, Continental stretching: An explanation of the Post-Mid-Cretaceous subsidence of the Central North Sea Basin: Journal of Geophysical Research, v. 85, p. 3711–3739.

SHACKLETON, N. J. AND KENNETT, J. P., 1975, Paleotemperature history of the Cenozoic and the initiation of Antarctic glaciation: Oxygen and carbon analyses in DSDP sites 277, 279: Washington, D.C., Initial Report DSDP 24, p. 653–659.

SHARMA, G. D., 1965, Formation of silica cement and its replacement by carbonates: Journal of Sedimentary Petrology, v. 35, p. 733–745.

SHAW, H. F. AND PRIMMER, T. J., 1991, Diagenesis of mudrocks from the Kimmeridge Clay Formation of the Brae Area, UK North Sea: Marine and Petroleum Geology, v. 8, p. 270–277.

SPRUNT, E. S., 1981, Causes of quartz luminescence colours: Chicago, Scanning Electron Microscopy, Part 1, SEM Inc. p. 525–535.

STOKER, S. J. AND BROWN, S., 1986, Coarse clastic sediments of the Brae field and adjacent areas, North Sea: a core workshop: Edinburgh, British Geological Survey, 11 p.

STOW, Ɔ. A. V., BISHOP, C. D., AND MILLS, S. J., 1982, Sedimentology of the Brae oilfield, North Sea. Fan models and controls: Journal of Petroleum Geology, v. 5, p. 129–148.

TURNER, C. C., COHEN, J. M., CONNELL, E. R., AND COOPER, D. M., 1987, A depositional model for the South Brae oilfield, *in* Brooks, J. and Glennie, K., eds., Petroleum Geology of North West Europe: London, Graham and Trotman, p. 853–864.

WILKINSON, M., CROWLEY, S. F., AND MARSHALL, J. D., 1992, Model for the evolution ofoxygen isotope ratios in the porefluids of mudrocks during burial: Marine and Petroleum Geology, v. 9, p. 98–105.

ZIEGLER, P. A., 1978, North-Western Europe: tectonics and basin development: Geologie en Mijnbouw, v. 57, p. 589–626.

ZINKERNAGEL, U., 1978, Cathodoluminescence of quartz and its application to sandstone petrology: Contributions to Sedimentology, v. 8, p. 1–69.

EARLY SIDERITE CEMENTATION AS A CONTROL ON RESERVOIR QUALITY IN SUBMARINE FAN SANDSTONES, SONORA CANYON GAS PLAY, VAL VERDE BASIN, TEXAS

SHIRLEY P. DUTTON, H. SCOTT HAMLIN, ROBERT L. FOLK, AND SIGRID J. CLIFT
Bureau of Economic Geology, The University of Texas at Austin, Austin, Texas 78713–7508

ABSTRACT: Early precipitation of siderite cement in Sonora Canyon sandstones (Wolfcampian) in the Val Verde Basin, southwest Texas strongly influenced later diagenesis and reservoir quality in these low-permeability gas reservoirs. Sandstones of the Sonora Canyon interval were deposited in water depths of 100 to 500 m in coalesced submarine fans basinward (southwest) of the northwest-trending shelf margin. Sonora Canyon sandstones are composed of hundreds of feet of fan-lobe turbidites and local channel-fill facies deposited on the continental slope and basin floor.

Sonora Canyon sandstones are fine-grained sublitharenites and litharenites (average composition $Q_{77}F_4R_{19}$). Grain-rimming siderite rhombs 1 to 2 μm long were the earliest major cement to precipitate, in volumes ranging from 0 to 38%. Siderite is concentrated in bedding-parallel layers 8 to 10 cm thick or in irregular patches 3 to 8 cm in diameter. Isotopic composition of the siderite falls in a narrow range, $\delta^{13}C$ averaging 2.4‰ (PDB) and $\delta^{18}O$ averaging 31.1‰ (SMOW). The isotopic data indicate that siderite cement formed in a methanogenic geochemical environment at a burial depth of about 300 to 600 m (27°C) from sea-water-derived pore fluids ($\delta^{18}O = 0‰$). Bacterial reduction of iron accompanying anaerobic bacterial methanogenesis increased the Fe^{+2} in the pore fluids and, in the absence of sulfide, siderite precipitated. Subspherical nannobacterial bodies (0.05 to 0.15 μm) are revealed by etching siderite in warm HCl. These bodies are locally abundant, ranging to 100 per μm² of siderite crystal surface; other parts of the crystals contain virtually no bodies. The bacteria presumably helped trigger siderite precipitation.

Abundant early siderite inhibited later porosity loss by compaction and quartz cementation; siderite-rich sandstones (containing ≥10% siderite) average 33% minus-cement porosity and 6% quartz cement. Siderite-poor sandstones (<10%), are extensively cemented by quartz (average = 11%) and are much more compacted (16% minus-cement porosity). Siderite-rich sandstones retain higher porosity (7.9%) and permeability (0.042 md) than do siderite-poor sandstones (average porosity = 6.4%, geometric mean permeability = 0.006 md). Best matrix reservoir quality in Sonora Canyon sandstones occurs in siderite-cemented zones.

INTRODUCTION

Although Sonora Canyon sandstones of the Val Verde Basin in southwest Texas contain abundant natural gas resources, the gas is difficult to extract economically because extensive diagenesis has reduced average permeability in these sandstones to less than 0.1 md. One of the most common diagenetic changes was precipitation of grain-rimming siderite cement, and, contrary to what might be expected, reservoir quality is best in sandstones that contain the most siderite. This paper (1) ascertains the geochemical and hydrologic conditions under which the siderite precipitated and (2) quantifies the effect of siderite cementation on reservoir quality. Because the geochemical conditions under which siderite can precipitate are tightly constrained (Berner, 1981), we can interpret the chemical conditions that developed in the Sonora Canyon sandstones quite precisely. Furthermore, because the depositional environment of the Sonora Canyon sandstones is well defined, such interpretation increases our knowledge of the spectrum of deposits in which siderite can form.

The Canyon gas reservoirs in the Val Verde Basin, lying at depths between 900 and 2,400 m, have produced 2.3 trillion cubic feet (Tcf) of gas (Hills, 1968; Bebout and Garrett, 1989). The Canyon reservoirs are designated "tight gas sandstones" because of their low permeability, which generally ranges between 0.001 and 0.03 md under reservoir pressure conditions (Dutton et al., 1993b). Understanding the diagenetic variability of Sonora Canyon sandstones can help us predict zones of higher permeability and identify economic production methods in this part of the Canyon gas play.

GEOLOGIC SETTING

This study focused on the Sonora Canyon, one of several Canyon sandstone intervals of Late Pennsylvanian to Early Permian age in the Val Verde Basin (Fig. 1). The lower to middle Wolfcampian Sonora Canyon sandstones (Hamlin et al., 1995), which were deposited in shelf-edge, slope, and adjacent basin-

floor settings, lie in the northern part of the Val Verde Basin adjacent to the Ozona Arch and the southwestern margin of the Eastern Shelf (Fig. 1). Desmoinesian–Virgilian carbonate-bank facies rim the Ozona Arch and Eastern Shelf (Rall and Rall, 1958; Holmquist, 1965), delineating shelf-margin positions that persisted during deposition of the overlying Sonora Canyon sandstones (Wolfcampian; Fig. 2). These shelf edges included both distinct shelf-slope topographic breaks and more irregular and topographically gradational ramps. In the basin, dark clay–

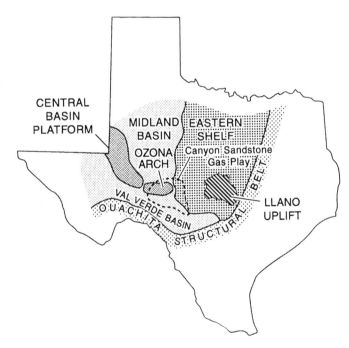

FIG. 1.—Principal Late Pennsylvanian–Early Permian tectonic elements in West Texas. The Canyon Sandstone gas play extends across the northern Val Verde Basin, the southwestern margin of the Eastern Shelf, and the Ozona Arch, which separates the Val Verde and Midland Basins.

shale and thin limestones overlie Desmoinesian carbonates and are overlain by Sonora Canyon sandstones. The Sonora Canyon is a wedge-shaped interval composed primarily of multiple-sourced, coalesced submarine fans forming submarine ramps and slope aprons (terminology of Reading and Richards, 1994; Fig. 2). Sonora Canyon sandstones grade basinward into deep-water mudstones and landward into shallow-shelf mudstones, isolated sandstones and thin limestones (Hamlin et al., 1992a). On the basis of stratigraphic reconstruction of the height of the shelf margin above the basin floor, we estimate that water on the slope and in the basin during Canyon deposition was 100 to 500 m deep (Hamlin et al., 1995). Deposition of the Sonora Canyon slope wedge prograded the slope and shelf margin, constructing a platform for overlying Wolfcampian–Leonardian shallow-shelf carbonate deposits (Fig. 2).

Submarine fans are turbidite systems (Shanmugam and Moiola, 1988; Mutti and Normark, 1991), and turbidites are the most common sedimentary features identified in Sonora Canyon cores (Fig. 3). In the Canyon submarine-fan depositional systems, the primary reservoir facies are channel-fill and fan-lobe sandstones (Mitchell, 1975; Berg, 1986; Hamlin et al., 1992a, b, 1995). Sonora Canyon channel-fill facies (3 to 30 m thick) are composed mainly of massive, locally conglomeratic sandstones. Although thin mudstone beds and laminations occur locally, bedding is irregular, and complete Bouma sequences (Bouma, 1962) are rare. Thick sequences of regularly bedded turbidites characterize Sonora Canyon fan-lobe facies (Fig. 3). Proximal fan-lobe turbidites are composed of massive to laminated sandstone beds 0.3 to 1.5 m thick and mudstone beds less than 0.2 m thick. Distal lobe-fringe turbidites have sandstone–siltstone beds less than 0.3 m thick interbedded with thicker mudstones. All gradations between these two end-member bedding styles exist in medial positions on fan lobes. The sandstone depocenters in the middle of the Sonora Canyon slope wedge are composed largely of hundreds of feet of fan-lobe turbidites and local channel-fill facies.

Sediment source areas of Sonora Canyon sandstones lay mainly to the north and east (Hamlin et al., 1992a). Sediment from rising Ouachita highlands was transported across the broad Eastern Shelf (Fig. 1) during lowstand of sea level (Ham-

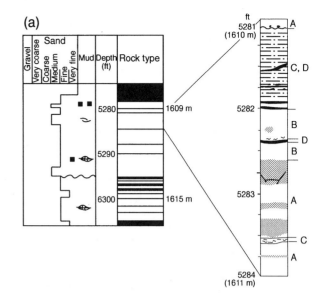

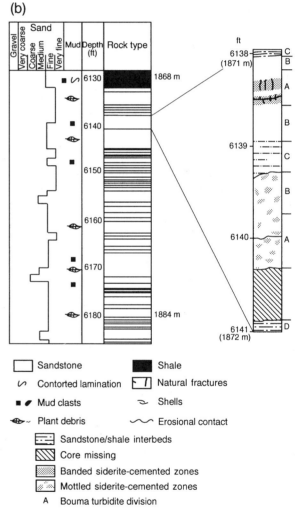

FIG. 3.—Core description from the Enron Sawyer 144A No. 5 well, Sutton County, Texas. (A) Sandstone between 1,609 and 1,614 m (5,278 and 5,294 ft) is probably a small channel fill. The underlying thin sandstones are turbidites. (B) Interbedded sandstones and mudstones are thick- and thin-bedded turbidites. Detailed insets show rock types and location of siderite-cemented zones. Modified from Hamlin et al. (1992a) and Marin et al. (1993).

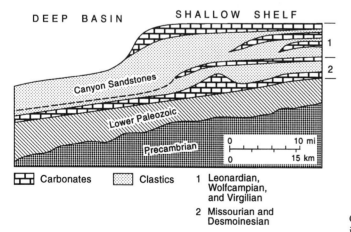

FIG. 2.—Schematic cross section showing Late Pennsylvanian–Early Permian depositional topography, northeastern Val Verde Basin. Modified from Rall and Rall (1958).

lin et al., 1995). Depositional patterns suggest that coarse clastic sediment from the southern part of the Ouachita Structural Belt was trapped in the adjacent, rapidly subsiding foredeep and did not reach the eastern or northern parts of the Val Verde Basin (Fig. 1). The part of the Ouachita Structural Belt that flanks the Eastern Shelf, however, was an important source area of the Sonora Canyon.

METHODS

Fifty-one thin sections were examined from cores from two wells, the Phillips Ward "C" No. 11 and the Enron Sawyer 144-A No. 5, in Sonora Canyon gas fields in Sutton County, Texas. Thin sections were taken from a total of 146 m of core that gave good coverage of the entire 400-m-thick Sonora Canyon interval (Hamlin et al., 1992a). Core depths range from 1,608 to 2,005 m below ground surface. Matrix-free sandstones were sampled preferentially because cementation was more extensive in them compared with sandstones having detrital clay matrix.

Composition of Sonora Canyon sandstones was determined by standard thin-section petrography and scanning electron microscopy (SEM) using an energy-dispersive X-ray spectrometer. Thin sections were stained for K-feldspar and carbonates. Point counts (200 points) of thin sections from representative samples from the two cores were used to determine mineral composition and porosity. Counting error varies with the percentage of the constituent. A constituent that is 50% of the sample (whole-rock volume) has the maximum error of $\pm 3.6\%$, whereas a constituent that is 10% of the sample has an error of $\pm 2.1\%$, and one that is 2% of the sample has an error of $\pm 0.9\%$ (Folk, 1974). Grain size and sorting were determined by grain-size point counts (50 points).

Siderite cement was analyzed for stable-isotope ratios of carbon and oxygen by allowing carbonate samples to react with phosphoric acid and release CO_2 gas. The gas preparation procedure used in the stable-isotope measurements applied reaction rates and $\delta^{18}O$ variations for carbonates and phosphoric acid that were described by Walters et al. (1972) for rocks having mixed carbonate mineral assemblages. The CO_2 recovered in 1 hour at 25°C, considered to represent the calcite fraction in each sample, was discarded. The gas evolved after 2 hours at 150°C represented the siderite in each sample. We used the equilibrium fractionation factor ($\alpha = 1.00771$) for siderite-phosphoric acid of Rosenbaum and Sheppard (1986). The siderite CO_2 was then analyzed for carbon and oxygen stable isotopes using a VG SIRA 12 triple collection mass spectrometer. Results of oxygen isotopic analyses are reproducible to $\pm 0.2‰$ and results of carbon analyses to $\pm 0.1‰$. Oxygen isotope ratios determined at 150°C were corrected to 25°C by linear interpolation (Rosenbaum and Sheppard, 1986), and all values are reported in per mil (‰) deviations relative to the CO_2 liberated from the PDB standard at 25°C by reaction with more than 100% phosphoric acid. Delta values were corrected for ^{13}C and ^{17}O variations according to Craig (1957).

CANYON SANDSTONE PETROGRAPHY

Sonora Canyon sandstones are typically composed of very fine to medium-grained sandstone, although coarser, locally conglomeratic sandstone is present in the channel-fill facies. In general, upper Sonora Canyon sandstones are finer grained than are lower Sonora Canyon sandstones. Sorting ranges from good to poor, but most clean sandstones (defined as containing $\leq 2\%$ clay matrix) are moderately well sorted.

Framework Grain Composition

Sonora Canyon sandstones are mineralogically immature and are classified as sublitharenites and litharenites in the sandstone classification of Folk (1974). Average composition of the essential framework grains (normalized to 100%) is 77% quartz, 4% feldspar, and 19% rock fragments ($Q_{77}F_4R_{19}$). Plagioclase is the most abundant feldspar, composing from 0 to 8% of the whole-rock volume; orthoclase is rare. Sedimentary rock fragments, including chert, sandstone, siltstone, shale, and mud rip-up clasts, are the most abundant lithic grains. Fine-grained, low-rank metamorphic rock fragments, such as slates or phyllites, also are common. The composition of the rock fragments is consistent with the sediment source being from the Ouachita Structural Belt, which shed detritus from older sedimentary rocks and their metamorphosed facies into the adjacent basin. Detrital organic matter, such as wood fragments, occurs in many of the sandstones and mudstones in volumes ranging from 0 to 10%.

Cements and Replacive Minerals

Authigenic cements and replacive minerals constitute between 7 and 52% of the whole-rock volume in the sandstone samples. Authigenic quartz, siderite, ferroan dolomite, calcite, chlorite, illite, kaolinite, and pyrite all occur in Sonora Canyon sandstones. Major diagenetic events in the burial history of Sonora Canyon sandstones were (1) siderite and chlorite cementation, (2) mechanical compaction, (3) quartz cementation, (4) feldspar dissolution and illite and kaolinite precipitation, and (5) iron-bearing calcite and ferroan dolomite precipitation.

Siderite.—

The volume of siderite cement in Sonora Canyon sandstones varies from 0 to 38% (median volume = 2%). Siderite, one of the first cements to precipitate, commonly rims detrital grains in Sonora Canyon sandstones (Figs. 4, 5). The siderite rims are formed of clusters of 1- to 2-μm-long siderite rhombs (Fig. 6). Many siderite rims contain flakes of authigenic chlorite, suggesting that chlorite and siderite both formed relatively early in the diagenetic history.

When an opaque white card is inserted beneath the thin section and the thin section is examined in strong, oblique, reflected light, the siderite rims and masses appear pale smoky brown. The brown color suggests that the siderite crystals contain disseminated organic matter. To determine the source of this coloration, we etched the siderite for 1/2 hour in 10% HCl at 50°C. SEM examination then revealed the presence of subspherical bodies (0.05 to 0.15 μm), which we interpreted as fossil nannobacteria (Folk, 1993). We presume that the brown color results from the presence of carbon-rich remnants of cell walls that are invisible at the SEM scale. Nannobacteria are locally abundant (Fig. 7A, B), ranging to 100 bodies per μm² on the siderite crystal surfaces; other parts of the crystals contain virtually no bodies. As we discuss in the section on siderite diagenesis, bacteria presumably helped trigger siderite precip-

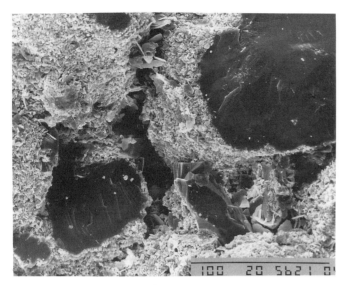

FIG. 4.—SEM photograph of siderite and chlorite cement rimming detrital quartz grains. Sample from a depth of 1,713 m in the Phillips Ward "C" No. 11 well. Scale bar is 100 μm. Photo by J. Mendenhall.

FIG. 6.—SEM photograph of a cluster of siderite rhombs. Sample from a depth of 1,713 m in the Phillips Ward "C" No. 11 well. Scale bar is 1 μm. Photo by J. Mendenhall.

FIG. 5.—SEM photograph showing close-up of contact between a detrital quartz grain (top of photo) and rhombs of grain-rimming siderite cement. Sample from a depth of 1,941 m in the Phillips Ward "C" No. 11 well. Scale bar is 10 μm. Photo by J. Mendenhall.

itation by consuming organic matter and reducing Fe^{3+} in the unconsolidated sediment and producing Fe^{2+} and HCO_3^-.

Most Sonora Canyon sandstones contain some siderite, but certain layers and patches are extensively cemented by siderite. Abundant siderite cement commonly occurs in bedding-parallel layers (Fig. 8) that average 8 to 10 cm in thickness and are rarely more than 25 cm. Abundant siderite also occurs in indistinct, irregular patches that are typically 3 to 8 cm in diameter (Fig. 9). Distribution of siderite can be estimated qualitatively by measuring the number of feet of sandstone that is stained red-brown by oxidation of siderite cement. An average of 30%

of all Sonora Canyon sandstone in the two cores is iron-oxide stained; in some intervals, as much as 47% of the sandstone is stained. The percentage of stained sandstone slightly increases upward. Siderite concretions and layers also occur in interbedded shales.

A comparison of siderite-rich and siderite-poor sandstones reveals similarities and differences between them (Table 1). Siderite-rich Sonora Canyon sandstones, arbitrarily defined as sandstones having ≥10% siderite cement, have grain size and sorting similar to that of siderite-poor sandstones (sandstones having <10% siderite). In each well, the average depth of occurrence of siderite-rich sandstones is shallower than that of siderite-poor sandstones. All of the siderite-rich samples occur in basal, Bouma A (massive or graded sandstone) turbidite divisions, whereas siderite-poor samples divide equally between Bouma A and Bouma B (parallel laminated sandstone) divisions. Point counts reveal that the siderite-rich samples contain less organic matter than do the siderite-poor samples (Table 1). In general, organic matter visible in core is absent or rare in most Bouma A facies but common to abundant in Bouma divisions B to E, the D (parallel laminated siltstone) and E (mudstone) divisions apparently being the most organic rich.

Oxygen and carbon isotopic compositions of siderite cements were determined for five samples that were selected because they represent a large depth range and contain abundant siderite but little calcite or ferroan dolomite. Siderite $\delta^{18}O$ values (corrected to 25°C) have a narrow range from $-0.66‰$ to $+1.16‰$ (PDB) (Table 2). The average $\delta^{18}O$ composition is $+0.3‰$ (PDB), which is equivalent to 31.1‰ (SMOW). Carbon-isotope $\delta^{13}C$ values range from $+0.56‰$ to $+3.95‰$ and average $+2.4‰$ (PDB). The average $\delta^{18}O$ and $\delta^{13}C$ values of Sonora Canyon siderite are heavier than most values of marine siderite reported in the literature (summarized in Mozley, 1992) but still within the range of reported values. Interpretation of carbon and oxygen isotopic composition of siderite cement is discussed in the section on siderite diagenesis.

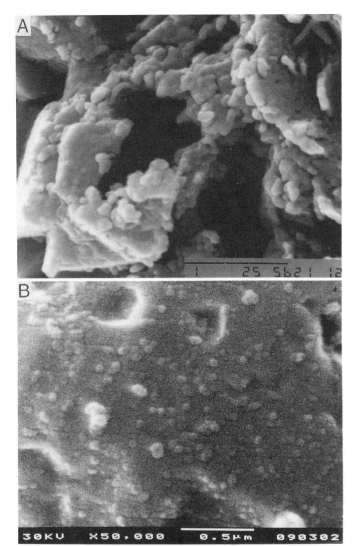

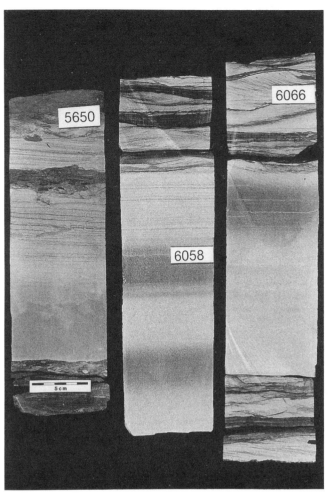

FIG. 8.—Core photo of typical siderite-cemented sandstone layers from the Phillips Ward "C" No. 11 well. Labels show depth in feet.

FIG. 7.—(A) SEM photograph of subspherical nannobacterial bodies in a siderite rhomb that was etched in warm HCl. Sample from a depth of 1,713 m in the Phillips Ward "C" No. 11 well. Scale bar is 1 μm. (B) SEM photograph of an etched surface of a siderite crystal, same locality, showing very abundant 0.03 to 0.06 μm nannobacterial bodies. Scale bar is 0.5 μm.

Other Cements.—

Although pyrite, not abundant in Sonora Canyon sandstones, can occur in volumes of as much as 5.5% (average = 0.3%), in most samples no pyrite was counted in 200 points. Pyrite apparently precipitated before siderite, because where it is most abundant, it forms the first rims around detrital grains and is overlain by siderite. Some detrital grains are partly surrounded by pyrite and partly by siderite, but the pyrite appears to have precipitated first. Pyrite also replaces organic particles and fossils.

Quartz cement precipitated after siderite. The volume of quartz cement in these sandstone samples varies from 1.5% to 20%. Quartz overgrowths are well developed in clean sandstones, their volume averaging 9%. Where quartz overgrowths are abundant, they completely fill some primary, intergranular pores (Fig. 10). An inverse relationship exists between quartz

and siderite cement; quartz cement is most abundant in samples containing little siderite cement (Fig. 11, Table 1). Where siderite is abundant, most quartz overgrowths are small, and they apparently nucleated where breaks occurred in the siderite rims (Fig. 12).

Authigenic chlorite, illite, and kaolinite occur in clean, matrix-free sandstones. Chlorite mainly occurs as platelets, whereas most illite has a fibrous morphology. Chlorite apparently precipitated early in the burial history; illite and kaolinite probably precipitated relatively late.

Iron-bearing calcite and ferroan dolomite cements also precipitated late in the diagenetic sequence; textural relationships show that they precipitated after quartz cement. They both replace framework grains (mainly feldspars) and fill intergranular pores that remained open throughout earlier phases of diagenesis. Ferroan dolomite volume ranges from 0 to 7% and calcite volume from 0 to 5%.

Porosity

Porosity observed in thin section varies from 0 to 5%; in clean sandstones, the average volume of primary porosity is

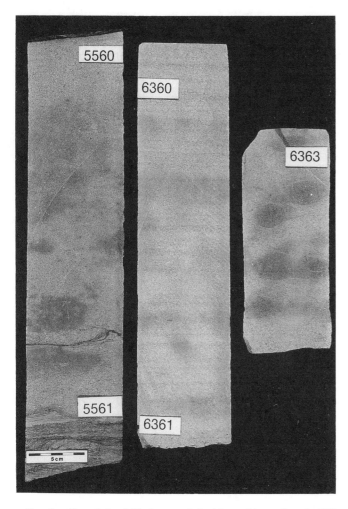

FIG. 9.—Core photo of siderite-cemented patches and layers from the Phillips Ward "C" No. 11 well. Labels show depth in feet.

TABLE 2.—ISOTOPIC COMPOSITION OF SIDERITE CEMENTS (PBD)*

Well	Depth (m)	δ18O	δ13C
Phillips Ward	1681.5	−0.34	0.56
Phillips Ward	1845.9	0.65	3.66
Enron Sawyer	1612.2	0.70	0.82
Enron Sawyer	1719.7	1.16	3.95
Enron Sawyer	1884.1	−0.66	2.76

*All values given in per mil (‰) relative to CO_2 liberated from PDB-1 at 25°C by dissolution using >100% phosphoric acid

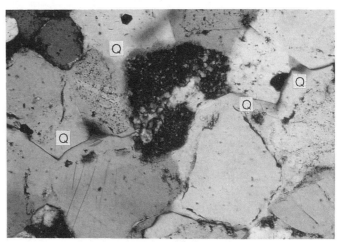

FIG. 10.—Siderite-poor sandstones experienced considerable quartz cementation. Quartz cement (Q) occludes much of the intergranular porosity. Sample from 1,945 m in the Phillips Ward "C" No. 11 well. Long dimension of photo is 0.65 mm. Cross-polarized light. From Hamlin et al. (1992a).

TABLE 1.—COMPARISON OF SIDERITE-RICH AND SIDERITE-POOR
SONORA CANYON SANDSTONES, SHOWING MEANS AND ONE
STANDARD DEVIATION

	Siderite-rich (≥10%)	Siderite-poor (<10%)
Average depth (m)*	1,720 ± 68	1,840 ± 109
Bouma division	A only	A or B
Framework-grain composition	$Q_{78}F_4R_{18}$	$Q_{75}F_4R_{21}$
Mean grain size (mm)	0.131	0.143**
Sorting (phi standard deviation)	0.66	0.62
Detrital clay matrix (%)*	0.1 ± 0.4	1.0 ± 1.8
Organic matter (%)*	0.1 ± 0.2	0.7 ± 1.3
Quartz cement (%)*	5.7 ± 3.1	10.7 ± 4.0
Siderite (%)*	24.0 ± 9.5	1.8 ± 2.2
Pyrite (%)	0.8 ± 1.5	0.1 ± 0.2
Total cement (%)*	32.3 ± 8.5	15.3 ± 4.8
Primary porosity (%)*	1.5 ± 1.1	0.4 ± 0.7
Minus-cement porosity (%)*	32.9 ± 8.5	14.8 ± 4.5
Porosimeter porosity (%)*	7.9 ± 1.3	6.3 ± 2.2
Geometric mean permeability (md)*#	0.042	0.006
Number of samples	14	30

*Means of the two populations are significantly different at the 95% confidence level as determined by t-test.

**Includes three samples of chert-pebble conglomerate and conglomeratic sandstone that were probably deposited in a channel or a slope canyon near the head of a submarine fan. If those three samples are omitted, mean grain size is 0.132 mm.

#Permeability measured at restored reservoir pressure.

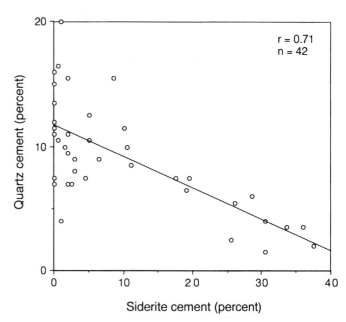

FIG. 11.—Quartz cement volume decreases with increasing volume of siderite cement. Relationship is significant at the 99% confidence level.

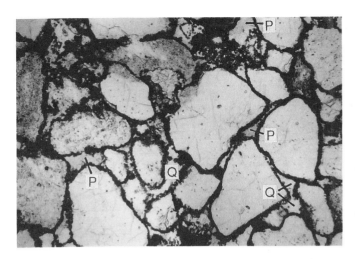

FIG. 12.—Thick, continuous siderite rims (dark) around quartz grains inhibited precipitation of quartz cement, thus preserving some intergranular porosity (P). Quartz cement (Q) that is in optical continuity with underlying grains presumably nucleated where breaks occurred in the siderite rims, probably out of the plane of the thin section. Sample from 1,713 m in the Phillips Ward "C" No. 11 well. Long dimension of photo is 0.65 mm. Plane-polarized light.

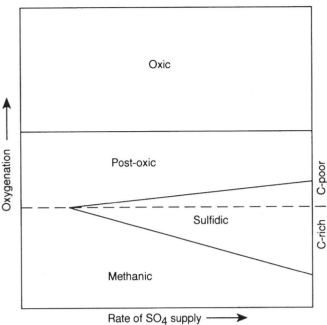

FIG. 13.—Classification of diagenetic environments (from Maynard, 1982; modified from Berner, 1981). Siderite can form in post-oxic and methanic environments; Canyon siderite is interpreted to have formed in a methanic environment.

0.8%, and average secondary porosity is 0.6%. Most secondary pores formed by dissolution of framework grains, mainly feldspar, chert, and clay clasts. Average porosity measured by porosimeter is 6.9% in clean sandstones. Thin-section porosity is generally lower than porosimeter porosity because micropores are present between clay flakes and within partly dissolved grains; the micropores cannot be accurately quantified in thin section. Although the total volume of micropores may be greater than that of macropores, microporosity contributes little to permeability. The difference between porosimeter-measured porosity and thin-section porosity provides an estimate of the volume of microporosity, which in clean Sonora Canyon sandstones averages 5.5%.

INTERPRETATION OF SIDERITE DIAGENESIS

Geochemical Conditions of Siderite Precipitation

Because siderite will precipitate only under certain narrow chemical conditions, the conditions under which the Sonora Canyon siderite formed can be closely constrained. The conditions necessary for siderite precipitation include low oxidation potential (low Eh), high partial pressure of CO_2 (high P_{CO_2}), very low sulfide concentration (low $[S^{-2}]$), and a high ratio of ferrous iron to calcium ion (high $[Fe^{+2}]/[Ca^{+2}]$) (Berner, 1971). Sulfide concentration must be low or pyrite will form, and the $[Fe^{+2}]/[Ca^{+2}]$ ratio must be high or calcite will form. The conditions necessary to precipitate siderite are met in two anoxic-nonsulfidic geochemical environments—the weakly reducing post-oxic environment and the highly reducing methanic environment (Fig. 13; Berner, 1981; Maynard, 1982).

Although siderite has long been recognized as a common authigenic mineral in freshwater environments because the sulfate concentration in meteoric water is low, many examples of early siderite cement that formed in marine environments have also been reported in the literature recently (see summary of references in Mozley, 1989). Berner (1981), Maynard (1982),

and Mozley and Wersin (1992) suggested that anoxic, nonsulfidic geochemical conditions may be more common in marine sediments than previously recognized. For post-oxic conditions to develop (Fig. 13), deep-sea sediments must contain sufficient metabolizable organic matter that aerobic bacteria consume all dissolved oxygen in the sediment during organic decomposition (Berner, 1981). However, organic matter must be insufficient to bring about sulfidic conditions, so that further bacterially mediated organic matter decomposition takes place by successive reduction of nitrate, manganese, and iron, but not sulfate (Berner, 1981). Froelich et al. (1979) called this anoxic-nonsulfidic geochemical environment the suboxic environment; Berner (1981) called it post-oxic (Fig. 13). Bacterial reduction of Fe^{3+} in this environment increases the Fe^{2+} in the pore fluids, and in the absence of sulfide, siderite precipitates.

The methanic environment forms under more strongly reducing conditions (Fig. 13). For methanic conditions to develop, organic matter must be sufficient that sulfate-reducing bacteria first reduce all sulfate in the pore waters to sulfide. If any reduced iron is present in the pore fluids of the sulfate-reduction zone, pyrite precipitates. Once all sulfate has been reduced, organic matter continues to decompose by microbial methanogenesis (Curtis, 1986), which produces methane and CO_2 (Fig. 14). Assuming a supply of reduced iron still exists, siderite precipitates.

Siderite that formed in a post-oxic environment can be distinguished from that which formed in a methanic environment by examining the isotopic composition of the carbon (Maynard, 1982). Carbon in CO_2 produced by oxidation of organic matter and corresponding bacterial reduction of nitrate, manganese, iron, and sulfate is isotopically light (−25‰ PDB), whereas CO_2 produced during methanogenesis is isotopically heavy

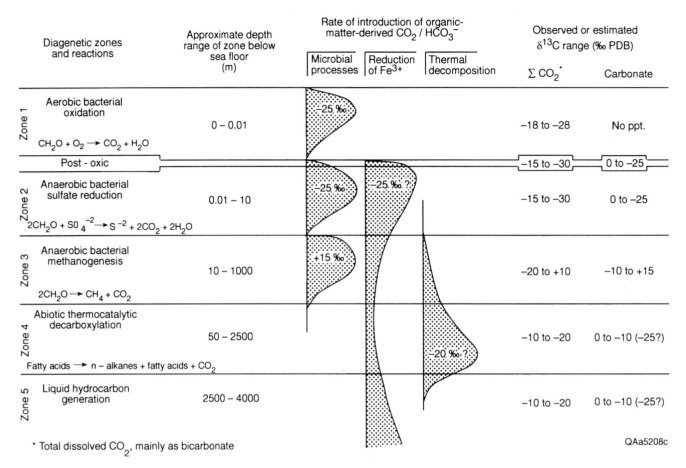

FIG. 14.—Zones of burial diagenesis of organic matter and resulting range of $\delta^{13}C$ values in dissolved carbon species (ΣCO_2) and of carbonates precipitated in the different zones (modified from Curtis, 1978; Garrison, 1981; Pisciotto, 1981; and after Mack, 1984). In this diagram, the post-oxic environment is represented by the narrow band between zones 1 and 2, in which microbial reduction of Fe^{+3} occurs but sulfate reduction has not begun.

(+15‰ PDB; Curtis, 1978, 1986; Fig. 14). Analyzing the $\delta^{13}C$ of the Sonora Canyon siderite should thus identify the geochemical zone in which it formed.

Interpretation of Canyon Siderite

The relatively heavy values of carbon (average of +2.4‰ PDB) indicate that most if not all of the Sonora Canyon siderite formed in a methanic environment. This interpretation is supported by the presence of pyrite in the siderite-rich samples (Table 1), which indicates that the sediments passed through the zone of sulfate reduction. Furthermore, the pyrite appears to have precipitated before the siderite. Textural evidence indicates that siderite cement in Sonora Canyon sandstones formed relatively early in the burial history of these deep-water marine sediments. Except where pyrite is present, siderite is the first cement around detrital grains, and the high minus-cement porosity of siderite-rich sandstones (33%) indicates that they were cemented before much compaction occurred. This interpretation of early siderite cementation is consistent with siderite precipitating at the shallower end of the depth range at which bacterial methanogenesis occurs, from approximately 10 to 1,000 m below the sediment-water interface (Curtis, 1978; Fig. 14). Our interpretation differs from that of Dutton et al. (1993a) because the isotopic data allowed us to determine that the sid-

erite precipitated in a methanic geochemical environment. The discussion following outlines our interpretation of how siderite precipited in Sonora Canyon sandstones.

Bioturbation is common in Sonora Canyon deposits, indicating that the bottom water in the basin could support organisms and thus was oxygenated. The oxidation zone in the sediment probably was quite thin, approximately 10^{-2} m (Fig. 14), and below that a zone of anaerobic sulfate reduction developed, in which pyrite precipitated. The Fe^{2+} in the pyrite probably came from microbial iron reduction of semiamorphous, colloidal particles of iron oxide and oxyhydroxide on larger detrital grains and clay flakes (Berner, 1981; Hesse, 1986; Coleman and Raiswell, 1993).

The relatively low volume of pyrite in the sandstones (average of 0.8% in siderite-rich sandstones) compared with that of siderite suggests that the supply of sulfide must have been limited. Because the supply of sulfate in the ocean water was limitless, diffusion of sulfate from the water column into the pore fluids of the sulfate-reduction zone must have been very limited to nonexistent, perhaps because rapid sedimentation of turbidite packages quickly isolated the sediment from sea water (Curtis and Douglas, 1993). The volume of pyrite that precipitated may thus have been limited to the supply of sulfate buried with the sea water in the pores. When all that sulfate was reduced to sulfide, no more pyrite precipitated. In general, rapid sedimentation rates move sediment quickly through a narrow sulfate-reduction zone, seldom more than 1 to 2 m deep, and into the methanogenesis zone (Curtis, 1986). Rapid sedimentation can thus limit the amount of pyrite that precipitates in a sediment. An alternative explanation of the low pyrite volume was given by Mozley and Carothers (1992), who also observed a paucity of pyrite associated with siderite. Following Berner (1985), they suggested that significant sulfate reduction can occur without pyrite forming, if H_2S gas escapes from the system.

Once sulfate reduction in the Sonora Canyon sandstones ended, the concentration of Fe^{2+} in the pore fluids began to rise because bacteria continued to reduce iron (Canfield, 1989; Fig. 14) but pyrite precipitation had ended. With sulfate depleted, organic decomposition continued by microbial methanogenesis in reactions of the general form $2CH_2O = CH_4 + CO_2$ (Irwin et al., 1977; Fig. 14). Simultaneous iron reduction increased the pore-water alkalinity (Coleman and Raiswell, 1993), so that in the presence of bicarbonate derived from methanogenesis, siderite precipitated.

The siderite has heavy carbon isotope values (average = 2.4‰ PDB) because light carbon fractionated into methane and the heavy carbon into CO_2 (Curtis, 1978). Carbon isotopic values of Sonora Canyon siderite, however, are not as heavy as the predicted value of +15‰ (Curtis, 1978), suggesting that some light carbon from CO_2 produced in the sulfate-reduction zone may also have precipitated in the siderite. Because no evidence exists of calcite precipitating in the sulfate-reduction zone, some of the light CO_2 generated there may have diffused into the underlying methanic zone (Gautier and Claypool, 1984). Claypool and Kaplan (1974) observed that in the early stage of methane production, the $\delta^{13}C$ of dissolved bicarbonate in deep-sea sediments most commonly shifts from about $-23‰$ to a moderately heavy $+5‰$. Garrison (1981) and Pisciotto (1981) stated that carbonates that precipitate in the methanogenic zone have $\delta^{13}C$ values ranging from $-10‰$ to $+15‰$ (Fig. 14). Finally, Gautier and Claypool (1984) noted that carbonates having $\delta^{13}C$ values as high as $+15‰$ are rare, and that lower values, such as those observed in Sonora Canyon siderite, may reflect "some commonly achieved dynamic balance between isotopic fractionation resulting from methane generation and the continued addition of isotopically light carbon through fermentation reactions" (p. 121).

Alternatively, the moderately positive $\delta^{13}C$ isotopic values may reflect a mixture of two generations of siderite that could not be separated when the samples were prepared for isotopic analyses. If some siderite formed in a post-oxic environment, it would be expected to have light carbon values (Maynard, 1982). If that siderite were analyzed along with siderite that formed later in the methanic zone, the resulting $\delta^{13}C$ values would represent a mixture of light carbon from the post-oxic siderite and heavy carbon from methanic siderite. Using differences in $\delta^{13}C$ composition and zonation of siderite crystals that are Fe-rich in the center and Mg-rich along the crystal margins, Mozley and Carothers (1992) identified two siderite generations that formed in post-oxic and methanic environments. Although zonation in the Sonora Canyon siderites has not been observed, the measured $\delta^{13}C$ isotopic values may nevertheless reflect mixtures of two generations of siderite. Some, if not all, of the

siderite must have precipitated in the methanic zone, however, because of the heavy $\delta^{13}C$ values.

Oxygen isotopic composition of the siderite also provides information about the conditions under which it precipitated. The fractionation equation between siderite and water of Carothers et al. (1988), $10^3 \ln \alpha = 3.13 \times 10^6 T^{-2} - 3.50$, relates isotopic composition of siderite to (1) temperature and (2) isotopic composition of the water from which the siderite precipitates. The possible combinations of temperature and water compositions that could have precipitated siderite having the observed average composition of 31.1‰ (SMOW) are plotted in Figure 15. Assuming that the $\delta^{18}O$ composition of the water from which the siderite precipitated was that of normal sea water, or 0‰ (SMOW), the temperature at which the siderite formed was about 27°C (Fig. 15). Using the present-day geothermal gradient of 29°C/km and assuming a temperature of 18°C at the sediment–water interface in water depths of about 200 m (Pickard, 1979), the siderite would have precipitated at a depth of about 300 m below the sea floor. If the temperature at the sediment-water interface were as cool as 10°C, the interpreted depth of siderite precipitation would be about 600 m. This range of interpreted depths of siderite precipitation is within the depths typical of the methanic zone, from 10 to 1,000 m below the sea floor (Curtis, 1978; Fig. 14). It is also similar to the depth of siderite formation, from about 120 to 1,200 m, in Cretaceous and Tertiary sediments on the continental rise off the east coast of the United States (Botz and von Rad, 1987). Sonora Canyon siderites thus have $\delta^{18}O$ values compatible with precipitation at relatively shallow burial depths from interstitial pore fluids derived from normal sea water. Precipitation of the siderite at 300 to 600 m is also consistent with the amount of compaction in the siderite-rich zones, judging from their minus-cement porosity values of about 33%. Experimental work by

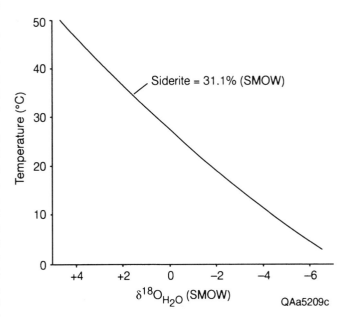

FIG. 15.—Loci of possible water temperatures and $\delta^{18}O$ compositions that could have precipitated Canyon siderite with the average $\delta^{18}O$ composition of 31.1‰ (SMOW). Equation relating temperature, $\delta^{18}O$-water, and $\delta^{18}O$-mineral is: $10^3 \ln a = 3.13 \times 10^6 \times T^{-2} - 3.50$ (Carothers et al., 1988).

Pittman and Larese (1991) predicts that well-sorted sandstones having a composition similar to that of the Sonora Canyon sandstones would reach a porosity of 33% after compaction to about 1,200 m. Because a decrease in sorting lowers porosity, the moderately sorted Sonora Canyon sandstones could have had porosity reduced to 33% at shallower depths, possibly by 300 to 600 m.

In contrast to the Sonora Canyon siderite, measured $\delta^{18}O$ values of many marine siderites reported in the literature are incompatible with the interpreted low temperature of formation or expected sea-water compositions (Gautier and Claypool, 1984; Mack, 1984; Carpenter et al., 1988; Mozley and Carothers, 1992; Thyne and Gwinn, 1994). Mozley and Carothers (1992) even suggested that low values of $\delta^{18}O$, typically 18 to 31‰, are "the norm" for marine siderites. One difference between many of those studies and this one is the water depth at which the sediments containing the siderite were deposited. Most of the formations having light $\delta^{18}O$ values, the Gammon Shale (Gautier and Claypool, 1984), Fox Hills Formation (Carpenter et al., 1988), Kuparuk Formation (Mozley and Carothers, 1992), and Cardium Formation (Thyne and Gwinn, 1994), were deposited in marine shelf environments that could potentially be exposed to meteoric-water incursion during lowstands of sea level. In contrast, because the Sonora Canyon sandstone was deposited in considerably deeper water on the slope and basin floor, it would be highly unlikely to have had any early exposure to meteoric water. The commonly observed low values of $\delta^{18}O$ in marine siderites may thus be explained by meteoric-water incursion, as each of these authors suggested. The Sonora Canyon sediments were deposited in water so deep that meteoric water did not penetrate them even during sea-level lowstands, and the early siderite cement precipitated from connate, sea-water-derived pore fluids. However, other factors in addition to meteoric-water penetration might cause the light $\delta^{18}O$ values observed in many marine siderites. Siderite in Pennsylvanian submarine fans deposited on the Eastern Shelf of the Midland Basin has light $\delta^{18}O$ values of 27.6 to 30.8‰ (SMOW; Mack, 1984), and early flushing by meteoric water would seem as unlikely in that setting as it would for the Sonora Canyon submarine-fan deposits.

Siderite Distribution in Canyon Sandstones

The factors that controlled the particular location of siderite layers and patches in Sonora Canyon sandstones are unclear. Whereas the horizontal layers of siderite may correspond to sediment horizons in which iron-reducing bacteria were particularly abundant, patches of siderite also may arise from bacteria flourishing at a point in the sediment, which triggered precipitation of siderite spherically around that point. Work on siderite concretions in salt-marsh sediments has shown that microbial populations are inhomogeneously distributed and that siderite concretions form in association with localized microbial populations (Coleman, 1993). Why bacteria may have flourished in particular layers or spots in the Sonora Canyon sandstone is unknown. Because the grain size and sorting of the siderite-rich zones are the same as those of the siderite-poor zones, a textural control on the distribution of the siderite layers seems unlikely. The presence of abundant organic matter apparently does not favor the formation of siderite either. One might expect sulfate

reduction and methanogenesis to be most extensive in organic-rich intervals, which would then develop the most abundant siderite, but instead siderite-rich zones contain less organic matter than do siderite-poor zones (Table 1). One possibility is that although the siderite-rich zones contained more organic matter at one time, it was consumed by bacteria during the methanogenesis reactions that led to siderite precipitation. The general paucity of organic matter in Bouma A divisions in Sonora Canyon core suggests, however, that, for some reason, the siderite developed preferentially in organic-poor zones. The explanation for this may not lie in the amount of organic matter, but in another factor that correlates negatively with organic matter. For example, variations in the amount of oxidized iron available to bacteria for reduction would influence the location of siderite layers (Coleman, 1993), and oxidized iron should be most abundant in organic-poor sediments. The preferential occurrence of siderite in Bouma A deposits, therefore, may be explained by the sorting of less dense organic matter into lower energy, laminated Bouma B-E facies during turbidite deposition and resulting higher levels of oxidized iron in the organic-poor Bouma A sequences.

An additional control on location and spacing of siderite layers may have been the distance over which diffusion was effective in the sediments. For example, if iron-reducing bacteria flourished in a layer having abundant oxidized iron and triggered the precipitation of siderite, the pore fluids in that immediate layer would become depleted in bicarbonate. Bicarbonate generated by microbial methanogenesis in the surrounding sediment would then diffuse through the pore fluids to that layer, where it would encounter the high iron concentration from continued microbial iron reduction and precipitate as siderite. Siderite layers are rarely more than 15 cm thick, and they are generally spaced at least 10 cm apart. Both the thickness and the spacing between layers may reflect the distance over which bicarbonate could diffuse to supply ions for siderite precipitation.

RESERVOIR QUALITY

Locating siderite-rich zones in the Sonora Canyon sandstone has economic importance because, as first reported by the National Petroleum Council (1980), the highest porosity in Canyon sandstones occurs in association with abundant siderite. Data from the two wells in this study support this observation. Siderite-rich sandstones average 1.5% primary porosity measured in thin section, whereas siderite-poor sandstones average 0.4% primary porosity (Table 1). Average porosimeter porosity of siderite-rich sandstones is 7.9%, compared with 6.3% in siderite-poor sandstones (Fig. 16, Table 1).

Siderite-cemented Canyon sandstones also have higher permeability than do sandstones without siderite, as the National Petroleum Council (1980), Berg (1986) and Huang (1989) previously noted, although they presented no supporting data. Siderite-rich sandstones in this study have a geometric mean permeability of 0.042 md, compared with siderite-poor sandstones having a mean permeability of 0.006 md (Table 1). Similarly the range of permeability is higher in siderite-rich than in siderite-poor sandstones (0.009 to 0.592 md versus 0.002 to 0.168 md, respectively; permeability measurements were made at calculated in situ overburden pressure). Furthermore, porosity and

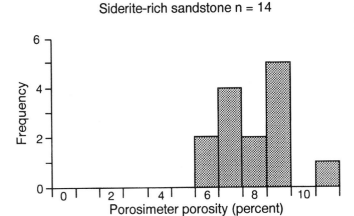

Siderite-rich sandstone n = 14

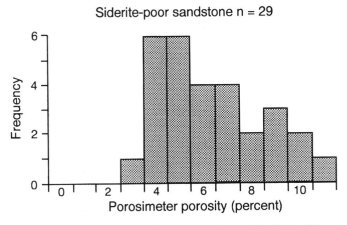

Siderite-poor sandstone n = 29

FIG. 16.—Histograms of porosimeter-measured porosity in siderite-rich and poor Canyon sandstones. The mean porosity of 7.9% of siderite-rich sandstones is significantly larger (at the 95% confidence level) than is the mean porosity of 6.3% of siderite-poor sandstones.

(Table 1). Assuming that the Sonora Canyon sandstones had a porosity of about 40% at the time of deposition, although siderite-rich sandstones lost only 7% porosity by compaction, siderite-poor sandstones lost considerably more porosity, an average of 25%, by compaction. Furthermore, as previously noted by Berg (1986) and Huang (1989), thick coats of siderite around detrital quartz grains inhibited some later nucleation of quartz cement (Fig. 10), thus preserving primary intergranular porosity. Siderite-rich sandstones contain an average of 6% quartz cement, but siderite-poor sandstones contain an average of 11% quartz cement (Fig. 10, Table 1). Even though the siderite itself fills primary porosity, therefore, the overall effect of the siderite is to preserve porosity.

CONCLUSIONS

Siderite cementation was a major diagenetic modification of Sonora Canyon sandstones, which were deposited in submarine fans on the slope and basin floor along the northeastern margin of the Val Verde Basin. Siderite precipitated fairly soon after deposition, at depths of about 300 to 600 m and temperatures of about 27°C from connate, sea-water-derived pore fluids having oxygen isotopic composition of 0‰ (SMOW). The siderite formed in a methanic geochemical environment in which bacterial processes controlled the chemistry of the pore fluids by organic-matter decomposition. Siderite layers and patches developed preferentially in organic-poor, Bouma A divisions of turbidites. The main control on the location of siderite layers probably was availability of oxidized iron. Once iron-reducing bacteria became established in a layer or spot and precipitation of siderite began, diffusion of bicarbonate from the surrounding sediment caused siderite precipitation to continue in that zone.

Siderite-rich layers retain the highest porosity and permeability in the overall low-permeability Sonora Canyon sandstones. Early siderite cementation preserved some intergranular porosity by inhibiting mechanical compaction and precipitation of quartz cement.

ACKNOWLEDGMENTS

This study was funded by the Gas Research Institute under contract no. 5082–211–0708, Stephen E. Laubach, principal investigator. We benefited from discussions with Bryan Bracken, Steve Fisher, Larry Mack, and Paul Wagner. Research Assistant Barbara Marin measured the thickness and location of every siderite band or patch in the cores, and Nina Baghai made grain-size point counts. Bruce Turbeville performed the isotopic analyses at the Bureau of Economic Geology Mineral Studies Laboratory under the direction of Steven W. Tweedy, Chemist-in-Charge. We appreciate the helpful reviews of an early version of this paper by Stephen E. Laubach and Raymond A. Levey and editorial comments from Lana Dieterich and Tucker F. Hentz. SEPM reviewers Andrew J. C. Hogg and an anonymous person suggested important improvements. Susan Lloyd did the final word processing, and Joel L. Larden and Richard L. Dillon drafted the figures. David Stephens photographed text illustrations. Publication authorized by the Director, Bureau of Economic Geology, The University of Texas at Austin.

permeability are higher in siderite-rich sandstones than in clean siderite-poor sandstones, those having ≤2% detrital matrix. Thus, the higher porosity and permeability in siderite-rich sandstones are not a function of differences in grain size, sorting, or volume of detrital clay. Core-analysis data from this study therefore confirm that the best matrix reservoir quality in Sonora Canyon sandstones apparently occurs in siderite-cemented zones.

On the basis of petrographic evidence, siderite-cemented sandstones seem to have the highest remaining porosity because they lost less intergranular porosity by compaction and quartz cementation than did sandstones without siderite. Siderite formed before significant burial and provided siderite-cemented sandstones with a rigid framework, so that loss of porosity by mechanical compaction during burial was arrested. Because sandstones having little siderite cement had no rigid framework, they were more intensely compacted during burial. Average minus-cement porosity, a measure of the amount of porosity remaining after compaction (Rosenfeld, 1949), is 33% in siderite-rich sandstones, compared with 15% in siderite-poor sandstones

REFERENCES

BEBOUT, D. G. AND GARRETT, C. M., JR., 1989, Upper Pennsylvanian and Lower Permian slope and basinal sandstone, in Kosters, E. C., Bebout, D. G., Seni, S. J., Garrett, C. M., Jr., Brown, L. F., Jr., Hamlin, H. S., Dutton, S. P., Ruppel, S. C., Finley, R. J., and Tyler, N., Atlas of Major Texas Gas Reservoirs: The University of Texas at Austin, Bureau of Economic Geology, p. 116–118.

BERG, R. R., 1986, Reservoir Sandstones: Englewood Cliffs, Prentice-Hall, 481 p.

BERNER, R. A., 1971, Principles of Chemical Sedimentology: New York, McGraw-Hill, 240 p.

BERNER, R. A., 1981, A new geochemical classification of sedimentary environments: Journal of Sedimentary Petrology, v. 51, p. 359–365.

BERNER, R. A., 1985, Sulphate reduction, organic matter decomposition and pyrite formation, in Eglinton, G., Curtis, C. D., McKenzie, D. P., and Murchison, D. G., eds., Geochemistry of Buried Sediments: London, The Royal Society, p. 25–38.

BOTZ, R. AND VON RAD, U., 1987, Authigenic Fe-Mn carbonates in Cretaceous and Tertiary sediments of the continental rise off eastern North America, Deep Sea Drilling Project Site 603: Initial Reports of the Deep-Sea Drilling Project, v. 93, p. 1061–1077.

BOUMA, A. H., 1962, Sedimentology of Some Flysch Deposits: a Graphic Approach to Facies Interpretation: Amsterdam, Elsevier, 168 p.

CANFIELD, D. E., 1989, Reactive iron in marine sediments: Geochimica et Cosmochimica Acta, v. 53, p. 619–632.

CAROTHERS, W. W., ADAMI, L. H., AND ROSENBAUER, R. J., 1988, Experimental oxygen isotope fractionation between siderite-water and phosphoric acid liberated CO_2-siderite: Geochimica et Cosmochimica Acta, v. 52, p. 2445–2450.

CARPENTER, S. J., ERICKSON, J. M., LOHMANN, K. C., AND OWEN, M. R., 1988, Diagenesis of fossiliferous concretions from the Upper Cretaceous Fox Hills Formation, North Dakota: Journal of Sedimentary Petrology, v. 58, p. 706–723.

CLAYPOOL, G. E. AND KAPLAN, I. R., 1974, The origin and distribution of methane in marine sediments, in Kaplan, I. R., ed., Natural Gases in Marine Sediments: Marine Science, v. 3, p. 99–140.

COLEMAN, M. L., 1993, Microbial processes: controls on the shape and composition of carbonate concretions: Marine Geology, v. 113, p. 127–140.

COLEMAN, M. L. AND RAISWELL, R., 1993, Microbial mineralization of organic matter: mechanisms of self-organization and inferred rates of precipitation of diagenetic minerals: Philosophical Transactions of the Royal Society of London A, v. 344, p. 69–87.

CRAIG, H., 1957, Isotopic standards for carbon and oxygen correction factors for mass-spectrometric analysis of carbon dioxide: Geochimica et Cosmochimica Acta, v. 12, p. 133–149.

CURTIS, C. D., 1978, Possible links between sandstone diagenesis and depth-related geochemical reactions occurring in enclosing mudstones: Geological Society of London Journal, v. 135, p. 107–117.

CURTIS, C. D., 1986, Mineralogical consequences of organic matter degradation in sediments: inorganic/organic diagenesis, in Leggett, J. K. and Zuffa, G. G., eds., Marine Clastic Sedimentology: London, Graham & Trotman, p. 108–123.

CURTIS, C. D. AND DOUGLAS, I., 1993, Catchment processes and the quantity and composition of sediment delivered to terminal basins: Philosophical Transactions of the Royal Society of London A, v. 344, p. 5–20.

DUTTON, S. P., CLIFT, S. J., FOLK, R. L, HAMLIN, H. S., AND MARIN, B. A., 1993a, Porosity preservation by early siderite cementation in Sonora Canyon sandstones,Val Verde Basin, southwest Texas (abs.): American Association of Petroleum Geologists 1993 Annual Convention Official Program, p. 95.

DUTTON, S. P., CLIFT, S. J., HAMILTON, D. S., HAMLIN, H. S., HENTZ, T. F., HOWARD, W. E., AKHTER, M. S., AND LAUBACH, S. E., 1993b, Major low-permeability-sandstone gas reservoirs in the continental United States: The University of Texas at Austin, Bureau of Economic Geology, Report of Investigations No. 211, 221 p.

FOLK, R. L, 1974, Petrology of Sedimentary Rocks: Austin, Texas, Hemphill, 182 p.

FOLK, R. L., 1993, SEM imaging of bacteria and nannobacteria in carbonate sediments and rocks: Journal of Sedimentary Petrology, v. 63, p. 990–999.

FROELICH, P. N., KLINKHAMMER, G. P., BENDER, M. L., LUEDTKE, N. A., HEATH, G. R., CULLEN, D., DAUPHIN, P., HAMMOND, D., HARTMAN, B., AND MAYNARD, V., 1979, Early oxidation of organic matter in pelagic sediments of the eastern equatorial Atlantic: suboxic diagenesis: Geochimica et Cosmochimica Acta, v. 43, p. 1075–1090.

GARRISON, R. E., 1981, Pelagic and hemipelagic sedimentation in active margin basins, in Douglas, R. G., Colburn, I. P., and Gorsline, D. S., eds., Depositional Systems of Active Continental Margin Basins: Los Angeles, Society of Economic Paleontologists and Mineralogists, Pacific Section, Short Course Notes, p. 15–38.

GAUTIER, D. L. AND CLAYPOOL, G. E., 1984, Interpretation of methanic diagenesis in ancient sediments by analogy with processes in modern diagenetic environments, in McDonald, D. A. and Surdam, R. C., eds., Clastic Diagenesis: Tulsa, American Association of Petroleum Geologists Memoir 37, p. 111–123.

HAMLIN, H. S., CLIFT, S. J., AND DUTTON, S. P., 1992a, Stratigraphy and diagenesis of Sonora Canyon deep-water sandstones, Val Verde Basin, southwest Texas: American Association of Petroleum Geologists, Southwest Section, Transactions, p. 209–220.

HAMLIN, H. S., CLIFT, S. J., DUTTON, S. P., HENTZ, T. F., AND LAUBACH, S. E., 1995, Canyon Sandstones—A geologically complex natural gas play in slope and basin facies, Val Verde Basin, southwest Texas: The University of Texas at Austin, Bureau of Economic Geology Report of Investigations No. 232, 74 p.

HAMLIN, H. S., MILLER, W. K., PETERSON, R. E., AND WILTGEN, N., 1992b, Results of applied research in the Canyon Sands, Val Verde Basin, southwest Texas: In Focus—Tight Gas Sands, v. 8, p. 1–32.

HESSE, R., 1986, Early diagenetic pore water/sediment interaction: modern offshore basins, in McIlreath, I. A. and Morrow, D. W., eds., Diagenesis: St. John's, Newfoundland, Geoscience Canada Reprint Series 4, p. 277–316.

HILLS, J. M., 1968, Gas in Delaware and Val Verde Basins, West Texas and southeastern New Mexico, in Beebe, B. W., ed., Natural Gases of North America, v. 2: Tulsa, American Association of Petroleum Geologists Memoir 9, p. 1394–1432.

HOLMQUIST, H. J., 1965, Deep pays in Delaware and Val Verde Basins, in Young, A. and Galley, J. E., eds., Fluids in Subsurface Environments: Tulsa, American Association of Petroleum Geologists Memoir 4, p. 257–279.

HUANG, F. F., 1989, Depositional environments, diagenesis and porosity relationships of the Canyon sands, Edwards and Sutton Counties, Texas: Unpublished Ph.D. Dissertation, Texas Tech University, Lubbock, 244 p.

IRWIN, H., CURTIS, C., AND COLEMAN, M., 1977, Isotopic evidence for source of diagenetic carbonates formed during burial of organic-rich sediments: Nature, v. 269, p. 209–213.

MACK, L. E., 1984, Petrography and diagenesis of a submarine fan sandstone, Cisco Group (Pennsylvanian), Nolan County, Texas: Unpublished M.S. Thesis, The University of Texas at Austin, Austin, 254 p.

MARIN, B. A., CLIFT, S. J., HAMLIN, H. S., AND LAUBACH, S. E., 1993, Natural fractures in Sonora Canyon sandstones, Sonora and Sawyer fields, Sutton County, Texas: Society of Petroleum Engineers 1993 Joint Rocky Mountain Regional Meeting and Low-Permeability Reservoirs Symposium, SPE Paper 25895, p. 523–531.

MAYNARD, J. B., 1982, Extension of Berner's "New Geochemical Classification of Sedimentary Environments" to ancient sediments: Journal of Sedimentary Petrology, v. 52, p. 1325–1331.

MITCHELL, M. H., 1975, Depositional environment and facies relationships of the Canyon sandstone, Val Verde Basin, Texas: Unpublished M.S. Thesis, Texas A&M University, College Station, 210 p.

MOZLEY, P. S., 1989, Relation between depositional environment and the elemental composition of early diagenetic siderite: Geology, v. 17, p. 704–706.

MOZLEY, P. S., 1992, Isotopic composition of siderite as an indicator of depositional environment: Geology, v. 20, p. 817–820.

MOZLEY, P. S. AND CAROTHERS, W. W., 1992, Elemental and isotopic composition of siderite in the Kuparuk Formation, Alaska: Effect of microbial activity and water/sediment interaction on early pore-water chemistry: Journal of Sedimentary Petrology, v. 62, p. 681–692.

MOZLEY, P. S. AND WERSIN, P., 1992, Isotopic composition of siderite as an indicator of depositional environment: Geology, v. 20, p. 817–820.

MUTTI, E. AND NORMARK, W. R., 1991, An integrated approach to the study of turbidite systems, in Weimer, P. and Link, M. L., eds., Seismic Facies and Sedimentary Processes of Submarine Fans and Turbidite Systems: New York, Springer-Verlag, p. 75–106.

NATIONAL PETROLEUM COUNCIL, 1980, Unconventional gas sources: Tight Gas Reservoirs, v. 5, part I, p. 1–222; part II, p. 10-1-19–24; and Executive Summary, Washington, D.C., National Petroleum Council, 32 p.

PICKARD, G. L., 1979, Descriptive Physical Oceanography: New York, Pergamon, 233 p.

PISCIOTTO, K. A., 1981, Review of secondary carbonates in the Monterey Formation, California, *in* Garrison, R. E. and Douglas, R. G., eds., The Monterey Formation and Related Siliceous Rocks of California: Los Angeles, Society of Economic Paleontologists and Mineralogists, Pacific Section, p. 273–283.

PITTMAN, E. D. AND LARESE, R. E., 1991, Compaction of lithic sands: experimental results and applications: American Association of Petroleum Geologists Bulletin, v. 75, p. 1279–1299.

RALL, R. W. AND RALL, E. P., 1958, Pennsylvanian subsurface geology of Sutton and Schleicher Counties, Texas: American Association of Petroleum Geologists Bulletin, v. 42, p. 839–870.

READING, H. G. AND RICHARDS, M., 1994, Turbidite systems in deep-water basin margins classified by grain size and feeder system: American Association of Petroleum Geologists Bulletin, v. 78, p. 792–822.

ROSENBAUM, J. AND SHEPPARD, S. M. F., 1986, An isotopic study of siderites, dolomites and ankerites at high temperatures: Geochimica et Cosmochimica Acta, v. 50, p. 1147–1150.

ROSENFELD, M. A., 1949, Some aspects of porosity and cementation: Producers Monthly, v. 13, p. 39–42.

SHANMUGAM, G. AND MOIOLA, R. J., 1988, Submarine fans: characteristics, models, classification, and reservoir potential: Earth-Science Reviews, v. 24, p. 383–428.

THYNE, G. D. AND GWINN, C. J., 1994, Evidence for a paleoaquifer from early diagenetic siderite of the Cardium Formation, Alberta, Canada: Journal of Sedimentary Research, v. A64, p. 726–732.

WALTERS, L. J., CLAYPOOL, G. E., CHOQUETTE, P. W., 1972, Reaction rates and $\delta^{18}O$ variation for the carbonate-phosphoric acid preparation method: Geochimica et Cosmochimica Acta, v. 36, p. 129–140.

QUANTIFYING COMPACTION, PRESSURE SOLUTION AND QUARTZ CEMENTATION IN MODERATELY- AND DEEPLY-BURIED QUARTZOSE SANDSTONES FROM THE GREATER GREEN RIVER BASIN, WYOMING

W. NAYLOR STONE AND RAYMOND SIEVER

Department of Earth and Planetary Sciences, Harvard University, 20 Oxford Street, Cambridge, Massachusetts 02138

ABSTRACT: We have quantified the effects of diagenesis on quartzose sandstones from a large range of locations and moderate- to great-burial depths in the Greater Green River Basin, southwest Wyoming. The quantification has allowed us to make basin-scale conclusions regarding mechanical compaction, intergranular pressure solution, and quartz cementation and changes in their relative importance to porosity decline during progressive burial to depths of 7000 m (22,966 ft). We use a point-counting technique which combines cathodoluminescence and light microscopy to quantify the abundance of quartz cement, intergranular volume (IGV = intergranular porosity + cements + matrix) and grain contact types (point vs. pressure solution contacts). We compare our data to published thermal history data to establish controls on porosity in quartzose sandstones with thermal maturities equivalent to vitrinite reflectance values of 0.4 to 2.1% R_o, which, on average, corresponds to present burial depths of 1500 m (4921 ft) to 7000 m.

The results suggest that most mechanical compaction and intergranular pressure solution has occurred at depths shallower than our sample range, which, correcting for erosion, means shallower than approximately 2000 m (6562 ft). The observation that IGV does not decrease with depth below 2000 m adds to a growing body of evidence that intergranular pressure solution is only rarely an important moderate-burial to deep-burial porosity reduction process. Compaction, except by stylolitization, in quartzose sandstones does not usually continue below shallow burial because the combined effects of mechanical compaction, pressure solution, and small amounts of shallow quartz cementation produce stable grain-packing arrangements. Mechanical compaction, that is, grain position rearrangement, sometimes aided by grain fracturing and rotation, results in IGV loss to approximately 30% and pressure solution further reduces IGVs to an average of 22.2%, with very little variability ($\sigma = 3.8\%$). Neither grain size nor abundance of grain-coating clays correlates with pressure solution abundance. Electron microscopic examination of materials along the surfaces of macroscopically-visible stylolites indicates that most of the material, which consists of carbonaceous matter and clay minerals, was deposited as a thin layer in the sandstone. This observation suggests that the amount of dissolution along a stylolite is best measured by the volume of stylolite cones rather than the thickness of stylolite seams.

The most important consideration for porosity prediction below 2000 m is the distribution of quartz cement, which generally increases in abundance with thermal maturity as measured by vitrinite reflectance and, to a lesser extent, with present depth. The average quartzose sandstone with a thermal maturity equivalent to 1.5% R_o, corresponding on average to maximum burial to ~5000 m (16,404 ft), has <2% porosity, 17.4% quartz cement and a few percent carbonate and clay cement. We calculate that 7.6% quartz cement can be provided internally to the average quartzose sandstone from intergranular pressure solution, stylolitization, feldspar dissolution, carbonate replacement of quartz and feldspar grains and clay mineral transformations. The remaining ~10% quartz cement observed in the average deep low-porosity sandstone is episodically imported to the sandstone during moderate to deep burial. We present data on aluminum abundance in quartz cement; these data support the hypothesis that quartz cement episodically precipitates over a large portion of a sandstone's burial history.

Given that a large amount of silica transport is required to import 10% quartz cement and that most cementation is occurring in the thermobaric hydrologic regime, we suggest that a significant proportion of the observed quartz cement may be imported during episodes of rapid fluid flow related to deep-basin mineral dehydration, tectonic forces, hydrocarbon generation and migration and the consequent breakage of overpressured compartments and flow along faults and fractures. The background quartz cementation is provided by internal silica-sources and compaction-driven flow, local density-driven flow and diffusion from silica-sourcing beds adjacent to the sandstones, where present. To predict deep porosity and permeability distribution, one can take advantage of the fact that most silica must be imported; where high flow rates are unlikely to have occurred and where adjacent beds have little silica-exportation potential, high porosity can occur at great depths.

INTRODUCTION

It is well-established that both compaction and cementation are important in sandstone lithification; however, the degree to which each process influences lithification of different kinds of sandstones over a large range of burial depths/temperatures has been incompletely quantified. Quantification of the effects of compaction and cementation on sandstone porosity and permeability is the first step to the formulation of working models of sandstone diagenesis and the distribution of porosity and permeability in sedimentary basins. For these models, we need to know how lithification varies over the range of framework grain compositions/textures, temperatures, pressures, fluid compositions, and hydrologic regimes (the diagenetic variables) found in sedimentary basins. Knowledge of the controls on, and the distribution of, porosity and permeability is essential for the construction of accurate sedimentary basin hydrologic models, the prediction of petroleum reservoir quality, the mapping of petroleum migration pathways and the characterization of deep hazardous waste injection sites. For basin hydrologic modeling, the cumulative volumes and average rates of fluid flow through sandstones during their burial history can be estimated by assessing the volume of silica-bearing fluid flow necessary to pro-

vide the observed quartz cement that could not be provided by internal processes such as pressure solution.

In this study, we quantify mechanical compaction, pressure solution and cementation in quartzose sandstones (mostly quartz arenites and subarkoses; a few sublitharenites). We sampled sandstones from a large range of moderate- to great-burial depths, thermal histories, and locations in the basin to study basin-wide diagenetic trends. We define moderate burial to be 2000 m (6562 ft) to 3500 m (11483 ft) and deep burial to be greater than 3500 m. We use a point-counting technique that combines cathodoluminescence (CL) and light microscopy to ensure accurate identification of intergranular pressure solution contacts and quartz cement, in addition to other components.

By restricting the study to quartzose sandstones, we are essentially holding constant one of the diagenetic variables (sandstone composition), thus allowing better evaluation of the effects of temperature and pressure history, fluid composition and paleohydrology. Other studies have quantified compaction in lithic-rich sandstones (Wilson and McBride, 1988; McBride et al., 1991).

Ever since microscopic examination of sandstones has been possible, quartz overgrowth cement and intergranular pressure solution have been recognized (Sorby, 1863, 1880). Since the time of this groundbreaking work, an enormous amount of research on sandstone diagenesis has been published; below we review only the literature pertaining to the interrelationships between pressure solution, quartz cementation and porosity decline, and the quantitative understanding of them through CL studies. Pressure solution was early suggested to be a quartz cement source by Waldschmidt (1941). Heald (1956) recognized that quartz cementation affected pressure solution also, due to the stabilization of packing arrangements through the enlargement of grain contacts. Progressively deeper drilling for petroleum in the early part of the century made geologists aware that sands lose porosity with increasing burial (e.g., Athy, 1930). Due to the difficulty in distinguishing quartz overgrowths from their detrital cores in light microscopy, it was hard to assess the relative roles of compaction and cementation in the observed porosity decrease and to assess the adequacy of pressure solution as a source for the observed cement. The arrival of electron microcopy and CL microscopes (Sippel, 1968) allowed better distinction of quartz cement from detrital grains, thereby providing the tool for answering these questions. In one of the first quantitative CL studies, Sibley and Blatt (1976) showed that outcrop samples of the Tuscarora Orthoquartzite had lost most of their porosity by cementation and a lesser amount by pressure solution. They estimated that only a third of the observed cement could have been provided by pressure solution. This quantification of the mass-balance problem confirmed what many previous petrographers had qualitatively established and dubbed the quartz cement problem.

Other CL studies concentrated on oil and gas reservoir sandstones that had experienced various degrees of lithification (Houseknecht, 1984, 1988; James et al., 1986). It was concluded from these studies that the fraction of total porosity that had been lost was eliminated mostly by compaction and pressure solution. These authors extrapolated their conclusions to deep burial, suggesting that pressure solution will continue with increasing burial until all porosity is lost. They also concluded that many porous reservoir sandstones displayed enough pressure solution to account for the observed quartz cement. Much of the discrepancy in conclusions between the latter authors and Sibley and Blatt (1976) arose as a result of the fact that Sibley and Blatt (1976) had evaluated sandstones which were representative of the end-product of diagenesis (orthoquartzites) whereas Houseknecht (1984, 1988) and James et al. (1986) had evaluated reservoir sandstones which had experienced various degrees of diagenesis. The relative importance of compaction, pressure solution, and cementation to the entire sandstone lithification process can be fully assessed only by evaluating sandstones which have completed diagenesis and approached zero porosity (i.e., orthoquartzites or quartzites). None of these studies adequately characterized the typical depths or temperatures over which each diagenetic process occurs, especially for deep burial.

When this research project started, these questions had not been resolved, as evidenced by the wide range of views in abstracts from the 1991 AAPG convention (e.g., Ajdukiewicz et al., 1991; Szabo and Paxton, 1991; Houseknecht et al., 1991; Lahann, 1991). The controversy in the literature on quartzose sandstone diagenesis continues, centered around three interrelated questions: (1) what diagenetic process, specifically com-

paction, pressure solution or cementation, is most important in the porosity decline process, and which process(es) should be modeled to predict the distribution of porosity; (2) at what depths and temperatures is each diagenetic process most active; and (3) to what degree does the typical quartzose sandstone supply silica to its own pore space from intergranular pressure solution, framework grain dissolution, stylolitization and other silicate mineral reactions; what are the remaining sources of quartz cement and how is the externally-provided silica transported to the sandstone?

SAMPLING APPROACH

The location map (Fig. 1) shows our broad, basin-scale sampling scheme. We sampled the deepest quartzose sandstones available at the U.S. Geological Survey Core Research Center and a number of quartzose sandstones from moderate burial depths, for comparison. We sampled each major sandstone facies and each major diagenetic zone (defined by degree of lithification, general cement type and general appearance of hand specimen) from the quartzose intervals cored in each well. The sampling of distinctly different diagenetic zones in each well allowed the evaluation of variation in diagenesis at any one location and depth interval. Two or more samples were pointcounted from each of 26 wells, for a total of 55 thin sections point-counted. Samples come from various positions relative to structure and hydrocarbon-water contacts; most are from natural gas wells, although a significant number are from oil wells

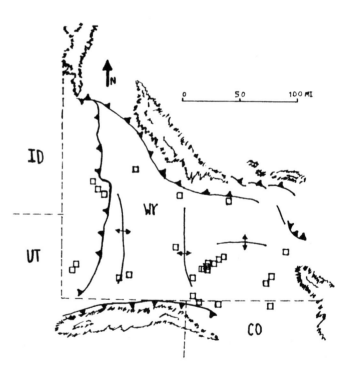

FIG. 1.—Greater Green River Basin and Wyoming Overthrust Belt. Samples were taken from wells marked by squares. Overthrust Belt is west of the thrust fault on the western edge of the basin, paralleling the Wyoming-Idaho and Utah borders. Marked by opposite pointing arrows on lines delineating trends of subsurface structure are, from west to east, the Moxa Arch, the Rock Springs Uplift, and the Wamsutter Arch, which divide the Greater Green River Basin into five smaller basins. Modified from Law et al. (1986).

and dry wildcat wells. No effort was made to distinguish the samples in regard to their structural position or hydrocarbon saturation.

The present depth and thermal maturity distribution of the samples is shown in Figure 2 (sources and explanation of vitrinite reflectance data are listed below in Methods). This figure illustrates the general correlation of present burial depth with vitrinite reflectance. Figure 2 also demonstrates the relatively even distribution of samples over a wide range of moderate to high thermal maturity. Differential uplift and unroofing, variable paleogeothermal gradients and measurement/method error are responsible for the scatter in the relationship. Vitrinite reflectance is a measure of integrated temperature-time history (primarily temperature) and therefore is most often a simple function of maximum burial depth (see review in Bostick, 1979). Because uplift/unroofing and paleogeothermal gradients were not consistent across the basin, present burial depth is not as good as vitrinite reflectance in the recording of temperature history, which is possibly the most important diagenetic variable.

The main units sampled were the Weber Sandstone (Upper Pennsylvanian), the correlative Tensleep Sandstone and the Nugget Sandstone (Lower Jurassic). We also sampled for comparison the Upper Cretaceous Frontier Formation and Ericson Sandstone. Figure 3 is a generalized stratigraphic column that excludes the great thicknesses of latest Upper Cretaceous and Tertiary strata and the small thicknesses of lower Paleozoic strata.

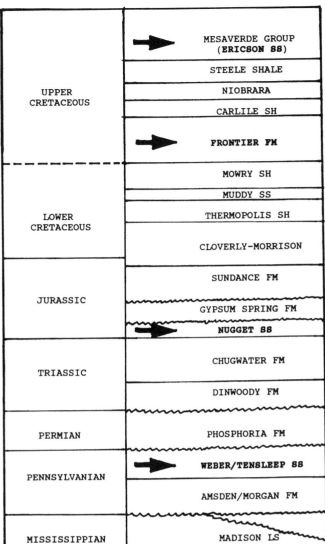

FIG. 3.—Generalized stratigraphic chart, excluding the lower Paleozoic section and the great thicknesses of latest Upper Cretaceous and Tertiary siliciclastics. Units sampled are marked by arrows.

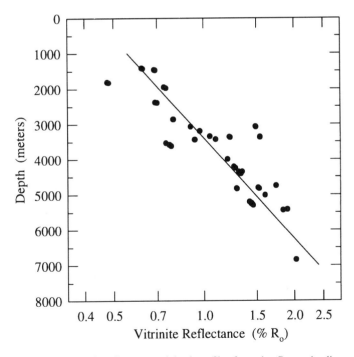

FIG. 2.—Vitrinite reflectance and depth profile of samples. Regression line is included to illustrate general increase in thermal maturity with present burial depth (log R_o = 1.06e^{-4} (depth) − .36). Scatter around line is related to differential uplift and erosion across the basin, in addition to measurement/method error and variable paleogeothermal gradients (r = .86). Sources of vitrinite reflectance data are listed in text.

STRUCTURAL AND STRATIGRAPHIC HISTORY

The Phanerozoic history of the Greater Green River Basin area includes several intervals of relatively rapid subsidence and sedimentation during regional uplift or thrusting in adjacent areas. The Weber and Tensleep Sandstones were deposited during the Ancestral Rocky Mountains epeirogeny, which occurred from the Late Mississippian to the Early Permian Periods (Maughan, 1990). They are quartzose sand erg deposits that prograded southwards across Wyoming starting in Middle Pennsylvanian time. In the Greater Green River Basin, they were deposited relatively conformably upon the shallow marine sandy dolomites, limestones, and mudstones of the Early Pennsylvanian Morgan Formation. In Permian time, following the deposition of the Weber and Tensleep, marine deposition returned, resulting in deposition of the Phosphoria Formation, which includes organic-rich shale, phosphorite, and chert (Edman and Surdam, 1984). The Nugget Sandstone, deposited in the Late Triassic and Early Jurassic, is also an erg deposit.

Beginning in Late Jurassic time, thrusting in the Idaho-Wyoming-Utah Overthrust Belt moved a series of thin sheets eastward along westward-dipping fault planes. Thrusting continued until early Eocene time, resulting in the eastward displacement of five major thrust sheets, the easternmost of which is the youngest. During thrusting, sedimentation occurred in the foreland, which consisted of the Greater Green River Basin, parts of adjacent basins, and strata now in the younger thrust sheets (Armstrong and Oriel, 1965; Royse et al., 1975; Wiltschko and Dorr, 1983; Burtner and Nigrini, 1994).

During the Laramide orogeny in the Late Cretaceous and Early Tertiary Periods, the foreland was segmented by the uplift of the Wind River and Uinta Mountains, creating the present configuration of the Greater Green River Basin. The greatest rates of subsidence and sedimentation in the Greater Green River Basin mostly occurred during the Laramide orogeny. Great thicknesses of Late Cretaceous and Early Tertiary strata were deposited, and Paleozoic and early Mesozoic strata in the centers of the individual basins were deeply buried. Also during this time, uplift of the Moxa Arch and the Rock Springs Uplift occurred, although there is some evidence that movement along the Moxa Arch began before the Laramide orogeny (Wach, 1977; Roehler, 1961). After the Early Tertiary, basin subsidence slowed and erosion removed variable amounts of sediment.

Comparison of data from the Weber (Tensleep), the Nugget, and the few samples of Ericson and Frontier shows that these units are similar in terms of their diagenetic products and their relationships with the diagenetic variables to be discussed in this paper. Therefore, data from the different units will not be distinguished in the following discussion.

METHODS

Our focus is on point-count data from a technique that combines CL with light microscopy. These data are supplemented by electron microprobe and scanning electron microscope (SEM) imaging and elemental data, oxygen and carbon isotopic data on carbonate cements, and fluid inclusion microthermometry. We use published data on the thermal, pressure, and fluid histories to establish controls on the observed diagenetic variability.

For accurate quantification of pressure solution contacts and quartz cement abundance, CL imaging is essential, because it allows accurate distinction between authigenic and detrital quartz. Our point-counting method, modified from Gale (1985), consists of the following steps: (1) take 18 CL photographs (1000 ASA Agfachrome™ color slide film, 35 sec exposures) on a Nuclide, Inc. Luminoscope™, each photograph containing approximately 10–15 grains; (2) project the CL photographs on a screen with 16 grid points; (3) locate the photographed areas on the thin section under the polarizing microscope; (4) identify the phase at each grid point using both the CL photograph and transmitted and reflected light; (5) measure the length of long and short axes and contact indices (see below) of each of three grains (at additional grid points) in each photograph; and (6) add up data, for a total of 288 compositional points and 54 grain size/contact index points. Figures 4A and 4B (see p. 97) are CL photographs that illustrate quartz cement and pressure solution contacts.

The advantages of using CL photographs in combination with a polarizing microscope are: (1) better CL resolution due to long photograph exposure times; (2) the ability to use high power objectives, a rotating stage, and reflected light on the polarizing microscope (most luminoscope setups are unable to achieve these advantages); (3) prevention of excessive eye fatigue from looking at dim CL in the luminoscope; and (4) reduction of electron beam damage (causing changes in intensity and color of luminescence) on the thin section. The major disadvantage of our CL-light microscopy point-counting method is that it is very time consuming.

The compositional variables counted for this study include: (1) all phases present, noting whether they are detrital, pore-filling, or grain-replacive, and, for clay minerals, whether they are grain-coating or pore-filling; and (2) porosity, noting whether it is primary or secondary, and intergranular or intragranular. We quantified compaction processes with the variables "intergranular volume" and contact indices. Intergranular volume (IGV), often called "minus-cement porosity," is the sum of intergranular cements, matrix, and intergranular porosity. Contact indices include the contact index (CI), which is the average number of contacts per grain, and the tight packing index (TPI), which is the average number of pressure solution contacts (long, sutured, or concave-convex) per grain. The contact index was first used by Taylor (1950) and the tight-packing index by Wilson and McBride (1988). Ductile grain contacts, although relatively rare, were counted separately but were added to pressure solution contacts for the TPI.

Cement textures were characterized by SEM and electron microprobe imaging. Major and trace element compositions of some authigenic minerals were determined using the energy dispersive spectrometers (EDS) and wavelength dispersive spectrometers (WDS) on an electron microprobe. One of the objectives of these analyses was to determine the nature of the materials along stylolite surfaces to refine methods for the determination of the amount of dissolution along stylolites. An objective of the WDS analyses was to determine whether the study of trace element compositions (especially aluminum) of authigenic quartz could help us understand the diagenetic environment of authigenic quartz precipitation.

Carbon and oxygen isotopic analyses of carbonate cements helped bracket the temperatures of carbonate precipitation and the precipitating pore fluid isotopic compositions. The isotopic data are presented in Stone (1995); conclusions from this work are used in Figure 5.

Microthermometry of aqueous fluid inclusions in authigenic quartz in overgrowths and in multi-grain fracture fills helped constrain temperatures of authigenic quartz precipitation. This data is presented in Stone (1995) but is mentioned in the following discussion.

Thermal history data were taken from the literature and consist primarily of vitrinite reflectance (measured in oil: R_o) data from Merewether et al. (1987), Pawlewicz et al. (1986), and unpublished data from GeoChem Laboratories, Inc. (G. Bayliss, pers. commun., 1994). Apatite fission track and organic maturation data from Burtner and Nigrini (1994) and Burtner et al. (1994) and structural data from Royse et al. (1975) and Dixon (1982) supplemented the vitrinite reflectance data from GeoChem Laboratories, Inc. in the Wyoming Overthrust Belt. In most cases, reflectance measurements were made on vitrinite from petroleum source strata (e.g., Phosphoria Formation) in the same or nearby wells. We extrapolated the vitrinite reflectance values from these strata to the depths of the sampled sandstones by using average vitrinite reflectance gradients reported in the references listed above.

Information on present-day pore fluid pressures was gained from Spencer (1987; pers. commun., 1991) and the Wyoming Oil and Gas Fields Symposium, Greater Green River Basin (Oil and Gas Fields Symposium Committee, Wyoming Geological Association, 1979). Pore water major and trace elemental data and isotopic data are from Burtner (1987; pers. commun., 1994).

SUMMARY OF PETROGRAPHY AND DIAGENESIS

The framework grain composition of the samples ranges from 80–99% quartz (% of total framework constituents), with an average of 91%; 0–17% feldspar (mostly K-feldspar), with an average of 6%; and 0–18% lithics (mostly sedimentary rock fragments, including chert), with an average of 2%. Therefore, most samples can be classified as quartz arenites or subarkoses; a few are sublitharenites.

The predominant cement is quartz overgrowth cement, which ranges from 0–26.7% and averages 11.2% of total rock volume. Carbonate cements include poikilotopic calcite spherical nodules (0.5–14 mms in diameter) and dolomite pore-filling and grain-replacive rhombs. Total carbonate ranges from 0–13.5% and averages 2.3%. There is an average of 0.5% feldspar overgrowths. Authigenic clay and clay matrix ranges from 0–7.3%, averages 1.3% and mostly consists of grain-coating smectite/illite and illite clay, but also pore-filling smectite/illite and illite, grain-coating and pore-filling chlorite and pore-filling and replacive kaolinite. Other authigenic materials include, in small amounts, iron-oxides, pyrite and hydrocarbon residue.

Total porosity ranges from 0–21.5% and averages 8.6%. This average includes 7.1% intergranular porosity and 1.5% intragranular secondary porosity. Most intragranular secondary porosity occurs in partially-dissolved feldspars but includes oversized pores in which there is no remnant of the dissolved grain. Evidence of intergranular secondary porosity from the dissolution of carbonate cement was present in one sample.

IGVs range from 15.3% to 31.3%, average 22.2%, and have a standard deviation of 3.8%. We infer that mechanical compaction reduced IGVs to an average of 30% (see below) and that pressure solution further reduced IGVs to the data average of 22.2%.

Petrographic, isotopic, fluid inclusion, and diagenetic product distribution data from our work and the literature allow us to construct the general sequence of diagenetic processes (paragenetic sequence) shown in Figure 5. Because the main concern of this paper is compaction, pressure solution and quartz cementation, we discuss carbonate and clay authigenesis only as they affect compaction, pressure solution and quartz cementation.

COMPACTION PROCESSES

Mechanical Compaction

The depositional IGVs of typical sands range from 40–50% (Pryor, 1973; Schenk, 1983; Atkins and McBride, 1992), although it has been reported by Dickinson and Ward (1994) that modern eolian sands from the Namib Desert have average depositional IGVs of 34%. Packing rearrangement, including grain fracturing and rotation, lumped under the term "mechanical compaction", reduces IGVs from depositional values during shallow burial. The relative importance to IGV decline of mechanical compaction versus pressure solution can be assessed from the functional relationship between IGV and TPI (Fig. 6), because TPI is a measure of the number of pressure solution contacts per grain and because IGV decline records the effects of both mechanical compaction and pressure solution. In Figure 6, we show that an extrapolation of a linear regression of TPI and IGV intersects the IGV abscissa at 30%. This indicates that pressure solution does not replace mechanical compaction as the dominant compaction process until an IGV of approximately 30%. This number is therefore also an estimate of the minimum IGV to which this type of natural sand mechanically compacts. We note below that a linear relationship may not be the best description of the relationship as TPI = 0 is approached. Also, a better constraint of the relationship between IGV and TPI would be gotten from a larger database which can be divided into different sorting and shape classes, because sorting and shape differences likely result in much of the scatter in Figure 6. As will be shown below, the establishment of 30% as the IGV at the onset of significant pressure solution is critical for models which estimate amounts of silica liberated during pressure solution to different IGVs.

Fracturing and rotation of individual grains contribute significantly to mechanical compaction in a few samples. Many of the fractured grains are subsequently healed by quartz cement, enabling easy identification in CL (Fig. 7, see p. 97). During point counting, we recorded the number of fractured/healed grains; Figure 8 shows the distribution of the average proportion of such grains in our samples. This number is only a fraction of the total fractured and rotated grains because some grains may have escaped subsequent healing and would not, therefore, be easily recognized in our CL images. Most samples have only a few percent fractured/healed grains; therefore, the contribution of this process to mechanical compaction is not very significant except in a few samples. Grain fracturing/healing has been documented in CL studies of many other sandstones (Sippel, 1968; Sibley and Blatt, 1976; Dickinson and Milliken, 1993; Dunn, 1993).

Due to the very low content of clay-rich lithic grains, ductile deformation is not a very significant compaction process in these samples.

Intergranular Pressure Solution

Variable amounts of intergranular pressure solution have occurred, further reducing IGVs from 30% to a minimum of 15.3%, with an average of 22.2%. As can be seen in the relatively tight distribution of IGVs around an average of 22.2% (Fig. 9), most samples have undergone some pressure solution, and no sample has escaped mechanical compaction (Note: we did not sample any of the few entirely poikilotopic calcite-cemented samples, which have IGVs close to depositional values due to early cementation. A few samples do, however, have some poikilotopic calcite-cemented patches that were point-counted, which results in a higher average IGV for these sam-

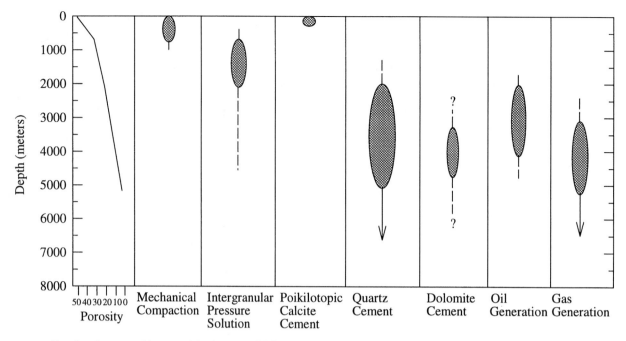

FIG. 5.—Summary of interpreted depth ranges of different diagenetic processes. See text for supporting data and conclusions.

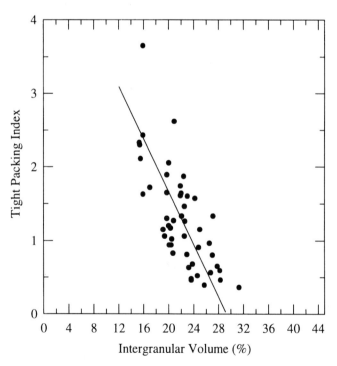

FIG. 6.—Crossplot of IGV versus TPI. Comparison of IGV, which measures the effects of both mechanical compaction and pressure solution, and TPI, which measures the effects of pressure solution only, establishes the relative importance of mechanical compaction and pressure solution to IGV decline. Extrapolation of the regression to TPI = 0 provides an estimate of the IGV at the onset of significant pressure solution (reduced major axis regression: TPI = −.1802 (IGV) + 5.26). Scatter mostly results from sorting and shape differences (r = −.71).

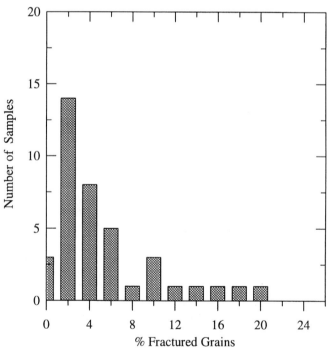

FIG. 8.—Histogram of percent of fractured/healed grains encountered during point counting.

ples). Given the large range of depths, locations in the basin, grain size and shape distributions and burial histories of the samples, there is very little variability in the distribution of

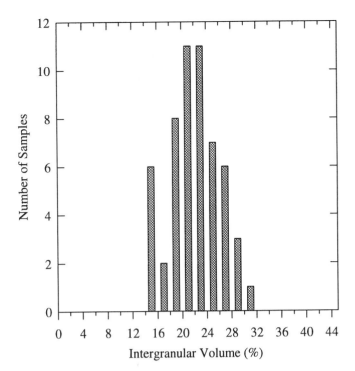

FIG. 9.—Histogram of IGV distribution of samples.

IGVs. One standard deviation of the IGVs is 3.8%; at least some of the variability results from measurement error. We attribute the low variability in mechanical compaction and pressure solution to: (1) the high efficiency and constancy of the mechanical compaction process; (2) the fact that at least some pressure solution occurs in most sandstones; and (3) the negative-feedback aspect of pressure solution; that is, it slows and becomes increasingly inefficient as grain contacts are enlarged by pressure solution and the consequent quartz cementation.

One of the most important results of the study is that there is a similar range of IGV throughout the depth/R_o range (Fig. 10) sampled. The graphs in Figure 10 demonstrate that over the depth and thermal maturity range sampled, there is no increase in extent of mechanical compaction or pressure solution with increasing depth or thermal maturity. Correcting for erosion, we conclude that both mechanical compaction and pressure solution operated primarily at depths of less than approximately 2000 m (6562 ft). Various workers have shown that compaction or pressure solution correlate with depth or thermal maturity over certain intervals (Taylor, 1950; Houseknecht and Hathon, 1987; Houseknecht, 1988, 1991; McBride, et al, 1991); however, these studies are of both lithic and quartzose sandstones and/or are of sandstones from a shallower depth range than our samples. Houseknecht (1991) shows data from Tertiary and Cretaceous, mostly lithic-rich, sandstones from a single well in the Greater Green River Basin that shows a slight correlation between IGV and depth (essentially the same range as our samples), although the statistical significance of this relationship is difficult to assess, because only a crossplot is shown. Houseknecht and Hathon (1987) could not demonstrate a consistent relationship between pressure solution and thermal maturity in a large data set from several petroleum fields with different thermal histories. McBride et al. (1991) showed in a study of Wilcox and Carrizo sandstones in the Gulf of Mexico Basin that compaction decreased porosities rapidly in the first 1200 m (3937 ft) but more slowly and variably at greater depths, essentially stopping at approximately 3000 m (9842 ft), just above depths of regional overpressuring. Unfortunately, there is not much data on mechanical compaction and pressure solution in presently-subsiding quartzose sands buried between 50 and 2000 m, which is quite possibly the range over which most mechanical compaction and pressure solution occur.

Several deep samples have relatively high porosity and IGV (e.g., 18% porosity at 4356 m (14291 ft); 15.6% at 4808 m (15774 ft)). These and most of the other units examined are not now regionally overpressured (Spencer, 1987, pers. commun., 1991; Oil and Gas Fields Symposium Committee, Wyoming Geological Association, 1979). In other basins there are sandstones with even higher porosities at greater depths than these. The Norphlet Formation in the Gulf Coast Basin contains some beds with porosities of greater than 20% at depths of greater than 5500 m (18044 ft), only some of which are in regions that are overpressured (Dixon et al., 1989; Ajdukiewicz et al., 1991; Schmoker and Schenk, 1994). We conclude that high fluid pressures are not necessary to prevent pressure solution in deep porous sandstones. The reason for this is that grain contacts are enlarged from pressure solution and small amounts of quartz cementation which occur at shallow depths (Fig. 11, see p. 97), resulting in packing arrangements which are stable to great lithostatic pressures. Stephenson et al. (1992) modeled the maximum porosity possible at different stress/temperature regimes by assuming that it is directly related to the area of the grain contacts, regardless of whether the contact enlargement was achieved by pressure solution or the precipitation of supporting (load-bearing) cement. According to their model, high porosity units can occur at great depths because compaction equilibrium exists in porous sandstones with sufficient supporting cement or grain-contact enlargement from pressure solution. Most deep sandstones have very low porosity because of the precipitation of large quantities of cement, far beyond the amount required for compaction equilibrium.

Using as evidence (1) our observation that IGVs do not decrease with depth below 2000 m in the Green River Basin, (2) a growing body of data indicating similar results (e.g., Ajdukiewicz et al., 1991; Szabo and Paxton, 1991; Land et al., 1987; McBride et al., 1991; Stone, 1995), and (3) the deep, anomalously-high-porosity sandstones in regions with near-hydrostatic fluid pressures, we conclude that mechanical compaction and intergranular pressure solution are not very important porosity-reduction processes at moderate to great depths (>2000 m). Where burial is too rapid for shallow packing stabilization to occur, pressure solution, grain fracturing and mechanical rearrangement will continue with burial (Houseknecht and Hathon, 1987). Also, in sandstones in which quartz cementation is inhibited (e.g., from grain-coating clays) and in which there are high enough fluid flow rates to remove liberated silica, pressure solution may continue with further burial, but it will become increasingly inefficient as grain contacts are enlarged.

Controls on Pressure Solution Variability

It has been suggested that intergranular pressure solution operates to a greater extent in finer grained sandstones (Housek-

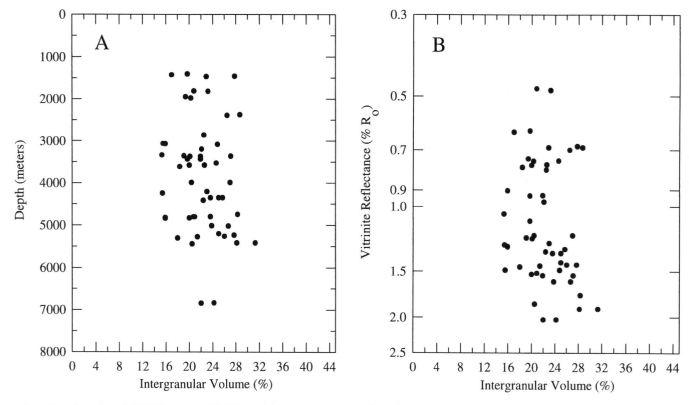

FIG. 10.—Crossplots of: (A) IGV vs depth. (B) IGV vs. vitrinite reflectance. Regardless of whether present burial depth or thermal maturity is used to examine distribution of IGV, there is no decrease in IGV with increasing depth or R$_o$. The same range in IGV exists at shallow depths as at great depths, indicating that compaction processes have operated primarily shallower than our samples depth range.

necht, 1984, 1988; James et al., 1986) and in sandstones with grain-coating clays (Heald, 1956; Weyl, 1959; Houseknecht, 1988). Houseknecht and Hathon (1987) summarized the results of quantitative CL studies on several different individual petroleum fields. From this work, they identified correlations of variable statistical significance between grain size and pressure solution (as measured by overlap quartz; see below for discussion of this method) at the scale of the individual fields; they did not demonstrate a significant correlation in the entire data set. There is not an obvious reason that otherwise similar fine sandstones should pressure dissolve to lower IGVs than coarse sandstones. However, within many units and at a local scale, finer sandstones often differ texturally (e.g., in sphericity, roundness) and compositionally (e.g., in clay or feldspar content) from coarser sandstones. These differences may result in differences in the amount of pressure solution and cementation, yet these relationships are complex and not well-understood. In our sample set, there is no relationship between the degree of compaction and grain size (Fig. 12). We conclude that grain size does not exert an important control on pressure solution variability at the basin scale. Relationships between grain size and compaction probably only exist locally, where texture and composition are a strong and consistent function of grain size.

Small amounts of grain-coating clays (0–3.5%) were observed in many samples; however, because most samples were deposited in eolian environments, there are not large amounts of depositional clay. In our sample set, there is not a significant correlation between the amount of grain-coating clay and the

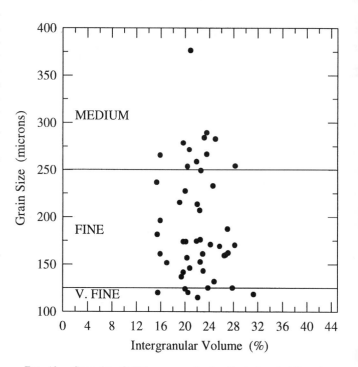

FIG. 12.—Crossplot of IGV versus grain size illustrating that there is no clear relationship between these variables. Other workers have identified field-scale relationships between pressure solution and grain size (see text); at the basin-scale, no trend exists.

degree of compaction (Fig. 13). The presence of grain-coating clay in samples that have not undergone significant pressure solution (Fig. 13) suggests that the multiple diffusion pathways that clay coats may provide along pressure-dissolving grain contacts (Weyl, 1959) may not always significantly enhance the pressure solution process. Likewise, the occurrence of sandstones that have been significantly pressure-dissolved yet contain no grain-coating clay (Fig. 13) demonstrates that grain-coating clay is not necessary for pressure solution to occur. Houseknecht and Hathon (1987) point out that some clay coats may be too thin to be identifiable, except by SEM. However, the relevance of such extremely thin clay coats to pressure solution is unknown, except that continuous diffusion pathways are probably not provided by thin coats. Bjorkum (1994) provides evidence that illitic clays often replace quartz grains at their surfaces, which leads to apparent, but not true, pressure solution textures. In sandstones with more abundant grain-coating clay than our samples, enough to inhibit quartz cementation, it is probable that greater amounts of pressure solution are necessary to stabilize the grain packing arrangement in the absence of supporting cement.

Due to the basin-scale sampling scheme of our study, it is difficult to establish whether regional variations in the amount of pressure solution occur. Only a few areas were sampled with a density suffiient to establish a robust distribution of IGV for that area. For instance, we sampled one well and one field more intensely than other wells or fields; a similar range in IGV was measured in the one well as was measured in the entire field and in our entire data set. This observation suggests that the same range in extent of compaction occurs at multiple scales (well, field, basin), from which we can infer that regional variability in pressure solution is not highly significant. Regional variations may, however, result in a few percent deviation from the basin-wide average IGV. For example, the Nugget Sandstone samples from the Overthrust Belt tend to have slightly lower than average IGVs, which is consistent with observations by Houseknecht (1988). Also, these samples have low quartz cement abundances given their moderate levels of thermal maturity. We can infer that the greater amount of pressure solution that has occurred in the Nugget in the Overthrust Belt is a result of the low abundance of quartz cement; compaction equilibrium from grain contact enlargement was achieved primarily by pressure solution. Knowledge of regional variations in the distribution of early, stabilizing quartz cement may help in understanding regional variations in extent of pressure solution.

Stylolitization

A third compaction process affecting some sandstones in the Greater Green River Basin is stylolitization. A stylolite is a surface, generally at or parallel to bedding, that is marked by a thin layer of dark material and has a characteristic interdigitate appearance in cross-section resulting from the cone-in-cone morphology of the surface of a stylolite (Fig. 14). Stylolites result from uneven dissolution along a surface in response to lithostatic loading or tectonic compression (Stockdale, 1922; Bates and Jackson, 1980). Stylolitization is a compaction process that does not affect IGV, rather, it affects only the thickness of the sandstone unit. Stylolites are present to a significant extent in only a few of the cores examined for this study. The extent to which the material along a stylolite surface was deposited as a lamina or concentrated as an insoluble residue of originally-dispersed material has been a source of considerable debate (Heald, 1955; Tada and Siever, 1989). The resolution of this problem has significant consequences for methods which estimate the amount of dissolution along stylolites. The two methods in question are: (1) measuring the amplitude of the stylolites, assuming that dissolution occurs irregularly along an originally-horizontal depositional laminae; and (2) measuring

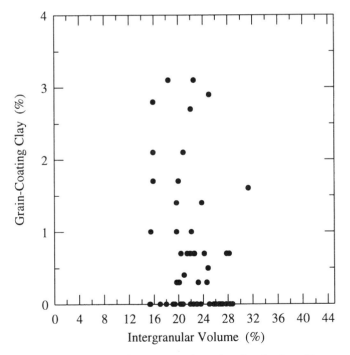

FIG. 13.—Crossplot of IGV versus grain-coating clay abundance illustrating that there is no clear relationship between these variables at the basin-scale. Grain-coating clay is not necessary for intense pressure solution, nor does the presence of grain-coating clay necessarily result in pressure solution.

FIG. 14.—Stylolite surface. Note the cone shape of the stylolite columns. The volume of the cones may be representative of the amount of material dissolved along the stylolite. Piece of core is approximately 10 cm (3.9 in) wide.

the thickness of the stylolite seam and comparing it to the abundance of the stylolitic material in the surrounding rock, assuming that all stylolitic material is an insoluble residue of material that was formerly dispersed.

To better evaluate the degree to which the material along stylolites is depositional versus an insoluble residue, we characterized the textures and compositions of the materials using a SEM and electron microprobe. These analyses demonstrate that the materials consist primarily of solid organic matter, including some identifiable woody tissues, and clay minerals which texturally appear to be of both detrital and authigenic origin. The clays include smectite/illite, illite, chlorite and kaolinite, many of which show crystal faces. Because these materials do not occur in significant quantities in the pore space or within the grains surrounding the stylolites in most samples, they must have been deposited as separate laminae. Dissolution of quartz and feldspar grains could, however, provide some of the elements for authigenic clay growth. Other materials identified along the stylolites include pyrite framboids, small crystals of titanium oxides, and quartz overgrowths growing from the edge into the stylolitic materials. The quartz overgrowths with unflawed terminations possibly indicate that pressure solution is not occurring along the entire surface of the stylolite. Their presence may also help mechanically support the surfaces on either side of the stylolitic material, possibly lithifying the stylolite. Pyrite precipitation is possibly a consequence of organic matter oxidation and sulfate reduction. The titanium oxides may be the only true insoluble residue along stylolites, because they could be concentrated from the dissolution of rutilated and/or titanium-rich quartz. A comparison of the abundance of titanium within the stylolites to the abundance within the surrounding material could provide an estimate of the amount of dissolution, assuming all titanium is immobile.

Other evidence for the depositional nature of the material along stylolites includes: (1) they are generally parallel to bedding surfaces, including cross-bedding, (2) thin, obviously depositional, laminae of clay minerals and organic matter often contain tiny incipient stylolites, and (3) intensely-stylolitized units often have intervals with obviously depositional clay- and organic-rich thin laminae (Fig. 15), along which tiny, incipient stylolites have formed.

Recrystallization and growth of detrital clays and crystallization of the various authigenic minerals observed may be an important factor in the dissolution along stylolite surfaces. Replacement by force of crystallization is thought to be important in many diagenetic processes (Maliva and Siever, 1988). Bjorkum (1994) demonstrated the importance of illite and muscovite replacement of quartz and suggested that replacement of quartz at grain contacts leads to apparent, but not true, pressure solution contacts. In the case of stylolitization, the clay- and organic-rich laminae may provide the ideal chemical and mineralogical environment for quartz dissolution and replacement. The presence of fine authigenic clay crystals and undeformed titanium and pyrite crystals in many stylolites suggests that pressure between quartz grains on either side of a stylolite may not be a factor in the dissolution. If replacement is fundamental to the stylolitization process, some of the dissolved components are retained in the growing crystals, and replacement stops as soon as the conditions favoring crystal growth stop. The latter suggestion implies that most of the time, stylolites are not actively dissolving, and the total amount dissolved over their history is constrained by crystal growth.

From all of the above evidence, we conclude that stylolites begin forming at clay- and organic-rich laminae. Therefore, the thickness of the material reflects depositional processes, not the amount of dissolution. Dissolution is enhanced at the depositional discontinuity and is perhaps related to the diagenesis (including recrystallization) of the clays and/or organic materials. We conclude that the volume of cones along the stylolite surface (Fig. 14) is representative of the amount of material dissolved; therefore, the amplitude of the stylolites (in a cross-sectional view) is a good estimate of the amount of dissolution.

Because several lines of evidence suggest the necessity of clay and/or organic-rich laminae for stylolization to occur, we can now explain the lack of abundant stylolites in the sandstones from this study: it is because they are predominantly eolian. Sandstones such as the Cotton Valley Sandstones of East Texas, which were deposited in lagoonal and shoreface environments, contain numerous clay- and organic-rich laminae and, therefore, numerous stylolites (Stone, 1995).

QUARTZ CEMENTATION

Abundance, Distribution, and Precipitation Temperatures

Quartz cement occurs as overgrowths on detrital quartz grains throughout the units studied. Its abundance varies from 0 to 26.7%, and, in general, it increases with depth and thermal maturity. It is the most abundant cement, averaging 11.2% in the entire data set and 17.4% in the deep, low-porosity samples. We have shown that there is little variability in the extent of compaction and pressure solution and that most occurred shallower than our sample range; thus, IGV is not correlated with porosity (Fig. 16). Figure 17 shows that quartz cement abundance is the primary control on porosity. Variability in abundance of carbonate, clay and other minor cements and some variability in IGV are responsible for the scatter in the relationship in Figure 17.

Figure 18 illustrates the relationship between quartz cement abundance and present burial depth. Comparison of Figures 18 and 19 shows that there is a slightly stronger relationship between quartz cement abundance and thermal maturity as measured by vitrinite reflectance. The latter relationship was also observed by Dutton and Hamlin (1992) in a study of Frontier Formation sublitharenites buried from moderate to great burial depths along the Moxa Arch in the Greater Green River Basin. We infer, therefore, that the maximum temperature and the maximum burial depth experienced by a quartzose sandstone are the most relevant parameters in determining quartz cement abundance. In Figures 20 and 21, we show correlations between porosity and thermal maturity (Fig. 20) and porosity and depth (Fig. 21). It has long been known that porosity generally decreases with depth (Athy, 1930). More recently, it has been suggested that thermal maturity is a better predictor of porosity than depth (Schmoker and Gautier, 1988). We have now established that the most important process controlling these relationships at depths of greater than 2000 m in the Green River Basin is quartz cementation.

As can be seen in Figure 20, most samples with thermal maturities greater than 1.5% R_o have less than 2% porosity. On average, 1.5% R_o corresponds to present depths of 4700 m

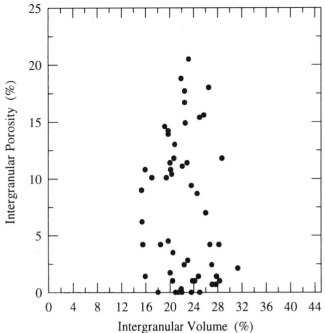

FIG. 16.—Crossplot of IGV versus intergranular porosity illustrating that variability in porosity is not related to variability in compactional processes (r = .09), for samples buried over the depth range of this study.

FIG. 15.—Depositional laminae of carbonaceous material and clay minerals, along which tiny, incipient stylolites have formed (not visible at this magnification). This piece of core is from a section that has a gradation from intervals with depositional laminae with no stylolites to intervals that are intensely-stylolitized. Nodule below lower arrow is pyrite.

(15,420 ft), but variability in amounts of erosion and uplift and in paleogeothermal gradients across the basin requires that thermal maturity be considered in cement abundance prediction rather than present depth. Given that there has been at least a few hundred meters of erosion everywhere across the basin, it is likely that near-zero porosity was attained during maximum burial to depths somewhat greater than 4700 m. Therefore, as a general rule, we conclude that quartz cementation was complete when near-zero porosity was attained at an R_o of approximately 1.5% and a depth of approximately 5000 m (16,404 ft), given average paleogeothermal gradients. Quartz cementation continues with increasing burial in sandstones which have remaining pore space at an R_o greater than 1.5%.

There are several possible explanations for the relationship between quartz cement abundance and thermal maturity: (1) kinetics: increasing rates of reaction with temperature, and/or threshold temperatures being reached for significant quartz nucleation rates; (2) relationship between silica availability and temperature: many diagenetic reactions which release silica must reach a temperature threshold before proceeding and may proceed at a faster rate at higher temperatures; (3) exponential increase in quartz solubility with temperature (see quartz solubility curve in Siever, 1962): if quartz is precipitating from the

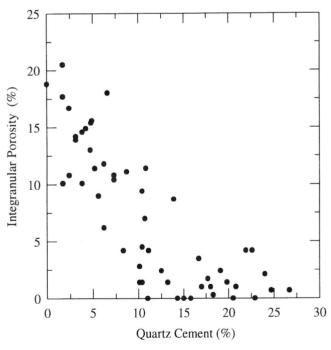

FIG. 17.—Crossplot of quartz cement versus intergranular porosity illustrating that most variability in porosity is a result of variability in quartz cement abundance (r = −.82). Variability in carbonate and clay cement abundances and IGV account for the scatter.

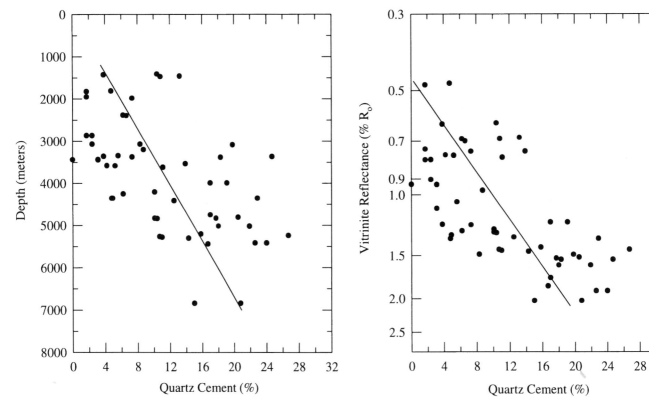

FIG. 18.—Crossplot of quartz cement versus depth illustrating the general increase in quartz cement abundance with present burial depth (r = .56). Comparison with Figure 19 shows that vitrinite reflectance is a better predictor of porosity than present depth. The regression can be used to predict the average quartz cement abundance at a given depth. (Qtz cem = $3.01e^{-3}$ (depth) − .144) (depth is independent variable).

FIG. 19.—Crossplot of quarz cement versus thermal maturity as measured by vitrinite reflectance (r = .66). Comparison with Figure 18 shows that vitrinite reflectance is a better predictor of porosity than present depth. The regression can be used to predict the average quartz cement abundance at a given thermal maturity. (Qtz cem = 29.36 ($\log R_o$) + 9.92) (R_o is independent variable).

cooling of hotter fluids, more quartz is precipitated per unit temperature decrease at higher temperatures; (4) thermal maturation of organic matter influencing silica solubility and transport, silica-providing silicate reactions, or basin hydrodynamics (as shown in Fig. 5, most quartz cementation corresponds to depths/temperatures of oil and wet gas generation); and (5) greater burial results in greater exposure to hot fluids derived from high-grade diagenetic and low-grade metamorphic reactions in the deeper portions and basement of sedimentary basins. To establish the relative importance of these mechanisms, we need better constraints on the timing and temperatures of quartz cementation and the potential silica sources. We discuss these issues below and then propose a model to explain the increase in quartz cement abundance with thermal maturity.

From the observed increase in abundance of quartz cement with increase in thermal maturity (Fig. 19), we infer that there is progressive cementation during burial of a sandstone; that is, small amounts of quartz cement are precipitated during shallow burial (with pressure solution being the most likely source for early cement), and with increasing burial, quartz cement is episodically added to a sandstone (see also Leder and Park, 1986). Walderhaug (1994) presents fluid inclusion data from his own studies and the literature (both mostly from samples from the North Sea Basin) which show temperatures of homogenization ranging mostly from 75–165°C, with a few as high as 190°C.

This range closely matches the temperature range implied by the relationship between vitrinite reflectance and quartz cement in our study (Fig. 19). Some deeply-buried samples from Walderhaug's (1994) study displayed a large range of homogenization temperatures, the maximum of which are at or near present-day formation temperatures. This suggests that quartz cementation in these samples occurred progressively over a large portion of the burial history. Fluid inclusion homogenization temperatures from a single deep sample from our study (present depth = 5227 m) range from 134–199°C, with a single measurement giving a temperature of 229°C (Stone, 1995). These measurements provide evidence that our deep samples have also undergone progressive cementation over a large portion of their burial history and evidence that hotter-than-ambient fluids may be involved in cementation (discussed below). Grant and Oxtoby (1992), on the other hand, showed relatively small ranges of homogenization temperatures in individual samples and an increase in the average and minimum fluid inclusion homogenization temperature with depth, suggesting that cementation in the deeper units did not begin until deep burial. Grant and Oxtoby (1992) further suggest that because the homogenization temperature averages and ranges at given depths mimic the present-day geothermal gradient, cementation occurred over a relatively short duration (<5 ma), synchronously throughout the entire sedimentary column. Haszeldine and Os-

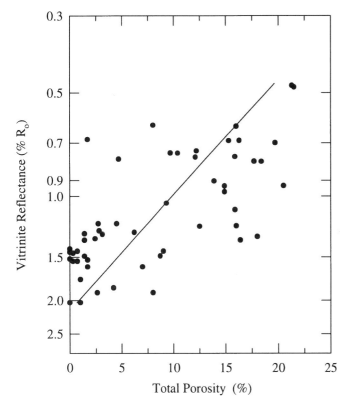

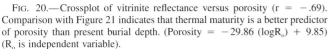

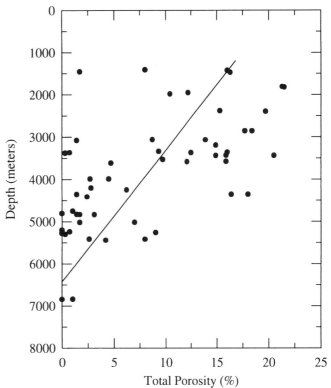

FIG. 20.—Crossplot of vitrinite reflectance versus porosity (r = −.69). Comparison with Figure 21 indicates that thermal maturity is a better predictor of porosity than present burial depth. (Porosity = −29.86 (logR_o) + 9.85) (R_o is independent variable).

FIG. 21.—Crossplot of depth versus porosity (r = −.61). Comparison with Figure 20 indicates that thermal maturity is a better predictor of porosity than present burial depth. (Porosity = −3.22 e^{-3} (depth) + 20.67) (depth is independent variable).

borne (1993) suggest that the latter observation results from stretching and re-equilibration of fluid inclusions during burial, whereas Walderhaug (1994) suggests that it results when rapid burial through the low-temperature portion of the quartz cement window does not give sufficient time for low-temperature inclusions to form. We argue against the Grant and Oxtoby (1992) conclusion because it implies the simultaneous initiation of silica-sourcing reactions throughout the sedimentary column and because it is not consistent with their observed increase in quartz cement abundance with depth.

We conclude from our observations in Figure 19 and from fluid inclusion data (compiled in Walderhaug, 1994) that quartz cementation in the typical deep sandstone occurred episodically over a large range of burial temperatures above 75°C, until near-zero porosity was attained, most often at a thermal maturity near 1.5% R_o (approximately equivalent to T_{max} = 200°C, Barker and Goldstein, 1990). Our data on the abundance of aluminum in quartz cement in Green River Basin sandstones sheds further light on the variability of temperatures and/or fluid compositions involved in the cementation process, and therefore, the duration of quartz cementation episodes.

Aluminum in Quartz Cement

Aluminum is a trace element in quartz. Its concentration in igneous or metamorphic quartz is generally very low (<100 ppm Al_2O_3), but in authigenic quartz precipitated at sedimentary basin temperatures, its concentration ranges from several hundred to a few thousand ppm. Variation in aluminum concentration in quartz cement is thought to cause variation in CL intensity and colors, resulting in observed CL zoning in quartz overgrowths (Henry et al., 1986; Ramseyer and Mullis, 1990; Holzwarth, W., pers. commun., 1991). To establish patterns of aluminum concentration in quartz cement, we analyzed multiple overgrowths in each of 9 samples with different thermal histories. The aluminum concentration was measured using the wavelength dispersive spectrometers on an electron microprobe. The results are reported as weight percent Al_2O_3. The detection limit is approximately 80 ppm and the measurement error is approximately 40 ppm. Figure 22 shows the results of these analyses. Each column of data points at a given vitrinite reflectance value represents measurements on randomly-selected quartz overgrowths (or quartz crystals in fracture fills) in a single polished thin section. The sample with the lowest reflectance value comes from a present burial depth of approximately 1200 m (3937 ft) and the sample with the highest value comes from a present burial depth of approximately 7000 m (22,966 ft).

Figure 22 shows that aluminum concentrations in quartz overgrowths of a given sample vary greatly and that the range is greater in most higher-R_o samples. The higher-R_o samples have approximately the same minimum aluminum concentration as the lower-R_o samples, but the mean and maximum aluminum concentrations are greater in most of the higher-R_o sam-

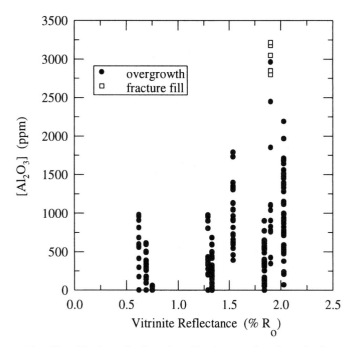

FIG. 22.—Aluminum abundance in authigenic quartz from 9 samples from a broad range of thermal maturities. The large range in aluminum concentration in most high-R_o samples suggests that quartz cement was precipitated from a large range of fluid compositions or over a large temperature range. The higher maximum and average concentration in most of the high-R_o samples suggests that aluminum was either in greater concentration in deeper pore waters or that there is more partitioning of aluminum into quartz at higher temperatures.

ples. The sample with the fracture-filling quartz, which is buried at 5418 m (17775 ft), has the largest range and highest single values of aluminum concentration. Fluid inclusion microthermometry work in the fracture-filling quartz gave homogenization temperatures of 190–225°C, which indicates that large amounts of aluminum are incorporated in quartz precipitated from aqueous fluids at relatively high temperatures. One quartz overgrowth outside of the fracture fill in the same sample has a similar aluminum concentration to that of the fracture fill, and a different overgrowth contains a fluid inclusion that homogenizes at 190°C. Both lines of evidence indicate that at least some of the overgrowth cementation occurred at the same high temperatures as the fracture fill cementation and that high aluminum concentrations correlate with high precipitation temperatures.

We can assume that aluminum is partitioned into authigenic quartz via a simple functional relationship between the aluminum concentration in the quartz, temperature and the aluminum concentration in the precipitating fluids. From the large amount of variability in aluminum concentrations in the higher-R_o samples, we conclude that there was a diversity of either fluid compositions or temperatures of precipitation, or both, responsible for quartz cementation in the higher-R_o sandstones. This supports the notion that quartz cementation occurs over a relatively large portion of the burial history of a sandstone and that deeper burial results in an incrementally larger amount of quartz cementation. The observed higher average and maximum concentrations in most of the higher-R_o samples suggest that either (1) aluminum is partitioned into quartz in greater amounts at higher

temperatures, or (2) aluminum concentrations in the formation fluids increase with depth/temperature in a basin. It has been suggested that the effect of pH on aluminum speciation in formation fluids may control the partitioning of aluminum into authigenic quartz (Ramseyer, pers. commun., 1993); if true, then pH changes with depth may explain part of the increase in average and maximum aluminum concentrations with increase in thermal maturity.

Internal Silica Sources

To better understand the controls on the quartz cement—thermal maturity relationship and to evaluate the implications of the observed cement abundances for basin hydrology, it is necessary to estimate the amount of cement derived from processes internal to the sandstone. These processes include intergranular pressure solution, stylolitization, silicate mineral dissolution and silicate mineral transformations.

The importance of intergranular pressure solution as a source of silica for quartz cementation has been debated for a number of years. This dispute has not been satisfactorily resolved due to the paucity of quantitative CL observations on the amounts of cement, mechanical compaction and pressure solution in samples from a broad range of depths and geographic locations. Sibley and Blatt (1976), Houseknecht (1984, 1988) and James et al. (1986) have quantified the amount of silica liberated during pressure solution by point-counting overlap quartz. This is defined as an area near a pressure solution contact that appears to have been formerly occupied by a portion of a quartz grain which has since been pressure dissolved. The overlap quartz method requires the petrographer to reconstruct original grain shapes, so it is petrographer-dependent and imprecise, especially in sandstones that did not originally have spherical and well-rounded grains. We have chosen instead to use pressure solution models from Rittenhouse (1971). With these models and our data on IGV and quartz cement abundances, we can estimate the importance of pressure solution as an internal source of the quartz cement observed in our samples.

Rittenhouse (1971) constructed geometric models of grain interpenetration during pressure solution for a range of packing geometries and grain shapes. These models give curves which estimate the amount of cement produced during IGV decline resulting from pressure solution. The shape of a curve is dependent upon the packing geometry used; for instance, if orthorhombic packing, rotated 30°, is used for equally-sized spherical grains, significantly more silica is liberated for a given IGV decline than if cubic packing, rotated 45°, is used. The different packing geometries will have different initial (prepressure solution) porosities as well. The tightest packing geometry possible for equally-sized spheres is rhombohedral packing, which results in a porosity of 25.9%. Variability in shape, sorting, and packing geometry typically result in initial porosities greater than 25.9%. As discussed above and shown in Figure 6, a comparison of the tight packing indices and the IGVs in our data set, which includes sandstones with grains of variable sorting and shapes (though mostly towards the well-sorted, well-rounded, and spherical end of the spectrum), indicates that mechanical packing has resulted in average prepressure solution porosities of approximately 30%.

After producing curves for different packing geometries and grain shapes, Rittenhouse (1971) constructed best estimate curves for natural sands which have a natural range of shapes, sortings, packing geometries and initial porosities. We have chosen to use his "maxima for any sand curve" from his Figure 6, which estimates the maximum amount of cement in any natural sand that can be produced during pressure solution to a given IGV (i.e., this curve has the maximum "cement to solution ratio"). In Figure 23, we show this curve and our data on IGV and quartz cement abundances, from which we estimate the maximum proportion of the observed cement that could have been provided by pressure solution. Because the relationship between TPI and IGV may not be linear near TPI = 0, that is, the initiation of pressure solution at point contacts (at which there is a high radius of curvature) may result in a relatively large amount of IGV decrease for a very small amount of pressure solution, we begin the model curve at 32%, rather than 30%. Also, small amounts of initial pressure solution may aid further mechanical compaction, because pressure solution at irregular point contacts can allow grains to slide past one another, thereby resulting in a relatively large IGV decrease for a small amount of pressure solution (Fuchtbauer, 1967). Lastly, we use a correction factor because very small initial pressure solution contacts are under-identified during point-counting of TPI. To prevent a bias towards counting small pressure solution contacts more frequently in coarser sandstones, in which small pressure solution contacts would be more easily identifiable

than in finer sandstones, we required that a pressure solution contact be at least 1/8 of the perimeter of the smallest of the two grains in contact. Because of this constraint, very small pressure solution contacts were counted as point contacts. By starting the Rittenhouse model at 32%, rather than 30%, we compensate for the above factors by slightly overestimating the amount of quartz cement produced at a given IGV.

The few data points that fall along the Rittenhouse model curve in Figure 23 contain the exact amount of quartz cement predicted to have been produced by pressure solution. It is important to note that the general trend of the data points is perpendicular to the Rittenhouse curve. This suggests that pressure solution does not simply redistribute silica from the grain contacts to the adjacent pore space; rather, most sandstones are either distinct importers or exporters of silica (Houseknecht, 1988), requiring silica transport at some scale greater than that of a single thin section. Points that fall below the curve have less quartz cement than was produced by pressure solution; these sandstones are silica exporters. As previously mentioned, the porous Nugget Sandstone samples from the Overthrust Belt have slightly lower than average IGVs and low quartz cement abundances; these samples are, therefore, mostly exporters (Fig. 23) confirming Housknecht's (1988) observations. Sandstones that fall above the line are importers with respect to intergranular pressure solution; however, because there are other internal sources (e.g., stylolites and silicate mineral dissolution/transformations; see below), sandstones that fall above the curve may not all be net silica importers. We can conclude only that intergranular pressure solution produces only a small proportion of the observed quartz cement for most of these samples. The Rittenhouse model predicts that pressure solution to the average IGV of 22.2% produces approximately 4% quartz cement. The average quartz cement abundance is 11.2%; therefore, it is necessary to call upon stylolites, silicate mineral dissolution/transformations and external silica sources to provide the remainder of the observed quartz cement: 7.2%.

Above we estimated the average imbalance between the observed cement abundances and the amount that could have been provided by pressure solution for the entire data set, which contains many samples with significant porosity. More relevant for our purposes is estimation of the proportion of observed quartz cement in the deep, entirely-cemented quartzose sandstones that is produced by pressure solution. That is, we should evaluate the sandstones that are representative of the end-product of diagenesis: quartzites, defined here as quartzose sandstones with less than 2% intergranular porosity. The imbalance will be much greater than the average of 7.2% for the entire data set when one considers that many samples have some porosity that would eventually be filled by quartz cement if they were to be buried deeper. Recall that pressure solution is significant only at relatively shallow depths, such that the deep, entirely-cemented samples have seen no more pressure solution than the shallower, uncemented samples. Also, for this reason, pressure solution provides only the relatively early quartz cement. In Figure 24, we plot on the IGV-quartz cement graph only the sandstones with less than 2% intergranular porosity. This graph shows that quartzites have a severe imbalance between the amount of observed quartz cement and the amount that could have been provided by pressure solution. The average IGV of the quartzites (22.7%) is essentially the same as the average

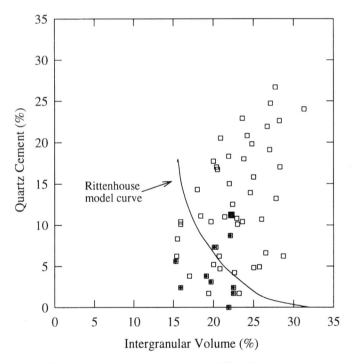

Fig. 23.—Crossplot of quartz cement versus IGV with a pressure solution model curve from Rittenhouse (1971). The model curve is Rittenhouse's "maxima for any sand" (i.e., highest cement-solution ratios) curve, starting at an IGV of 32% (see text). The hollow squares represent analyses from individual samples; the large, filled square is the average of all samples; the squares with dots in the center are samples from the Nugget Sandstone in the Overthrust Belt. Samples above the model curve are all importers of silica (with respect to pressure solution only); those below the curve are exporters. Note that the Nugget sandstones in the Overthrust Belt are mostly exporters. At the data average IGV of 22.2%, the model predicts that about 4 % quartz cement is produced by pressure solution.

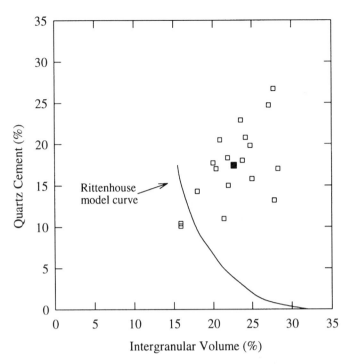

FIG. 24.—Crossplot of quartz cement versus IGV for samples with less than 2% porosity (quartzites), and a pressure solution model curve from Rittenhouse (1971). This graph illustrates that the end-product of diagenesis (a quartzite), which is the typical quartzose sandstone with $R_o > 1.5\%$, is typically a sandstone that has undergone much too little pressure solution to account for the observed cement abundance. At the quartzite average (large, filled square) IGV of 22.7%, the model predicts that less than 4% quartz cement is produced by pressure solution, yet an average of 17.4% quartz cement is observed.

mate that they provide an insignificant amount of the observed quartz cement in these cores. In the few cores where they are abundant, we must use a measure of their abundance and size to estimate the amount of silica liberated. We concluded above that because stylolites begin at thin clay- and organic-rich laminae, the thickness of the stylolite seam is not related to the amount of material dissolved. Thus, we will use the other proposed measure of the amount of quartz cement produced by stylolitization: the amplitude of the stylolite columns. We did not make measurements of abundances and amplitudes of the stylolites observed in the cores. Instead, we estimated the amount of silica liberated along stylolites in a hypothetical, intensely-stylolitized section of sandstone core. Figure 25 shows the stylolitized sandstone model; it was drawn to be representative of the maximum abundance and amplitude of stylolites in sandstones ever observed by the authors (the most intensely-stylolitized sandstone observed in the Cotton Valley Group of East Texas was not quite as stylolitized as the hypothetical model in Fig. 25). Measurement of the abundances and amplitudes of the stylolites in this model sandstone will therefore yield an estimate of the maximum amount of quartz cement provided by stylolitization. The sandstone was assumed to be composed entirely of quartz and have a porosity of 8%, requiring that both quartz grains and quartz cement be dissolved along the stylolite surface. The relatively low porosity was used because we assume that stylolites are active mostly in well-lithified sediments; this low-porosity assumption results in a higher estimate of the amount of silica liberated. The calculation resulted in the estimate that a maximum of 4% quartz cement can

IGV of the entire data set (22.2%). The Rittenhouse model predicts that at an IGV of 22.7%, less than 4% quartz cement is produced by pressure solution. The average quartz cement abundance of the quartzites is 17.4%. We therefore conclude that 13.4% quartz cement must be provided by other internal sources or by importation from external sources during moderate to deep burial.

We would like to emphasize that much of the disagreement over the importance of intergranular pressure solution as a source of observed quartz cement results from the analysis of sandstones at different stages of diagenesis. For instance, Houseknecht (1988) examined many sandstones which were relatively intensely pressure dissolved but not yet cemented (i.e., they were porous reservoir sandstones). Houseknecht (1988) therefore concluded that pressure solution provides more than the observed cement, which was true for the porous sandstones he observed but not for sandstones in general. Sibley and Blatt (1976), on the other hand, studied a sandstone (the Tuscarora Orthoquartzite) which has very low porosity, similar to our quartzites. They concluded from their overlap quartz point counts that pressure solution provided only a third of the observed quartz cement in the Tuscarora (7% quartz cement in sandstones which average 21%), which is slightly higher than our estimate based on the Rittenhouse model.

The most difficult internal silica source to quantify is stylolitization. Fortunately, because stylolites are not abundant or non-existent in most of the cores analyzed, we can safely esti-

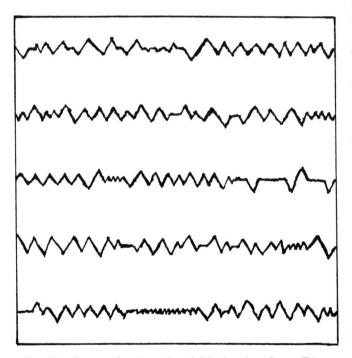

FIG. 25.—Cartoon of an intensely-stylolitized section of core. This was drawn to be representative of the most-intensely stylolitized sandstone ever encountered by the authors and to give an estimate of the maximum amount of cement produced by the stylolitization process. Because most sandstones in the Green River Basin contain no stylolites, the minimum (and typical) amount of cement produced by stylolitization is zero.

be provided by stylolitization. The estimated range in the Green River Basin sandstones is, therefore, 0% (for sandstones with no stylolites) to 4%, for the most intensely-stylolitized sandstone. We can now make the further assumption that the amount of silica produced from stylolitization in the average sandstone in the Green River Basin is at the midpoint between the least and most intensely stylolitized units: 2%. Because the majority of sandstones contain no stylolites and because our calculation method gives a liberal estimate of produced silica, this estimate of the average is probably high.

The other internal silica-producing silicate mineral reactions are dissolution of feldspar and unstable silicate rock fragments, replacement of quartz and feldspar by dolomite and calcite, and recrystallization of detrital smectite to illite. To estimate the amount of quartz cement provided by these reactions we use the observed average abundances of: (1) secondary porosity (1.52%), (2) carbonate replacement of quartz and feldspar (0.77%), and (3) illite/smectite clay (0.85%). We assume that all secondary porosity was produced by the dissolution of feldspars, that carbonate replacement is 50% of quartz and 50% of feldspar, and that all clay minerals were originally detrital smectite and are now well-crystallized illite (Note: illite/smectite clay includes all highly birefringent, micaceous, clay-size materials encountered during optical petrography.) Using the stoichiometry of these reactions (see Totten and Blatt, 1993, for summary of clay mineral silica-producing reactions) and assuming that all silica produced is reprecipitated as quartz cement, we estimate that an average a total of 1.6% quartz cement is produced by these reactions. Further recrystallization of illite to muscovite during higher grades of diagenesis and low-grade metamorphism will also liberate silica (Totten and Blatt, 1993). This reaction is not likely to be important for considering the *internal* silica sources in our quartzites, because most have not been buried to depths of this transition. Because some of the silica liberated during dissolution of feldspars could have been reprecipitated as clay minerals and because some of the illite/ smectite clay is authigenic and/or not recrystallized, the estimate is probably high.

Table 1 summarizes the internal quartz cement budget. Details of the calculations are in Stone (1995). Keeping in mind that we have used assumptions which yield the maximum amount of silica by each process, we find that the typical low-porosity sandstone, a quartzite averaging 17.4% quartz cement, must have imported enough silica to produce about 10% quartz cement. Such a large amount of silica importation requires either large amounts/rates of regional fluid flow or proximity of the sandstone to external silica sources. We can take advantage of the silica importation requirement to both constrain paleohydrological models and predict where variations in paleohydrology or proximity to external silica sources have resulted in deviation from the relationship between quartz cement abundance and thermal maturity.

Model to Explain Quartz Cement Abundance Increase with Thermal Maturity

Using the evidence discussed above, we propose the following model to explain the observed increase in quartz cement abundance with thermal maturity: (1) for kinetic reasons of quartz nucleation and precipitation rates, a minimum tempera-

TABLE 1.—SUMMARY OF INTERNAL SOURCES AND SINKS OF SILICA IN THE TYPICAL DEEP, LOW-POROSITY SANDSTONE: A QUARTZITE

Quartz Cement Internal Budget (average "quartzite")		
SOURCES:		
	pressure solution:	4 %
	stylolitization:	2 %
SINKS:	silicate reactions:	1.6 %
	quartz cement:	17.4 %
	fracture fills:	?
NET REQUIRED TO IMPORT:		approx. 10 %

ture of approximately 75°C must be reached before cementation at geologically significant rates can occur. For this reason, shallow pressure solution is often not accompanied by quartz cementation in the adjacent pore space. Evidence for 75°C as the lower limit of quartz cementation is available from the fluid inclusion studies (compiled in Walderhaug, 1994) and oxygen isotopic studies (compiled in Aplin and Warren, 1994). Additional evidence is provided by the low-R_o sandstones in our data set which have undergone substantial pressure solution but no quartz cementation, except in grain fractures, where a fresh quartz surface has allowed nucleation and growth of quartz cement at low temperatures. (2) Above the threshold temperature (note: we use threshold temperature loosely, to signify the onset of significant cementation), small amounts of quartz cement begin to precipitate from the formation fluids that contain silica from pressure solution. This early quartz cement further slows pressure solution by further enlarging grain contacts. (3) Quartz cement is added episodically during further burial, with the rate determined primarily by the rates of external silica-providing reactions and by the rates of fluid flow and diffusive transport, which bring silica into the sandstone from distant external sources and proximate external sources, respectively. A background rate of quartz cementation is related to compaction-driven flow, ongoing internal reactions and diffusion from adjacent beds, but much cementation occurs during short-duration episodes of unusually rapid fluid flow, related to mineral dehydration, tectonic forces, rapid subsidence, hydrocarbon generation and migration, and the consequent breakage of overpressured compartments and flow along faults and fractures. (4) Both the internal and external silicate reactions which provide silica to the flowing formation fluids occur episodically, because threshold temperatures must be reached before extensive reaction can occur (e.g., smectite-illite transformation). Sandstones that have been buried deeply have had more opportunity to be exposed to formation fluids supplied with silica from a variety of these temperature-threshold dependent silicate reactions, including deeper (i.e., >5000 m), high-grade diagenetic and low-grade metamorphic dehydration reactions. The dehydration of minerals during high-grade diagenesis and low-grade metamorphism provides water and silica for focusing into shallower sandstones with remaining porosity. (5) Deeper sandstones have been buried through a larger portion of the thermal gradient, which, because quartz solubility exponentially increases with temperature (Siever, 1962), requires that more silica be released (per unit volume of fluid) from the upward-flowing fluids which have been saturated with silica from deeper-basin reactions. When hot fluids rapidly travel through faults/fractures to a cooler sandstone, a relatively large amount of silica is released during cooling to ambient temperatures.

IMPLICATIONS FOR BASIN HYDROLOGY

We have shown that most quartz cement observed in the deep quartzites of the Greater Green River Basin was externally-sourced and that this cement importation occurred mostly over a R_o range of 0.5% to 1.5% (the R_o at which the typical sandstone has <2% porosity). These R_o values correspond, on average, to maximum burial depths of approximately 2000 m to 5000 m. Using the classification of sedimentary basin hydrologic regimes from Galloway (1984), we see that most quartz cementation occurs at greater depths than the meteoric regime (i.e., flow under elevation head) and the shallow, most-active part of the compactional regime. Therefore, quartz cementation occurs in the deeper parts of the compactional regime and in the thermobaric regime (see Harrison and Tempel, 1993, for an outline of the characteristics of the hydrologic regimes in different types of basins). Due to the very low solubility and solubility gradient of quartz at sedimentary basin temperatures, an enormous amount of silica transport is required in portions of the basin that have been thought to be relatively stagnant. The mechanism of silica transport (molecular diffusion vs. advection or convection) and the scale of advective/convective transport (between adjacent beds vs. long-distance) depend upon the proximity of a sandstone to potential silica-source beds. Evaluation of external silica-sourcing is beyond the scope of this paper. Because external sourcing is critical to the issue of silica transport, we limit our discussion below to a general outline of the likely silica transport mechanisms. For a more thorough discussion of external silica-sourcing and silica transport, see Stone (1995).

Where potential silica-source beds are proximate (e.g., shales interbedded with sandstones), silica transport occurs by molecular diffusion in addition to advective or convective fluid transport. Diffusion has been proposed to be important in the transport of silica from shales to adjacent sandstones (Fuchtbauer, 1967; Wood and Surdam, 1979; McBride, 1989), yet it is unknown whether diffusion can operate over longer distances. If adjacent shales are important sources of silica, diffusion combined with short-distance advective transport driven by shale compaction and mineral or hydrocarbon phase changes, could provide substantial amounts of quartz cement. Also, local density and thermal variability can result in small-scale convectional recirculation (Wood and Hewett, 1982, 1984) between nearby silica-source beds and sandstones. As discussed below, local time-transient fluid pressure variability can drive fluids between silica-source beds and sandstones in adjacent pressure compartments.

Where external silica sources are distant from the sandstone undergoing cementation, silica importation must occur by fluid transport. The precipitation of quartz cement from advecting or convecting fluids likely occurs primarily in response to the cooling of fluids flowing across thermal gradients (Siever, 1959; Leder and Park, 1986; Land et al., 1987). Simple calculations of the amount of fluid required to import quartz cement in the absence of molecular diffusion (e.g., von Englehardt, 1967; Land and Dutton, 1978; Bjorlykke, 1979, 1983; Bjorlykke and Egeberg, 1993) suggest that enormous volumes of fluid are required. McBride (1989) summarizes these calculations, indicating that most calculations require 10^4 to 10^5 pre-cement pore volumes of water to provide 5 to 15% quartz cement. Because this large volume of water is not thought to be available in the compactional and thermobaric hydrologic regimes, recirculation between silica-sourcing beds and sandstones must occur and/or fluids must be focused into the sandstones. These volume calculations are, however, based upon the cooling of fluids at low temperatures, such as occur in the meteoric hydrologic regime. Because quartz solubility exponentially increases with temperature (Siever, 1962), the cooling of fluids at higher temperatures (i.e., those at the actual temperatures of precipitation) results in more precipitation of quartz (per unit volume of fluid) than the same amount of cooling of fluids at lower temperatures; therefore, most of these calculations have overestimated the volume of fluids required. Nonetheless, the required volumes of fluids are still orders of magnitude more than the volumes believed to be present in the deep compactional and thermobaric regimes. The required volumes are less, however, if other factors control quartz solubility (e.g., organic acid complexing; Bennet and Siegel, 1987) or hotter-than-ambient fluids are involved in the cementation.

In addition to a fluid volume problem, current knowledge of fluid flow rates in the compactional and thermobaric regimes (e.g., in Harrison and Summa, 1991) suggests that flow rates are orders of magnitude slower than required for significant cement importation. Our knowledge of flow rates is, however, very poor. Stone (1995) calculates average paleo-flow rates required to import the large volumes of observed, externally-sourced quartz cement. These calculations indicate that flow rates might be much higher than suggested by modeling (e.g., Harrison and Tempel, 1993). Fluid convection models (e.g., Wood and Hewett, 1982, 1984; Leder and Park, 1986) use sufficient flow rates to successfully redistribute silica from warmer to cooler areas within homogeneous sand bodies with high porosities and permeabilities. Because our work shows that most silica must come from outside the sand bodies (i.e., mostly from fine, low-permeability units), convection may not be a plausible silica importation mechanism. Little is known about the efficiency of convective transport across highly-variable permeability zones (e.g., in a mixed sandstone-shale sequence) or large-scale convective transport from basement and near-basement silica sources to sandstones at the depths of active cementation. However, it is unlikely that convection cells, especially large-scale cells, can be maintained where permeabilities are low and highly variable.

Flow within the thermobaric regime, which is driven primarily along faults and fractures by mineral and hydrocarbon reactions, tectonic stresses, and thermal and density variability, has not been adequately studied, either by observation (fluid inclusion, isotopic and elemental analyses of fluids and cements) or numerical modeling. The time scales of excess fluid pressure generation and release are unknown. If fluid pressures change rapidly (at geologic time scales), significant rates of fluid flow and fluid recirculation could occur as fluids move from areas of active excess fluid pressure generation to areas in which excess fluid pressure generation is slow or in which excess fluid pressures were previously released. The generation and migration by buoyancy of petroleum fluids can result in the displacement and movement of aqueous, mineralizing fluids, yet little is understood about petroleum migration, much less the effects of it upon aqueous flow. It may be no coincidence that the temperatures of active oil and wet gas generation overlap with those of most active quartz cementation; rather, the effect of petroleum generation and migration on basin hydrology, and, perhaps, on quartz solubility, may be important in silica transport. The above flow mechanisms, in addition to local convection, may be important in fluid recirculation at the depths of active cementation, which also overlap with the smectite-illite transition, a silica source and fluid flow drive. In deep basins such as the Greater Green River Basin, there are numerous higher-grade diagenetic and low-grade metamorphic dehydration reactions occurring in sedimentary rocks at greater temperatures and depths (i.e., below 5000 m) than those of most active quartz cementation and smectite-illite transition. There is little doubt that during high-grade diagenesis and low-grade metamorphism of pelitic rocks, a large amount of silica and water are released (see Stone, 1995). Because of the low permeabilities of these deep-basin rocks, silica-bearing aqueous fluids are expelled upwards along fractures and faults and focused into sandstones with remaining porosity and permeability. In this manner, sandstones with remaining porosity and permeability serve as conduits for waters released from the dehydration of all deeper rocks.

If hot fluids from the deep-basin travel rapidly through faults and fractures to cooler sandstones, a large amount of silica is precipitated from a relatively small volume of fluid during cooling to ambient temperatures, thereby lowering the required flow rates and volumes necessary to import the large quantities of observed cement. This results from the exponential increase in quartz solubility with temperature (Siever, 1962). Burley et al. (1989) present fluid inclusion evidence that indicates that hot fluids episodically traveling up faults are involved in sandstone cementation. Our fluid inclusion work on quartz overgrowths and authigenic quartz in adjacent fractures suggests that hotter-than-ambient fluids were responsible for the fracture fill cementation and a portion of the overgrowth cementation (Stone, 1995).

SUMMARY OF CONCLUSIONS

Based primarily upon CL/light microscopy point-count data from quartzose sandstones from a broad range of locations and moderate to deep burial in the Greater Green River Basin and Wyoming Overthrust Belt, the following conclusions have been made about the controls on porosity decline at various depths and thermal maturities:

(1) Mechanical compaction, pressure solution and quartz cementation have all contributed to the reduction of porosity of the average quartzose sandstone sample from 40–50% at the time of deposition to less than 2% at a thermal maturity of 1.5% R_o, which, on average, corresponds to a maximum burial depth of approximately 5000 m (16404 ft). All three processes were important in total porosity decline, but their relative importance was different at different depths. Mechanical compaction reduced porosities of the average quartzose sandstone to approximately 30%, and pressure solution further reduced porosities to around 22%, mostly at depths shallower than 2000 m (6562 ft). Quartz cementation was the primary control on porosity-depth trends below 2000 m of burial. The reason that compaction did not continue to be significant in porosity decline below 2000 m is that grain packing arrangements were stabilized by grain-contact enlargement from the combined effects of pressure solution and small amounts of early quartz cementation.

(2) The best predictor of the average and range of quartz cement abundance and, therefore, porosity, below 2000 m, is the thermal maturity of that sandstone, as measured by vitrinite reflectance. Present burial depth also correlates with quartz cement abundance and porosity. Most quartz cementation occurred over a thermal maturity range of 0.5–1.5% R_o, which corresponds, on average, to maximum burial depths of 2000 to 5000 m. Below 5000 m, the typical sandstone has <2% porosity (i.e., is a quartzite), due to the precipitation of an average 17.4% quartz overgrowths.

(3) An evaluation of internal sources of silica for the average quartzite leads to the conclusion that intergranular pressure solution provided enough silica for 4% quartz overgrowths. Stylolitization (although only locally present) provided an average of 2% quartz overgrowths. Feldspar dissolution, carbonate replacement and clay mineral transformations together provided an additional 1.6%. The internal silica sources thus produced a total of 7.6% quartz overgrowths. Because the typical quartzite contains 17.4% quartz cement, an average of 9.8% quartz cement must have been provided by external silica sources.

(4) The best model to explain the observed increase in quartz cement abundance with thermal maturity that fits: (1) our observations of the distribution of quartz cement, (2) our knowledge of silica sources and general basin hydrology, (3) the observed variations in aluminum content, and (4) our fluid inclusion data and fluid inclusion data from the literature, is as follows: after a threshold temperature, approximately 75°C., for significant rates of quartz nucleation is reached, quartz cement is episodically added to a sandstone during progressive burial. A steady, background rate of quartz cementation occurs from the reactions of internal silica-sources and molecular diffusion and fluid transport from external silica sources in adjacent beds (where present). The remaining silica for quartz cementation is transported to the sandstones during short-duration episodes of unusually rapid fluid flow, related to mineral dehydration, tectonic forces, rapid subsidence, hydrocarbon generation and migration, and the consequent breakage of overpressured compartments and flow along faults and fractures. The dehydration of minerals during higher-grade diagenesis (i.e., >5000 m) and low-grade metamorphism provides water and silica for focusing into shallower sandstones with remaining porosity. As sandstones reach higher tempera-

tures during burial, more silica is released from the upward-flowing fluids (per unit volume of fluid, because of exponential solubility increase) which have been saturated with silica from a variety of deep-basin silicate reactions. When hot fluids rapidly travel through faults/fractures to a cooler sandstone, a relatively large amount of silica (per volume fluid) is released during cooling to ambient temperatures.

(5) A large amount of solute transport is required to import the observed volumes of quartz cement; most of this transport occurs at depths well below the meteoric regime and much of the compactional regime. As our knowledge of thermobaric regime flow improves through a combination of observation (fluid inclusion, isotopic and elemental data, fracture/fault imaging) and numerical modeling, we may reach the point at which we can sufficiently characterize the fluid flow history of a given sandstone unit at a given thermal maturity, to predict the volume of quartz cement. This necessarily involves analysis of stratigraphy and structure to characterize the distance of a sandstone to potential external silica-source beds and to characterize both local and regional fluid flow paths. We can base deep porosity distribution models on the simple observation that porosity decline in the moderate and deep portions of sedimentary basins is primarily a result of the importation of large quantities of cement: *where high flow rates are unlikely to have occurred and where adjacent beds have little silica-exportation potential, high porosity can occur at great depths in quartzose sandstones.*

ACKNOWLEDGMENTS

This paper reports the results of the first portion of a multi-basin study of deep sandstone diagenesis supported by NSF Grant # EAR-9017748 to R. Siever. Additional support was provided by BP Exploration Operating Company, Ltd., USGS Branch of Petroleum Geology and Harvard University's Daly Fund. The samples were taken from the USGS Core Research Center; we thank their staff for help during sampling. The study greatly benefited from help by various personnel at the USGS Branches of Petroleum Geology and Sedimentary Processes, most notably Bob Burruss, Ted Dyman, Don Gautier, Sam Johnson, Bill Keighin, Ed Maughan, Mark Pawlewicz, Janet Pitman, Rich Pollastro, Chris Schenk, Jim Schmoker and Chuck Spencer. Geoffrey Bayliss, Roger Burtner and Larry Tedesco helped obtain unpublished thermal maturity data for the Overthrust Belt. We appreciate helpful reviews by Roger Burtner, Mark Cooper and David Houseknecht.

REFERENCES

AJDUKIEWICZ, J. M., PAXTON, S. T., AND SZABO, J. O., 1991, Deep porosity preservation in the Norphlet Formation, Mobile Bay, Alabama (abs.): American Association of Petroleum Geologists Bulletin, v. 75, p. 533.

APLIN, A. C. AND WARREN, E. A., 1994, Oxygen isotopic indications of the mechanisms of silica transport and quartz cementation in deeply buried sandstones: Geology, v. 22, p. 847–850.

ARMSTRONG, F. C. AND ORIEL, S. S., 1965, Tectonic development of the Idaho-Wyoming thrust belt: American Association of Petroleum Geologists Bulletin, v. 49, p. 1847–1866.

ATHY, L. F., 1930, Density, porosity, and compaction of sedimentary rocks: American Association of Petroleum Geologists Bulletin, v. 14, p. 1–24.

ATKINS, J. E. AND MCBRIDE, E. F., 1992, Porosity and packing of Holocene river, dune, and beach sands: American Association of Petroleum Geologists Bulletin, v. 76, p. 339–355.

BARKER, C. E., AND GOLDSTEIN, R. H., 1990, Fluid-inclusion technique for determining maximum temperature in calcite and its comparison to the vitrinite reflectance geothermometer: Geology, v. 18, p. 1003–1006.

BATES, R. L. AND JACKSON, J. A., eds., 1980, Glossary of Geology: Falls Church, American Geological Institute, 749 p.

BENNET, P. C. AND SIEGEL, D. I., 1987, Increased solubility of quartz in water due to complexing by organic compounds: Nature, v. 326, p. 684–686.

BJORKUM, P. A., 1994, How important is pressure in causing dissolution of quartz in sandstones? (abs.): American Association of Petroleum Geologists Annual Convention Official Program, v. 3, p. 105.

BJORLYKKE, K., 1979, Cementation of sandstones—a discussion: Journal of Sedimentary Petrology, v. 49, p. 1358–1359.

BJORLYKKE, K., 1983, Diagenetic reactions in sandstones, in Parker, A. and Sellwood, B. W., eds., Sediment Diagenesis: Dordrecht, Riedel, NATO Advanced Study Institute, Series C, v. 115, p.277–286.

BJORLYKKE, K. AND EGEBERG, P. K., 1993, Quartz cementation in sedimentary basins: American Association of Petroleum Geologists Bulletin, v. 77, p. 1538–1548.

BOSTICK, N. H., 1979, Microscopic measurement of the level of catagenesis of solid organic matter in sedimentary rocks to aid exploration for petroleum and to determine former burial temperatures—a review, in Scholle, P. A. and Schluger, P. R., eds., Aspects of Diagenesis: Tulsa, Society of Economic Paleontologists and Mineralogists Special Publication 26, p. 17–44.

BURLEY, S. D., MULLIS, J., AND MATTER, A., 1989, Timing diagenesis in the Tartan Reservoir (UK North Sea): constraints from combined cathodoluminescence microscopy and fluid inclusion studies: Marine and Petroleum Geology, v. 6, p. 98–120.

BURTNER, R. L., 1987, Origin and evolution of Weber and Tensleep formation waters in the Greater Green River and Uinta-Piceance basins, northern Rocky Mountain area, U.S.A.: Chemical Geology, Isotope Geoscience Section, v. 65, p. 255–282.

BURTNER, R. L. AND NIGRINI, A., 1994, Thermochronology of the Idaho-Wyoming Thrust Belt during the Sevier Orogeny: a new, calibrated, multi-process thermal model: American Association of Petroleum Geologists Bulletin, v. 78, p 1586–1612.

BURTNER, R. L., NIGRINI, A., AND DONELICK, R. A., 1994, Thermochronology of Lower Cretaceous source rocks in the Idaho-Wyoming Thrust Belt: American Association of Petroleum Geologists Bulletin, v. 78, p. 1613–1636.

DICKINSON, W. W. AND MILLIKEN, K. L., 1993, Sandstone compaction by interrelated brittle deformation and pressure solution, Etjo Sandstone, Namibia: Geological Society of America Abstracts with Programs, v. 25, p. A-65.

DICKINSON, W. W. AND WARD, J. D., 1994, Low depositional porosity in eolian sands and sandstones, Namib Desert: Journal of Sedimentary Research, v. A64, p. 226–233.

DIXON, J. S., 1982, Regional structural synthesis, Wyoming salient of the Western Overthrust Belt: American Association of Petroleum Geologists Bulletin, v. 66, p. 1560–1580.

DIXON, S. A., SUMMERS, D. M., AND SURDAM, R. C., 1989, Diagenesis and preservation of porosity in the Norphlet Formation (Upper Jurassic), southern Alabama: American Association of Petroleum Geologists Bulletin, v. 73, p. 707–728.

DUNN, T. L., 1993, Early quartz grain fracturing and annealing: an important brittle deformation mechanism of early compaction in first cycle orogenic sandstones, revealed by scanning electron cathode luminesence microscopy (abs.): Geological Society of America Abstracts with Programs, v. 25, p. A-335.

DUTTON, S. P. AND HAMLIN, H. S., 1992, Interaction of burial history and diagenesis of the Upper Cretaceous Frontier Formation, Moxa Arch, Green River Basin, Wyoming: Wyoming Geological Association Guidebook, Fourty-third Field Conference, p. 37–50.

EDMAN, J. D. AND SURDAM, R. C., 1984, Diagenetic history of the Phosphoria, Tensleep, and Madison Formations, Tip Top Field, Wyoming, in McDonald, D. A. and Surdam, R. C., eds., Clastic Diagenesis: Tulsa, American Association of Petroleum Geologists, Memoir 37, p. 317–346.

FUCHTBAUER, H., 1967, Influence of different types of diagenesis on sandstone porosity: Seventh World Petroleum Congress, v. 2, p. 353–369.

GALE, P. E., 1985, Diagenesis of the Middle to Upper Devonian Catskill facies sandstones in southeastern New York State: Unpublished M.A. Thesis, Harvard University, Cambridge, 241 p.

GALLOWAY, W. E., 1984, Hydrogeologic regimes of sandstone diagenesis, in McDonald, D. A. and Surdam, R. C., eds., Clastic Diagenesis: Tulsa, American Association of Petroleum Geologists Memoir 37, p. 3–14.

GRANT, S. M. AND OXTOBY, N. H., 1992, The timing of quartz cementation in Mesozoic sandstones from Haltenbanken, offshore mid-Norway: fluid inclusion evidence: Journal of the Geological Society of London, v. 149, p. 479–482.

HARRISON, W. J. AND SUMMA, L. L., 1991, Paleohydrology of the Gulf of Mexico Basin: American Journal of Science, v. 291, p. 109–176.

HARRISON, W. J. AND TEMPEL, R. N., 1993, Diagenetic pathways in sedimentary basins, in Horbury, A. D. and Robinson, A. G., eds., Diagenesis and Basin Development: Tulsa, American Association of Petroleum Geologists Studies in Geology 36, p. 69–86.

HASZELDINE, R. S. AND OSBORNE, M., 1993, Fluid inclusion temperatures in diagenetic quartz reset by burial: implications for oil field cementation, in Horbury, A. D. and Robinson, A. G., eds., Diagenesis and Basin Development: Tulsa, American Association of Petroleum Geologists Studies in Geology 36, p. 35–46.

HEALD, M. T., 1955, Stylolites in sandstones: Journal of Geology, v. 63, p. 101–114.

HEALD, M. T., 1956, Cementation of Simpson and St. Peter sandstones in parts of Oklahoma, Arkansas, and Missouri: Journal of Geology, v. 64, p. 16–30.

HENRY, D. J., TONEY, J. B., SUCHECKI, R. K., AND BLOCH, S., 1986, Development of quartz overgrowths and pressure solution in quartz sandstones: evidence from cathodoluminescence/backscattered electron imaging and trace element analysis on the electron microprobe (abs.): Geological Society of America Abstracts with Programs, v. 18, p. 635.

HOUSEKNECHT, D. W., 1984, Influence of grain size and temperature on intergranular pressure solution, quartz cementation, and porosity in a quartzose sandstone: Journal of Sedimentary Petrology, v. 54, p. 348–361.

HOUSEKNECHT, D. W., 1988, Intergranular pressure solution in four quartzose sandstones: Journal of Sedimentary Petrology, v. 58, p. 228–246.

HOUSEKNECHT, D. W., 1991, Use of cathodoluminescence petrography for understanding compaction, quartz cementation, and porosity in sandstones, in Barker, C.E. and Kopp, O.C., eds., Luminescence Microscopy: Quantitative and Qualitative Aspects: Tulsa, Society of Economic Paleontologists and Mineralogists Short Course Notes 25, p. 59–66.

HOUSEKNECHT, D. W., DINCAU, A. R., FREEMAN, C. W., 1991, Sandstone compaction: basis for porosity predictive capabilities (abs.): American Association of Petroleum Geologists Bulletin, v. 75, p. 597–598.

HOUSEKNECHT, D. W. AND HATHON, L. A., 1987, Petrographic constraints on models of intergranular pressure solution in quartzose sandstones: Applied Geochemistry, v. 2, p. 507–521.

JAMES, W. C., WILMAR, G. C. AND DAVIDSON, B. G., 1986, Role of quartz type and grain size in silica diagenesis, Nugget Sandstone, south-central Wyoming: Journal of Sedimentary Petrology, v. 56, p. 657–662.

LAHANN, R. W., 1991, A kinetic model of pressure solution/compaction for porosity prediction (abs.): American Association of Petroleum Geologists Bulletin, v. 75, p. 616.

LAND, L. S. AND DUTTON, S. P., 1978, Cementation of a Pennsylvanian deltaic sandstone: isotopic data: Journal of Sedimentary Petrology, v. 48, p. 1167–1176.

LAND, L. S., MILLIKEN, K. L., AND MCBRIDE, E. F., 1987, Diagenetic evolution of Cenozoic sandstones, Gulf of Mexico Sedimentary Basin: Sedimentary Geology, v. 50, p. 195–225.

LAW, B. E., POLLASTRO, R. M., AND KEIGHIN, C. W., 1986, Geologic characterization of low-permeability gas reservoirs in selected wells, Greater Green River Basin, Wyoming, Colorado, and Utah, in Spencer, C. W. and Mast, R. F., eds., Geology of Tight Gas Reservoirs: Tulsa, American Association of Petroleum Geologists Studies in Geology 24, p. 253–269.

LEDER, F. AND PARK, W. C., 1986, Porosity reduction in sandstone by quartz overgrowth: American Association of Petroleum Geologists Bulletin, v. 70, p. 1713–1728.

MALIVA, R. G. AND SIEVER, R., 1988, Diagenetic replacement controlled by force of crystallization: Geology, v. 16, p. 688–691.

MAUGHAN, E. K., 1990, Summary of the Ancestral Rocky Mountains epeirogeny in Wyoming and adjacent areas: Washington, D.C., United States Geological Survey Open-File Report OF 90–447, 12 p.

MCBRIDE, E. F., 1989, Quartz cement in sandstones: a review: Earth Science Reviews, v. 26, p. 69–112.

MCBRIDE, E. F., DIGGS, T. N., AND WILSON, J. C., 1991, Compaction of Wilcox and Carrizo sandstones (Paleocene-Eocene) to 4420 m, Texas Gulf Coast: Journal of Sedimentary Petrology, v. 61, p. 73–85.

MEREWETHER, E. A., KRYSTINIK, K. B., AND PAWLEWICZ, M. J., 1987, Thermal maturity of hydrocarbon-bearing formations in southwestern Wyoming and northwestern Colorado: Washington, D.C., United States Geological Survey, Miscellaneous Investigations Series, 1831.

OIL AND GAS FIELD SYMPOSIUM COMMITTEE, WYOMING GEOLOGICAL ASSOCIATION, 1979, Wyoming Oil and Gas Fields Symposium, Greater Green River Basin: Casper, Wyoming Geological Association, 428 p.

PAWLEWICZ, M. J., LICKUS, M. R., LAW, B. E., DICKINSON, W. W., AND BARCLAY, C. S. V., 1986, Thermal maturity map showing subsurface elevation of 0.8 percent vitrinite reflectance in the Greater Green River Basin of Wyoming, Colorado, and Utah: U.S. Geological Survey, Miscellaneous Field Studies Map, 1890.

PRYOR, W. A., 1973, Permeability-porosity patterns and variations in some Holocene sand bodies: American Association of Petroleum Geologists Bulletin, v. 57, p. 162–189.

RAMSEYER, K. AND MULLIS, J., 1990, Factors influencing short-lived blue cathodoluminescence of α-quartz: American Mineralogist, v. 75, p. 791–800.

RITTENHOUSE, G., 1971, Pore space reduction by solution and cementation: American Association of Petroleum Geologists Bulletin, v. 55, p. 80–91.

ROEHLER, H. W., 1961, The Late Cretaceous—Tertiary boundary in the Rock Springs Uplift, Sweetwater County, Wyoming: Wyoming Geological Association, 16th Annual Field Conference Guidebook, p. 96–100.

ROYSE, F., WARNER, M. A., AND REESE, D. L., 1975, Thrust belt structural geometry and related stratigraphic problems, Wyoming-Idaho-northern Utah, in Bolyard, D. W., ed., Deep Drilling Frontiers of the Central Rocky Mountains: Denver, Rocky Mountain Association of Geologists, p. 41–54.

SCHENK, C. J., 1983, Textural and structural characteristics of some experimentally formed eolian strata, in Brookfield, M. E., and Ahlbrandt, T. S., eds., Eolian Sediments and Processes: Amsterdam, Developments in Sedimentology 38, Elsevier, p. 41–49.

SCHMOKER, J. W. AND GAUTIER, D. L., 1988, Sandstone porosity as a function of thermal maturity: Geology, v. 16, p. 1007–1010.

SCHMOKER, J. W. AND SCHENK, C. J., 1994, Regional porosity trends of the Upper Jurassic Norphlet Formation in Southwestern Alabama and vicinity, with comparisons to formations of other basins: American Association of Petroleum Geologists Bulletin, v. 78, p. 166–180.

SIBLEY, D. F. AND BLATT, H., 1976, Intergranular pressure solution and cementation of the Tuscarora Orthoquartzite: Journal of Sedimentary Petrology, v. 46, p. 881–896.

SIEVER, R., 1959, Petrology and geochemistry of silica cementation in some Pennsylvanian sandstones, in Ireland, H.A., ed., Silica in Sediments: Tulsa, Society of Economic Paleontologists and Mineralogists, Reprint Series 7, p. 55–79.

SIEVER, R., 1962, Silica solubility, 0°–200°C, and the diagenesis of siliceous sediments: Journal of Geology, v. 70, p. 127–150.

SIPPEL, R. F., 1968, Sandstone petrology: evidence from luminescence petrography: Journal of Sedimentary Petrology, v. 38, p. 530–554.

SORBY, H. C., 1863, On the direct correlation of mechanical and chemical forces: Proceedings of the Royal Society of London, v. 12, p. 583–600.

SORBY, H. C., 1880, On the structure and origin of non-calcareous stratified rocks: Quarterly Journal of the Geological Society of London, v. 37, p. 49–92.

SPENCER, C. W., 1987, Hydrocarbon generation as a mechanism for overpressuring in Rocky Mountain Region: American Association of Petroleum Geologists Bulletin, v. 71, p. 368–388.

STEPHENSON, L. P., PLUMLEY, W. J., AND PALCIAUSKAS, V. V., 1992, A model for sandstone compaction by grain interpenetration: Journal of Sedimentary Petrology, v. 62, p. 11–22.

STOCKDALE, P. B., 1922, Stylolites: their nature and origin: Indiana University Studies, v. 9, p. 1–97.

STONE, W. N., 1995, Diagenesis and paleohydrology of deeply-buried quartzose sandstones: Anadarko, Greater Green River, East Texas, and Appalachian Basins: Unpublished Ph.D. Thesis, Harvard University, Cambridge, 372 p.

SZABO, J. O. AND PAXTON, S. T., 1991, Intergranular volume (IGV) decline curves for evaluating and predicting compaction and porosity loss in sandstones (abs.): American Association of Petroleum Geologists Bulletin, v. 75, p. 678.

TADA, R. AND SIEVER, R., 1989, Pressure solution during diagenesis: Annual Reviews in Earth and Planetary Science, v. 17, p. 89–118.

TAYLOR, J. M., 1950, Pore-space reduction in sandstones: American Association of Petroleum Geologists Bulletin, v. 34, p. 701–716.

TOTTEN, M. W. AND BLATT, H., 1993, Alterations in the non-clay-mineral fraction of pelitic rocks across the diagenetic to low-grade metamorphic transition, Ouachita Mountains, Oklahoma and Arkansas: Journal of Sedimentary Petrology, v. 63, p. 899–908.

VON ENGELHARDT, W., 1967, Interstitial solution and diagenesis in sediments, in Larsen, G. and Chilingarian, G. V., eds., Diagenesis in Sediments: Amsterdam, Developments in Sedimentology 8, Elsevier, p. 503–522.

WACH, P. H., 1977, The Moxa Arch, an overthrust model?: Wyoming Geological Association, Annual Field Conference Guidebook 29, p. 651–654.

WALDERHAUG, O., 1994, Temperatures of quartz cementation in Jurassic sandstones from the Norwegian Continental Shelf—evidence from fluid inclusion: Journal of Sedimentary Research, v. A64, p. 311–323.

WALDSCHMIDT, W. A., 1941, Cementing materials in sandstones and their influence on the migration of oil: American Association of Petroleum Geologists Bulletin, v. 25, p. 1839–1879.

WEYL, P. K., 1959, Pressure solution and the force of crystallization, a phenomenological theory: Journal of Geophysical Research, v. 64, p. 2001–2025.

WILSON, J. C. AND MCBRIDE, E. F., 1988, Compaction and porosity evolution of Pliocene sandstones, Ventura Basin, California: American Association of Petroleum Geologists Bulletin, 72, p. 664–681.

WILTSCHKO, D. V. AND DORR, J. A., JR., 1983, Timing of deformation in overthrust belt and foreland of Idaho, Wyoming, and Utah: American Association of Petroleum Geologists Bulletin, v. 67, p. 1304–1322.

WOOD, J. R. AND HEWETT, T. A., 1982, Fluid convection and mass transfer in porous sandstone—a theoretical model: Geochimica et Cosmochimica Acta, v. 46, p. 1707–1713.

WOOD, J. R. AND HEWETT, T. A., 1984, Reservoir diagenesis and convective fluid flow, *in* McDonald, D. A. and Surdam, R. C., eds., Clastic Diagenesis: Tulsa, American Association of Petroleum Geologists Memoir 37, p. 99–110.

WOOD, J. R. AND SURDAM, R. C., 1979, Application of convective-diffusion model to diagenetic processes, *in* Scholle, P.A. and Schluger, P.R., eds., Aspects of Diagenesis: Tulsa, Society of Economic Paleontologists and Mineralogists Special Publication 26, p. 93–141.

EARLY DIAGENESIS AND PALEOSOL FEATURES OF ANCIENT DESERT SEDIMENTS: EXAMPLES FROM THE PERMIAN BASIN

ARIEL MALICSE AND JIM MAZZULLO

Department of Geology and Geophysics, Texas A&M University, College Station, Texas 77843–3115

ABSTRACT: The Queen Formation is a sequence of carbonates, evaporites and siliciclastics that was deposited across the back-reef shelves of the Permian Basin during the late Permian (Guadalupian) time. Its upper clastic member was deposited in desert fluvial sandflat, eolian and sabkha environments during relative sea-level lowstands. During these lowstands, some of the exposed sediments on the shelves were subjected to soil forming processes. The presence of paleosol features has been documented in some of the carbonate units; however, their presence in the equivalent siliciclastic units has been commonly overlooked.

Most pedogenetic criteria were developed using examples from humid and vegetated regions; consequently, these criteria are absent or poorly developed in the soils of ancient deserts. Petrographic, SEM and core analyses of the upper clastic member of the Queen Formation from four fields in the Northwest shelf and Central Basin platform reveal subtle soil features that are common to arid regions. These features include: deformed structures (the destruction of the primary sedimentary structures by intra-sedimentary growth and collapse of salts), cutans (the mechanical infiltration and deposition of clays on the grain surface) and subsoil lamellae (the formation of subsurface lamellae by clay illuviation independent or controlled to some degree by the inherited stratification). Soil horizons are rare but have been identified in some of the cores. Superimposed on these pedogenetic features are early diagenetic events, which include: labile grain dissolution, hematitization and precipitation of authigenic species and precipitation of major pore-filling cements.

The pedogenetic features presented in this paper are alternative criteria that can be used to recognize subtle soil features in ancient sediments. These features also indicate subaerial exposures, and therefore, important to the sequence stratigraphic reconstruction of eustatically controlled shelves.

INTRODUCTION

Desert deposits occur throughout geologic time and were widespread during the Permo-Triassic periods, when large portions of the continents lay between the Horse Latitudes. These deposits usually form red beds that are interbedded with carbonate and evaporite units. Some of these deposits were exposed subaerially for long periods of time and were subjected to soil forming processes. The criteria for recognizing most paleosol features were developed using examples from vegetated or humid landscapes (Blodgett, 1988; Retallack, 1990); thus, many of these criteria are poorly developed or missing in ancient desert sediments.

The need to restrict the definition of paleosol is necessary to distinguish other surficial processes (e.g., sedimentary) from those that are strictly soil forming. However, problems arise when these criteria are applied to arid and hyper-arid regions or to a geologic time when vegetation was sparse or absent. For example, the destruction of sedimentary structures by vegetation or burrows is considered a paleosol feature (Blodgett, 1988), whereas the destruction of sedimentary structures by the surficial precipitation of evaporites is not. The criteria to distinguish paleosol features should therefore reflect the balance among various factors such as time, climate, and scarcity of vegetation. Surficial processes in arid to hyper-arid deserts are characterized by prolonged cycles of drying that favor the formation of poorly vegetated and poorly drained saline soils (salorthids); (Minashina, 1968; Driessen and Schoorl, 1973; Szalbolcs, 1989). Under these conditions, pedogenic processes may either lead to the formation, disruption and/or prevention of soil profile.

During late Permian (Guadalupian) time, the vast arid shelves of the Permian Basin were periodically exposed and subjected to the progradation of siliciclastics (Jacka et al., 1972; Garber et al., 1989; Malicse and Mazzullo, 1990; Mazzullo et al., 1991, 1992). The upper clastic member of the Queen Formation exemplifies some of these prograding units. It was deposited in desert fluvial sandflat, eolian, and sabkha environments, and contains pedogenetic features that are unique to the saline soils

of arid regions (Malicse and Mazzullo, 1990; Mazzullo et al., 1991, 1992). Data on pedogenic features were collected from the analyses of 27 cores and 265 thin sections from four fields in the Northwest shelf and Central Basin platform (Figs. 1, 2). The primary purpose of this paper is to characterize the various paleosol features in the upper clastic member of the Queen Formation.

The desert shelf deposits of the Permian Basin were also subjected to early diagenetic alterations shortly after their deposition. Many of these early diagenetic features were preserved in the red bed sequence and can be observed in thin sections

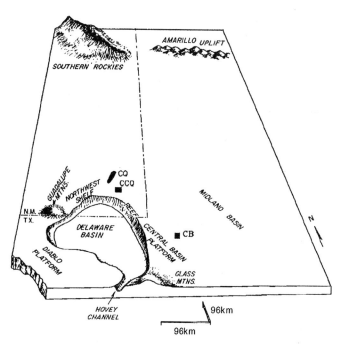

FIG. 1.—Permian Basin of West Texas, showing the locations of the study areas: Caprock Queen (CQ), Central Corbin Queen (CCQ) and Concho Bluff and Concho North (CB) fields.

Siliciclastic Diagenesis and Fluid Flow: Concepts and Applications, SEPM Special Publication No. 55

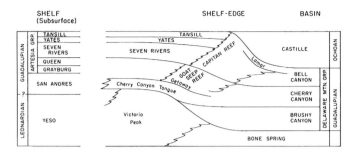

FIG. 2.—Stratigraphic cross-section of the Permian Basin (modified from Ward et al., 1986).

and SEM/EDS analyses. The secondary purpose of this paper is to document the early diagenetic features of the siliciclastics and establish a model showing their early diagenetic history.

PALEOGEOGRAPHIC AND PALEOCLIMATIC SETTING

The Guadalupian (late Permian) marks a major change in the large-scale configuration of the Permian Basin of west Texas and southern New Mexico (Oriel et al., 1967; Ward et al., 1986). The Delaware Basin, bordered on its landward margin by shelf-edge reefs, persisted as a deep-marine basin, while the Midland Basin gradually became filled and assumed the configuration of the surrounding shelves (Oriel, et al., 1967). Climatic conditions in the Permian Basin were arid to semi-arid for two reasons: the paleo-equator position of the Permian Basin, which lay along the path of the dry-subtropical northeasterly trade winds; and the presence of mountain ranges at the contact of Gondwana and Laurasia, which blocked moist equatorial easterlies from entering this region (Ziegler, et al., 1979; Mintz, 1981; Ross and Ross, 1988; Scotese and McKerrow, 1990). The flat, featureless shelves that surrounded the Delaware Basin were slowly subsiding and were constantly inundated by the cyclic influx of an epicontinental sea. During sea-level rise, the shelves became vast lagoons where carbonates and bedded evaporites accumulated. During relative sea-level fall, the shelves were subaerially exposed and were converted to desert coastal plains that stretched landward to the ancestral Rocky Mountains and the Amarillo Uplift (Mazzullo et al., 1991). Large volume of siliciclastics were deposited in various desert environments during these lowstands.

Contrary to earlier ideas that deserts are largely eolian, the landscapes of the Permian Basin deserts were vast fluvial sandflats and sabkhas (Fig. 3; Handford, 1981; Mazzullo et al., 1991, 1992). Eolian sand sheets were present but were confined downwind of the alluvial plains and sabkhas. The climate in the desert was arid to semi-arid punctuated by sporadic torrential rainfalls, which were more frequent along the distal mountain ranges. The sporadic rainfalls not only accelerated the weathering of the granitic rocks along the ancestral Rocky Mountains and Amarillo Uplift but also facilitated the transport of the resulting arkosic sediments to the downdip desert coastal plains. However, aridity is the principal factor controlling the depositional process, the lack of vegetation and the absence of a well-defined soil profile on the Permian Basin shelves.

The upper clastic member of the Queen Formation typifies many of the desert siliciclastics that prograded across the Per-

mian Basin shelves during lowstands. It was deposited in fluvial sandflat, eolian, and sabkha environments, and their lithofacies can be distinguished on the basis of sediment lithology and sedimentary structures.

Fluvial Sandflat and Delta Facies

Desert fluvial sandflats are broad poorly-vegetated sheetflood flats that are dissected by few small channels. They commonly result from torrential flashfloods that periodically inundate deserts. Fluvial sandflat deposits are usually narrow and elongate in their geometry except where they cross small topographic depressions (such as dry lakes) and widen into fan deltas. The channels in the fluvial sandflats and in the fan-delta plains are typically wide and shallow and tend to overflow in unconfined sheetfloods. These sheetfloods are considered the principal sedimentary agent in fluvial sandflats and fan-delta plains. In cores, these deposits consist of ripple cross-bedded fine to very fine sandstones, wavy to planar laminated sandy siltstones, siltstones and silty mudstones. Channel deposits are rare and are characterized by cross-bedded to planar-laminated medium to very fine sandstone that overlie scour surfaces.

Eolian Sand Sheet Facies

The vast desert plains of the back-reef shelves of the Permian Basin were subjected to long dry periods, which led to the formation of low-relief eolian sand sheets or ergs. Eolian sand sheets vary in size from small isolated patches, which form within fluvial sandflats, to widespread erg plains downwind from the source (Kocurek and Nielson, 1986).

The eolian sand sheet deposits of the Queen Formation are generally composed of planar-laminated fine to very fine sandstones, cross-bedded fine to very fine sandstones, very thin coarse to medium-grained sand deflation lags and no fossils. They are further distinguished by their high degree of sorting and a lack of silt and clay, which are effectively winnowed out by the wind (e.g., Kocurek, 1981; Fryberger et al., 1983).

Clastic-Sabkha Facies

Clastic-dominated sabkhas are poorly vegetated low-lying coastal plains that are adjacent to the fluvial sandflats, eolian sand sheets and exposed lagoonal deposits. Deposition in the sabkhas are dominated by fluvial and eolian sediments; thus, their lithologies are similar to the fluvial sandflats and eolian sand sheets. However, they are distinguished by: (1) the presence of displacive anhydrite nodules, which were precipitated at or near the water table as a result of the capillary evaporation of ground water brines (Kinsman, 1969; Sonnenfeld, 1984); and (2) their haloturbated appearance, which resulted from the distortion of the primary sedimentary structures associated with the expansive growth and dissolution of the evaporites (Smith, 1971).

The clastic-sabkha facies generally form red beds that are composed of deformed, wavy and planar laminated sandstone and siltstone beds. Their lithologies are characterized by laminated beds of sandstone and silty mudstone containing microfolds, fluid escape structures, evaporite nodules and no fossils (Mazzullo et al., 1991). The deformed beds are the most common sedimentary structures in the clastic sabkhas. They form

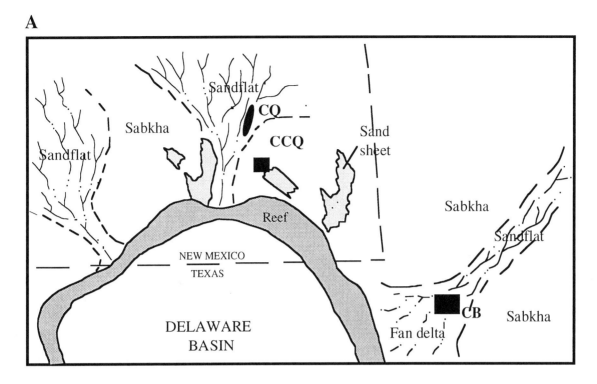

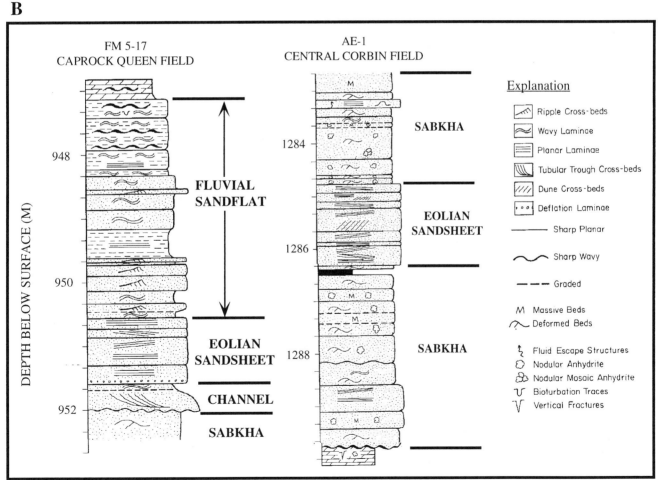

Fig. 3.—(A) Depositional model showing the progradation of the upper clastic member of the Queen Formation during relative sea-level lowstands. (B) Typical stratigraphic profile in the two fields on the Northwest Shelf.

thin to very thick (0.6–5.1 m) bedsets, and sometimes are interbedded with thin nodular-mosaic evaporite beds and laminated playa deposits. The laminated beds, which are less common, are typically thin (8–20 cm) and consist of interlaminated sandstone and silty mudstone with small evaporite nodules. The microfolds or salt-ridge structures within these laminated beds occur as minute chevron-shaped folds and as broad, gentle folds. Evaporite nodules are minor but distinct components of the sabkha facies. They are largely anhydrite, as discrete nodules and nodular-mosaic, and less commonly halites, as hopper crystals and secondary void-fillings.

PALEOSOL FEATURES

The relative sea-level lowstands during Guadalupian time resulted in the development of paleosol features in the subaerially exposed back-reef carbonates and siliciclastics of the Permian Basin. The association of pisolites and calcretes in some of the carbonates (Thomas, 1968; Dunham, 1969; Garber et al., 1989) indicates that they were subjected to soil forming processes during the lowstands. Despite the aridity, soil forming processes were also active during the siliciclastic deposition but were relatively inefficient to form a well defined soil profile. The terrestrial conditions in the Permian Basin deserts were characterized by long dry cycles interrupted by spasmodic rainfalls. The long dry spells not only retarded chemical weathering near the surface but also diminished the translocation of leached materials in the subsurface. This condition prevented the accumulation of argillic soil (Bt) horizon; (Allen, 1985). The formation of salt crusts also disrupted and prevented the development of a soil profile (e.g., Minashina, 1968; Driessen and Schoorl, 1973; Szalbolcs, 1989) and inhibited the growth of vegetation (e.g., Hanna and Stoops, 1976). The lack of vegetation limited the amount of residue available for soil organic matter production and explained the absence of root traces in the siliciclastics.

In summary, the soils of the exposed Guadalupian shelves were very poor to poorly drained and their surfaces were covered by salt crusts. Paleosols features therefore reflect processes that may lead either to the formation, disruption, and/or prevention of soil profile. The study on the upper clastic member of the Queen Formation shows that various micro- and macro-morphological features related to the development and/or destruction of soil profiles can be used to indicate soil-forming processes. These features include: (1) deformed structures, the destruction of primary sedimentary structures by intra-sedimentary growth and collapse of soluble salts; (2) cutans, the mechanical infiltration and deposition of clay on the grain surface; (3) subsoil lamellae, the formation of subsurface lamellae by clay illuviation; and (4) soil horizons, the formation of a soil profile by eluviation and illuviation processes.

Deformed Structures

In sabkha regions where ground water is shallow, salts precipitate within the surficial sediments to produce a blister-like surface referred to as puffy ground (Neal, 1972). Puffy grounds are seasonal features and they are readily destroyed by runoffs and rain (Cochran et al., 1988). The growth and collapse of puffy grounds destroy the primary sedimentary structures near or at the surface, and produce chaotic or deformed laminations, referred to as haloturbation (Fig. 4A). This mixing of the upper

sedimentary layers also disrupts any soil horizons that may have formed, a process synonymous to the proisotropic pedoturbation described by Johnson et al. (1987).

Haloturbation features are common in the argillic sabkha sediments of the upper Queen Formation where they occur in isolation from other pedogenetic features (e.g., soil horizons and root traces). They are readily distinguished from bioturbation features by their association with evaporite nodules and red beds, and the lack of defined shapes commonly associated with trace fossils. Thin sections of a haloturbated interval reveal sand with disrupted, dark reddish brown argillic laminae and few anhydrite nodules or halite hoppers.

Cutans

The infiltration of clays into desert sands is the most effective mechanism in the formation of cutans and in the accumulation of argillic subsurface (Bt) horizons (Walker, 1976; Pye, 1983; Retallack, 1990). Thick argillic horizons (greater than 5 cm) are generally lacking in the upper Queen Formation. Instead, there are detrital reddened cutans, clay coatings of varied thicknesses discontinuously draped on the grain surface. The cutans occur as grain and link cappings (coating on top of one or more grains), as coatings around detrital grains, and as bridging two or more grains (Brewer, 1960; Bullock et al., 1985; Figs. 4B-4E). They appear as clay platelets oriented parallel to the grain surface in SEM analyses. Common in all the terrestrial facies of the upper Queen Formation, the hematitic cutans occur either as an isolated feature or associated with gypsum nodules, subsoil lamellae, and with the rare soil (Bt) horizon.

The presence of cutans alone can not be used as an evidence of pedogenesis (Mack and James, 1992), particularly in humid temperate regions where a soil profile is readily formed. However, soil profiles on deserts are slow to developed, and if formed, they are subjected to water and wind erosion and haloturbation (Minashina, 1968; Radwanski, 1968; Johnson et al., 1987; USDA, 1988; Gordon, 1991). Therefore, cutans may persist as the only primary evidence of pedogenesis in the absence of a defined or preserved soil profile.

Subsoil Lamellae

Subsoil lamellae are defined as lamellae originating by clay illuviation that may be independent from or may be controlled to some degree by inherited stratification (Dijkerman et al., 1967). Although common in the sandy soils of humid temperate regions, subsoil lamellae are present in the fluvial and sabkha facies of the upper Queen Formation. They are distinguished from lamellae of sedimentary origin by the absence of ordered stratification (e.g., cross bedding and ripple marks) and the lack of bedding orientation (Figs. 4F–4G). In some cores, the lamellae may follow a crack that once transported mud-rich water during infiltration (Fig. 4H).

Subsoil lamellae may form through various processes, among them: (1) rhythmic efflorescence, (2) textural stratification and (3) flocculation of free iron oxides (e.g., Folks and Riecken, 1956; Wurman et al., 1959; Dijkerman et al., 1967). The subsoil lamellae in upper Queen Formation formed through a combination of these processes. The process of rhythmic efflorescence operates similarly as haloturbation where the original stratification is mixed incoherently within the profile by the formation

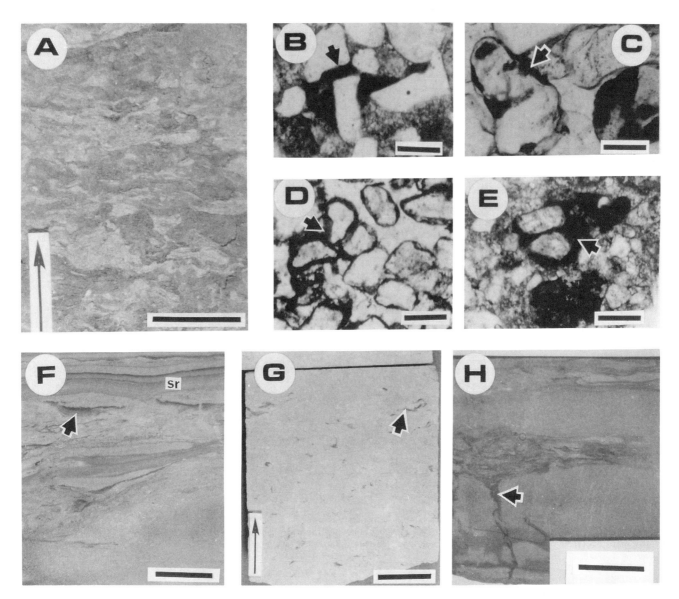

FIG. 4.—Macro- and micro-pedogenic features in the upper clastic member of the Queen Formation: (A) deformed structure in the argillic sabkha facies, (B) clay cutans as link capping on multiple grains, (C) clay cutans as coating on a single grain, (D) clay cutans as coating around detrital grains, (E) clay cutans as bridging of two grains, (F) subsoil lamellae (arrow) controlled to some degree by the inherent stratification (sr), (G) subsoil lamellae (arrow) independent of stratification, and (H) subsoil lamellae (arrow) following a crack or a void. Scale bars in the core pictures (A, F, G, H) equal 2.54 cm (1 inch). Scale bars in the photomicrographs (B-E) equal 50 microns.

and collapse of the puffy ground. Subsoil lamellae may also form through textural stratification process by halting the movement of a percolating muddy water. This occurs when the wetting front of the fine pore sediments comes in contact with the zone of coarser pore sediments (Dijkerman et al., 1967). If insufficient water is present, such as in deserts, the wetting front may be stopped indefinitely. Through the process of evaporation and gravitational settling, the clay from the water accumulates as lamellae. The formation of subsoil lamellae can also be accomplished by rhythmic movement and deposition of free iron oxide in suspension or solution with subsequent flocculation of migrating clay upon contact with precipitated iron (Wurman et al., 1959; Dijkerman et al., 1967).

Soil Horizons

Soil horizons are present in some of the cores at Corbin field in west Texas. The profile, which developed on a coastal dune facies, is generally characterized by 0.46-m-thick bleached white "E" horizon (eluvial horizon) grading downward to a 0.7-m-thick mottled red and white "B" horizon (illuvial horizon) and finally grading down to the unaltered "C" horizon (Fig. 5).

Mineralogical analysis between the E and B horizons show similar amounts of detrital quartz and feldspars (largely K-feldspars). The greatest difference between the two horizons is in the amount of grains with clay cutans, the amount of labile

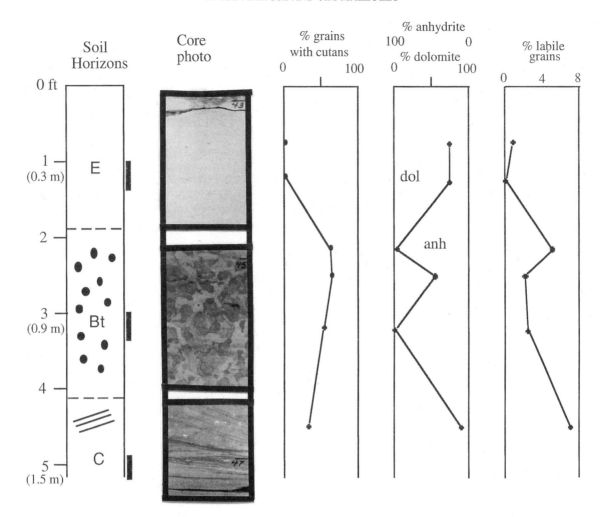

FIG. 5.—Soil horizons developed on the dune facies at Central Corbin Queen field. Plotted beside the profile are the variation in percent grains with clay cutans, relative ratio of dolomite and anhydrite cements and percent labile grains at the three horizons. Also shown is a representative core photo of each horizon. The thick solid lines on the right side of the soil profile show where the core photos were taken.

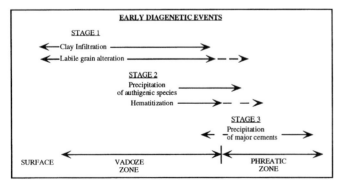

FIG. 6.—Model showing the early diagenetic sequence for the non-reservoir, red bed units of the upper clastic member of the Queen Formation. Stage 1 represent the earliest event and stage 3 the latest event.

opaque grains (largely Fe- and Ti-oxides), and the type of cement. The bleached E horizon lacked grains with clay cutans, cemented predominantly by dolomite, and contained few labile opaque grains. In contrast, the mottled B horizon generally contained grains with clay cutans, cemented predominantly by poikilotopic anhydrite, and contained more labile opaque grains.

The development of soil horizons along the coastal dune facies of the upper Queen Formation is marked by the progressive increase in depth through time of the elluvial E horizon. The E horizon developed by in-situ weathering of labile grains, flushing of early formed hematitic clays, and precipitation of authigenic carbonates. The depth of the E horizon may have fluctuated with the water table where the translocated iron oxide and clays are continuously being remobilized down the profile by continuous flushing of carbonate-rich water. The downward

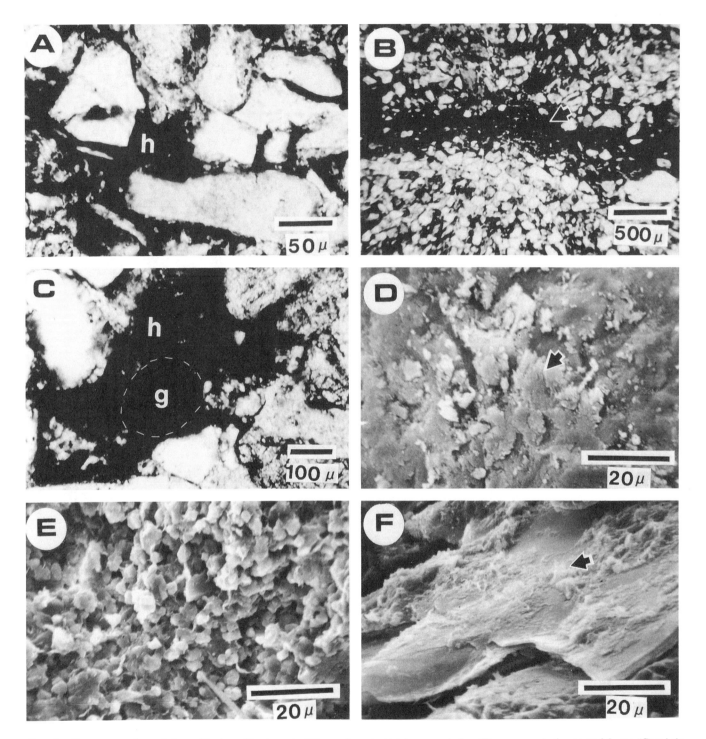

FIG. 7.—Occurrence and morphology of the hematitic pigments: (A) as an irregular blotchy concentration, (B) as a concentration on mud drapes, (C) as halo (h) around decomposing iron-bearing grain (g), (D) as irregularly-shaped platelets plastered on a detrital grain, (E) as clumped globules replacing an altered iron-bearing grain, and (F) as non-crystalline coatings (arrow) on micaceous minerals in a mud drape.

movement of the translocated materials is impeded by the water table subsequently forming the mottled B horizon. The predominance of anhydrite in the lower B horizon indicates that the water table is probably enriched with respect to $CaSO_4$.

EARLY DIAGENETIC FEATURES

The desert sediments of the Permian Basin shelves are arkosic to subarkosic and they are generally cemented by anhydrite or dolomite. They form red beds, although their color may

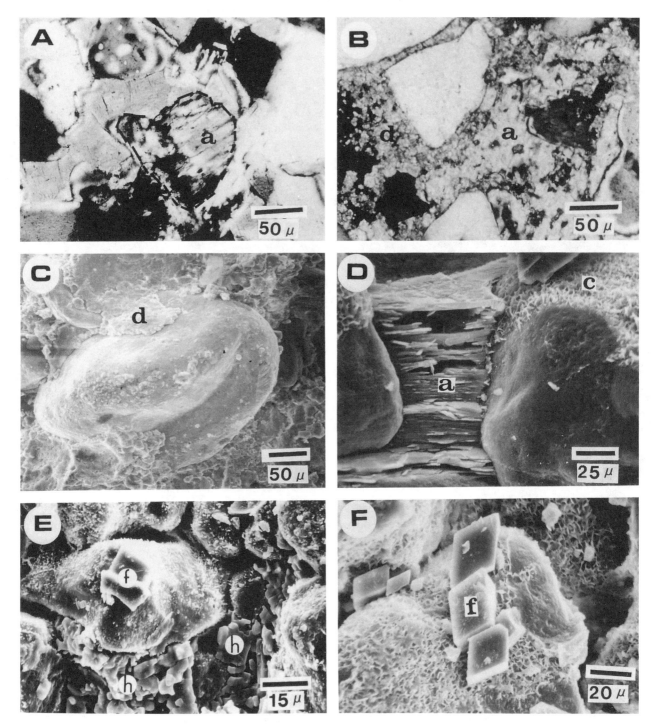

FIG. 8.—Major and minor authigenic minerals: (A) detrital feldspar grains partially replaced by anhydrite (a), (B) pore-filling anhydrite (a) and dolomite (d) cements, (C) SEM picture of an anhedral, pore-filling dolomite (d), (D) SEM picture of a pore-filling anhydrite cement (a) and grain-lining corrensite (c), (E) SEM picture of anhedral halite cement and grain-lining corrensite, and (F) authigenic K-feldspar rhombs (f) and grain-lining corrensite.

alter to pale yellowish brown during the late diagenetic migration of hydrocarbons. Many of the features observed in the red beds were inherited from their early diagenetic history. This sequence of early diagenetic events included the early alteration of labile grains, infiltration of clay, precipitation of hematite and other minor authigenic minerals, and precipitation of major

authigenic cements (anhydrite, dolomite, and halite; Fig. 6). The alteration of labile grains and clay infiltration, which are considered as the earliest diagenetic stage, took place near the surface, where the sediments were still permeable. The dissolution of labile grains through hydrolysis initiated along lines of grain weaknesses, such as the cleavage and boundary of the

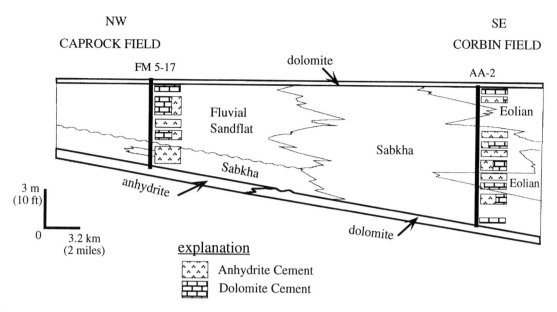

FIG. 9.—Diagramatic cross-section showing the correlation of facies at Caprock and Corbin fields on the Northwest Shelf and the relative amount of anhydrite and dolomite cements in these facies. The locations of the two fields are shown in Figure 1.

grain. The evidence for this early framework-grain alteration includes the pore-filling cements that have completely or partially filled up voids left by the dissolved grain. The mechanical infiltration of clays also took place at the surface and within the vadose zone. This process subsequently lead to the formation of clay cutans, which is considered here as a pedogenic feature.

Some of the materials removed during the first stage alteration were later precipitated as hematite (which gave the sediments their reddish brown color), corrensite (mixed-layer smectite and chlorite) and feldspar overgrowth. Most authigenic species were precipitated at the vadose zone, while the sediments were still permeable and uncemented, but probably continued until after the precipitation of the pore-filling cements. The last phase of the early diagenesis was the precipitation of anhydrite and dolomite, which reduced the primary porosity, hindered movements of fluids between sediments, and terminated the precipitation of additional authigenic species.

Post-Depositional Reddening

The siliciclastics along the Permian Basin shelves are generally reddish brown due to the presence of post-depositional hematite and hematitic clays. These hematitic pigments usually appear in thin section as: (1) irregular blotchy concentrations, (2) fine dissemination in the mud drapes, (3) clay cutans, (4) red halos in decomposing iron-oxide grains and (5) dark matrix encasing detrital grains (Figs. 7A–7C). Hematitic pigments in the form of irregular blotchy concentration and as fine dissemination in mud drapes are common in the silty and muddy facies (e.g., argillaceous sabkhas and fine grained fluvial sandflats). In contrast, the clay cutans, halos and dark matrix are associated with the sandy facies. The hematitic pigments appear as pseudohexagonal to irregularly-shaped platelets, clumped globules, and as non-crystalline coatings on micaceous minerals (Figs. 7D–7F) when examined by SEM analysis.

The presence of these pigments as red halos around decomposing iron-rich grains and surrounding cements indicates an in-situ reddening of the siliciclastics. The iron necessary to form hematite pigments was apparently derived largely from the alteration of ferriferous minerals in the vadose and phreatic zones shortly after deposition, because it could not have survived the rigors of transport. The iron released during alteration was then subsequently oxidized under the desert alkaline conditions (Walker, 1976).

Precipitation of Major Cements

Much of the siliciclastics on the Permian Basin shelves were cemented by dolomite and/or anhydrite, and to a lesser extent halite (e.g., Handford, 1981; Garber et al., 1989). The anhydrite typically occurred as pore-filling poikilotopic and blocky cement and as a grain-filler, whereas the dolomite occurred as fine to medium grained, subhedral to anhedral pore-filling cement and grain replacer (Figs. 8A–8D). The less common halite typically occurred as isolated patches of partially dissolved subhedral to euhedral pore-fillers and as individual cubes (Fig. 8E).

The mineralogy of the cements is primarily a function of the salinity of the ground water which is, in turn, controlled by the progressive evaporation of the interstitial water derived from the lagoon and the influence of meteoric waters (Bush, 1976; Patterson and Kinsman, 1977). The composition of the lagoon and its distance landward also control the type of cement that will precipitate. For example, the siliciclastics directly below or above a lagoonal facies contain cements with the same composition (Fig. 9). With increasing distance from the lagoonal facies contacts, the preference for either dolomite or anhydrite cement disappear. With the exception of the sabkha facies, which were cemented largely by anhydrite, the fluvial sandflat and eolian facies show no preference for either dolomite or anhydrite.

Although dolomite and anhydrite could precipitate during deep burial diagenesis, (e.g., Glennie et al., 1978), most of the cements in this study were early diagenetic features. The deep burial features of the same units are associated with the dissolution of these cements prior to the migration of hydrocarbons (see Malicse and Mazzullo, 1990; Mazzullo et al., 1991).

Precipitation of Minor Authigenic Minerals

Minor authigenic minerals are also present in the siliciclastics. Although occurring in only trace amounts, their presence is important because it explains the fate of the other ions removed during the early dissolution of some labile grains. These minor authigenic species include grain-lining corrensites and feldspar overgrowths. In SEM analysis, the authigenic corrensites formed boxwork and cornflake structures (Figs. 8D–8F). They are chemically composed of silicon, aluminum, iron and magnesium using Energy Dispersive Spectral analysis (EDS). In contrast, feldspar overgrowths occurred as discrete rhombs oriented parallel to the grain surface (Fig. 8F), and they are chemically composed of silicon, potassium and aluminum. The presence of grain-lining corrensites and feldspar overgrowths in the tightly cemented sandstones indicate that they were precipitated before the major cements, while the sediments were still permeable.

SUMMARY

The upper clastic member of the Queen Formation contains subtle pedogenetic features that are common to saline soils of arid regions. The recognition of these features are important because they represent subaerial exposures of eustatically controlled Permian Basin shelves. The pedogenetic features include: (1) deformed structures, distorted laminations resulting from haloturbation; (2) cutans, clay coatings on detrital grains resulting from clay illuviation; (3) subsoil lamellae, lamellae formed at the subsurface by prolonged clay illuviation; and (4) rare soil horizons. The upper clastic member was also subjected to other early diagenetic events. These diagenetic events are divided into three stages: stage 1, the earliest of the events, is characterized by early labile grain dissolution and translocation of secondary materials; stage 2 is characterized by the hematitization and precipitation of previously dissolved species; and finally, stage 3 is characterized by the precipitation of the pore-filling anhydrite and dolomite cements.

REFERENCES

ALLEN, B. L., 1985, Micromorphology of aridisols, *in* Soil Micromorphology and Soil Classification: Madison, Soil Science Society of America, p. 197–216.

BLODGETT, R. H., 1988, Calcareous paleosols in the Triassic Dolores Formation, southwestern Colorado: Washington, D. C., Geological Society of America Special Paper 216, p. 103–121.

BREWER, R., 1960, Cutans: their definition, recognition, and interpretation: Soil Science, v. 11, p. 280–292.

BULLOCK, P., FEDOROFF, N., JONGERIUS, A., STOOPS, G., TURSINA, T., AND BABEL, U., 1985, Handbook for Soil Thin Section Description: Wolverhampton, Waine Research Publications, 152 p.

BUSH, A., 1976, Some aspects of the diagenesis history of the sabkha in Abu Dhabi, Persian Gulf, *in* Purser., B. H., ed., The Persian Gulf-Holocene Carbonate Sedimentation and Diagenesis in a Shallow Epicontinental Sea: New York, Springer-Verlag, p. 395–407.

COCHRAN, G. F., MIHEVC, T. M., TYLER, S. W., AND LOPES, T. J., 1988, Study of salt crust formation on Owens (dry) Lake, California: Los Angeles Desert

Research Institute, Water Resources Center, Los Angeles Department of Water and Power, 103 p.

DIJKERMAN, J. C., CLINE, M. G., AND OLSON, G. W., 1967, Properties and genesis of textural subsoil lamellae: Soil Science, v. 104, p. 7–16.

DRIESSEN, P. M. AND SCHOORL, R., 1973, Mineralogy and morphology of salt efflorescence on saline soils in the Great Konya Basin, Turkey: Soil Science, v. 24, p. 436–442.

DUNHAM, R. J., 1969, Vadose pisolite in the Capitan Reef (Permian), New Mexico and Texas, *in* Friedman, G. M., ed., Depositional Environments in Carbonate Rocks: Tulsa, Society of Economic Paleontologists and Mineralogists Special Publication 14, p. 812–191.

FOLKS, H. C. AND RIECKEN, F. F., 1956, Physical and chemical properties of some Iowa soil profiles with clay-iron bands: Soil Science Society of America, v. 20, p. 575–580.

FRYBERGER S. G., AL-SARI, A. M., AND CLISHAM, T. J., 1983, Eolian dune, interdune, sandsheet, and siliciclastics sabkha sediments of an offshore prograding sand sea, Dharan area, Saudi Arabia: American Association of Petroleum Geologists Bulletin, v. 67, p. 280–312.

GARBER, R. A., GROVER, G. A., AND HARRIS, P. M., 1989, Geology of the Capitan shelf margin-subsurface data from the northern Delaware Basin, *in* Harris, P. M., and Grover, G. A., eds., Subsurface and Outcrop Examination of the Capitan Shelf Margin, Northern Delaware Basin: Tulsa, Society of Economic Paleontologists and Mineralogists Core Workshop 13, p. 3–269.

GLENNIE, K. W., MUDD, G. C., AND NAGTEGAAL, P. J. C., 1978, Depositional environments and diagenesis of Permian Rotliegendes sandstones in Leman Bank and Sole Pit areas of the UK southern North Sea: Journal of the Geological Society of London, v. 135, p. 25–34.

GORDON, N. S., AND GUTTERMAN, Y., 1991, Soil disturbance by a violent flood in Wadi Zin in the Negev Desert highlands of Israel: Arid Soil Research and Rehabilitation, v. 5, p. 251–260.

HANDFORD, C. R., 1981, Coastal sabkha and salt pan deposition of the lower Clear Fork Formation (Permian), Texas: Journal of Sedimentary of Petrology, v. 51, p. 761–778.

HANNA, F. S., AND STOOPS, G. J., 1976, Contribution to the micromorphology of some saline soils of the north Nile Delta in Egypt: Pedologie, v. 26, p. 55–73.

JACKA, A., THOMAS, C. M., BECK, R. H., WILLIAMS, K. W., AND HARRISON, S. C., 1972, Guadalupian depositional cycles of the Delaware Basin and Northwest Shelf, *in* Elam, J. and Chuber, S., eds., Cyclic Sedimentation in the Permian Basin (second edition): Tulsa, Society of Economic Paleontologists and Mineralogists, p. 151–173.

JOHNSON, D. L., WATSON-STEGNER, D., JOHNSON, D. N., AND SCHAETZL, R. J., 1987, Proisotropic and proanisotropic processes of pedoturbation: Soil Science, v. 143, p. 278–292.

KINSMAN, D. J. J., 1969, Modes of formation, sedimentary associations, and diagnostic features of shallow-water supratidal evaporites: American Association of Petroleum Geologists Bulletin, v. 53, p. 830–840.

KOCUREK, G., 1981, Significance of interdune deposits and bounding surfaces in eolian sand dunes: Sedimentology, v. 33, p. 795–816.

KOCUREK, G. AND NIELSON, J., 1986, Conditions favorable for the formation of warm-climate eolian sand sheets: Sedimentology, v. 33, p. 795–816.

MACK, G. H., AND JAMES, W. C., 1992, Paleosols for Sedimentologists: Boulder, Geological Society of America Short Course Notes, 127 p.

MALICSE, A. AND MAZZULLO, J., 1990, Reservoir properties of the desert Shattuck Member, Caprock Field, New Mexico, *in* Barwis, J. H., McPherson, J. G., and Studlick, J. R. J., eds., Sandstone Petroleum Reservoirs: New York, Springer-Verlag, p. 133- 152.

MAZZULLO, J., MALICSE, A., AND SIEGEL, J., 1991, Facies and depositional environments of the Shattuck Sandstone on the Northwest Shelf of the Permian Basin: Journal of Sedimentary Petrology, v. 61, p. 940–958.

MAZZULLO, J., MALICSE, A., NEWSOM, D., HARPER, J., MCKONE, C., AND PRICE, B., 1992, Facies, depositional environments, and reservoir properties of the Upper Queen Formation, Concho Bluff and Concho Bluff North Fields, Texas, *in* Mruk, D. H. and Curran, B. C., eds., Permian Basin Exploration and Production Strategies: Applications of Sequence Stratigraphic and Reservoir Characterization Concepts: Midland, West Texas Geological Society Symposium, Publication 92–91, p. 67–78.

MINASHINA, N. G., 1968, Soil formation and salt migration in the Murgab River delta: 9th International Congress Soil Science Transaction, v.1, p. 425–435.

MINTZ, L. W., 1981, Historical Geology (3rd ed.): Columbus, Charles E. Merrill Publishing Company, p. 374–400.

NEAL, J. T., 1972, Playa surface features as indicators of environment, *in* Reeves, C.C., ed., Playa Lake Symposium Proceeding: ICASALS Publication No. 4, p. 107–132.

ORIEL, S. S., MYERS, D. A., AND CROSBY, E. J., 1967, Paleotectonic investigations of the Permian System in the United States: Chapter C. West Texas Permian Basin Region: Washington, D.C., United States Geological Survey Professional Paper 515-C, p. 21–60.

PATTERSON, R. J. AND KINSMAN, D. J., 1977, Marine and continental ground water sources in a Persian Gulf coastal sabkha, *in* Frost, S. H., Weiss, M. P., and Sanders, J. B., eds., Reef and Related Carbonates—Ecology and Sedimentology: Tulsa, American Association of Petroleum Geologists Studies in Geology No. 4, p. 381–397.

PYE, K., 1983, Post-depositional reddening of late Quaternary coastal dune sands, north-eastern Australia, *in* Wilson, R. C. L., ed., Residual Deposits: Surface Related Weathering Processes and Materials: London, Blackwell Scientific Publications, p. 117–129.

RADWANSKI, S. A., 1968, Field observations of some physical properties in alluvial soils of arid and semi-arid regions: Soil Science, v. 106, p. 314–316.

RETALLACK, G. J., 1990, Soils of the past: an introduction to paleopedology: Boston, Unwin Hyman, 520 p.

ROSS, C. A., AND ROSS, J. R. P., 1988, Late Paleozoic transgressive-regressive deposition, *in* Wilgus, C. K., Ross, C. A., Posamentier, H., Van Wagoner, J., and Kendall, C. G. St. C., eds., Sea-level Changes: An Integrated Approach: Tulsa, Society of Economic Paleontologists and Mineralogists Publication 42, p. 217–226.

SMITH, D. B., 1971, Possible displacive halite in the Permian Upper Evaporite Group of northeast Yorkshire: Sedimentology, v. 17, p. 221–232.

SONNENFELD, P., 1984, Brines and Evaporites: Orlando, Academic Press Inc., 613 pp.

SCOTESE, C. R., AND MCKERROW, W. S., 1990, Revised world maps and introduction, *in* McKerrow, W. S. and Scotese, C. R., eds., Paleozoic Paleogeography and Biogeography: London, The Geological Society, p. 1–24.

SZABOLCS, I., 1989, Salt-affected Soils: Boca Raton, CRC Press Inc., 274 p.

THOMAS, C. M., 1968, Vadose pisolites in the Guadalupe and Apache mountains, West Texas, *in* Silver, B. A., ed., Guadalupian Facies, Apache Mountain Area, West Texas: Midland, Permian Basin Section Society of Economic Paleontologists and Mineralogists Guide Book, Publication 68–11, p. 32–35.

UNITED STATES DEPARTMENT OF AGRICULTURE SOIL CONSERVATIVE SERVICE, 1988, Soil Taxonomy: A Basic System of Soil Classification for Making and Interpreting Soil Surveys: Malabar, Krieger Publishing Company, 754 p.

Walker, T. R., 1976, Diagenetic origin of continental red beds, *in* Falke, H. and Dordrecht, D., eds., The Continental Permian in Central, West, and South Europe: London, Reidel Publishing Company, p. 240–282.

WARD, R. F., KENDALL, C. G. ST. C., AND HARRIS, P. M., 1986, Upper Permian (Guadalupian) facies and their association with hydrocarbons-Permian Basin, West Texas and New Mexico: American Association of Petroleum Geologists Bulletin, v. 70, p. 239–262.

WURMAN, E., WHITESIDE, E. P., AND MORTLAND, M. M., 1959, Properties and genesis of finer textured subsoil bands in some sandy Michigan soils: Soil Science Society of America, v. 23, p. 135–580.

ZIEGLER, A. M., SCOTESE, C. R., MCKERROW, W. S., JOHNSON, M. E., AND BAMBACH, R. K., 1979, Paleozoic paleogeography: Annual Review of Earth Planetary Science, p. 473–502.

ORGANIC AND AUTHIGENIC MINERAL GEOCHEMISTRY OF THE PERMIAN DELAWARE MOUNTAIN GROUP, WEST TEXAS: IMPLICATIONS FOR THE CHEMICAL EVOLUTION OF PORE FLUIDS

PHILLIP D. HAYS

United States Geological Survey, Water Resources Division, 401 Hardin, Little Rock, Arkansas 72211

SUZETTE D. WALLING AND THOMAS T. TIEH

Department of Geology and Geophysics, Texas A&M University, College Station, Texas 77843–3115

ABSTRACT: Aqueous species derived from first-order degradative reaction of organic material can modify Eh and pH conditions in the late burial environment, regulate solubility of minerals and dissolved constituents in the rock/water system and release organically bound metals into pore water as thermal stress associated with burial progresses. This study evaluates the role of organic matter alteration in controlling late burial diagenetic processes in the Delaware Mountain Group, a sequence of rocks that presents a unique opportunity for the study of coupled organic-inorganic diagenesis. The Delaware Mountain Group lies within the Permian Basin of west Texas and southern New Mexico and includes, in descending order, the Bell Canyon, Cherry Canyon and Brushy Canyon Formations. Characterization of mineral and organic geochemistry and study of natural geochemical tracers in the Delaware Mountain Group provide evidence of the organic drive for diagenetic processes in these rocks. Delaware Mountain Group siltstone organic matter is of a sufficient abundance, of suitable type and of a sufficiently advanced state of thermal maturity to have had the potential to generate organic acids and other thermogenic products capable of controlling pore-fluid chemistry and impacting diagenetic processes. The organic matter has yielded fluids associated with thermal stress; geochemical fossil and stable isotope correlation of siltstone organic matter with oil in Delaware Mountain Group sandstone reservoirs shows that the siltstones were the source of much of the oil. Mineralogic evidence of the role that organically-derived compounds played in late diagenetic processes includes the presence of abundant authigenic titanium oxides in these sandstones. The formation of authigenic titanium oxides indicates that titanium mobility was elevated by formation of organometallic titanium complexes—the only natural mechanism for enhancing titanium solubility. Finally, natural isotopic (carbon isotopes) and inorganic tracers (Mn, Zn, Ni, Co, V and Cr) establish a source/sink relation between the organic matter in the siltstones and authigenic minerals in the sandstones. The relation is manifest in the isotopic and minor element content of late stage authigenic products that formed in pore fluids whose chemistry was controlled by organic matter degradation. The various lines of evidence indicate that fluids carrying the products of organic matter degradation moved from the organic-rich siltstones into the sandstones where authigenic products formed and preserved a record of ambient pore-water chemistry.

INTRODUCTION

Sandstones of the Guadalupian Delaware Mountain Group in the Permian Basin of West Texas are important hydrocarbon reservoirs situated in a unique geologic and geochemical setting. A cursory inspection of these rocks bears out their uniqueness, revealing a complex detrital and authigenic mineralogy wrought by changing and potentially baffling conditions of pore-water chemistry. The rocks pose a two-fold question regarding (1) the mineralogic and diagenetic changes experienced by the Delaware Mountain Group sandstones during burial and (2) the impetus for these diagenetic occurrences, that is, the source and nature of fluids responsible for driving diagenetic processes and creating secondary porosity—the predominant porosity type—in these rocks.

The fluids that were responsible for driving diagenetic processes appear to have been derived from alteration of organic matter within Delaware Mountain Group siltstones. Originating in the siltstones, fluids bearing organic acids and other products of thermal degradation of organic matter have controlled late-burial dissolution and authigenesis in the sandstone. A diagram outlining this conceptual model is presented in Figure 1. A three-tined argument is used to demonstrate the role of organic matter diagenesis in controlling pore-fluid chemistry and affecting the overall diagenetic outcome in the sandstones. The arguments applied are based on data from (1) petrologic and inorganic geochemical study, (2) organic geochemical study and (3) study of natural tracers present in the system. The synthesis of the three arguments provides a conceptual model through which the diagenetic evolution of these sandstones can be best understood. This rock sequence presents a unique and valuable opportunity to study the role of organic diagenesis and porosity development in sandstones for several reasons: the abundance and organic-rich nature of the siltstones in the Del-

aware Basin; the lack of shale; the absence of meteoric flow in the basin deep; and the thick, uniform character of Delaware Mountain Group rocks. Whereas the potential of organic matter degradation to influence pore-water chemistry and contribute to porosity development has been generally accepted and well demonstrated experimentally, data on actual field examples of organic/rock interaction during diagenesis are limited; results of this study give a valuable assessment of the degree of diagenetic control possible in an organic-rich system.

Siliciclastic Diagenesis

In a hypothetical sediment body under static fluid conditions, chemical diagenetic alteration would be minimal. Diagenetic processes would be dependent primarily on temperature and pressure, serving only to maintain equilibrium between rock and water during burial. Such a system would have a rather simple diagenetic history and display a limited and predictable suite of authigenic minerals. Most sandstones exhibit diverse authigenic assemblages indicative of complex diagenetic histories that result from interplay of fluids of varying physiochemical make-up within the rock. These fluids, excluding those associated with magmatic processes, may be derived from several general sources including: meteoric recharge of a basin (Galloway, 1984; Longstaffe, 1984; James and Choquette, 1984), compactional dewatering of a basin (Galloway, 1984), brine migration (Hanor, 1987; Denham, 1992), mineral transformation (McMahon, 1989; Boles and Franks, 1979; Hower et al., 1976; Burst, 1969), or degradation of organic matter (Curtis and Coleman, 1986; Franks and Forester, 1984; Surdam et al., 1984; Crossey, 1985). These fluids can operate singly or in combination during burial to control pore-fluid chemistry within sandstones. Interactions between these fluids and sandstones may result in substantial redistribution of material and

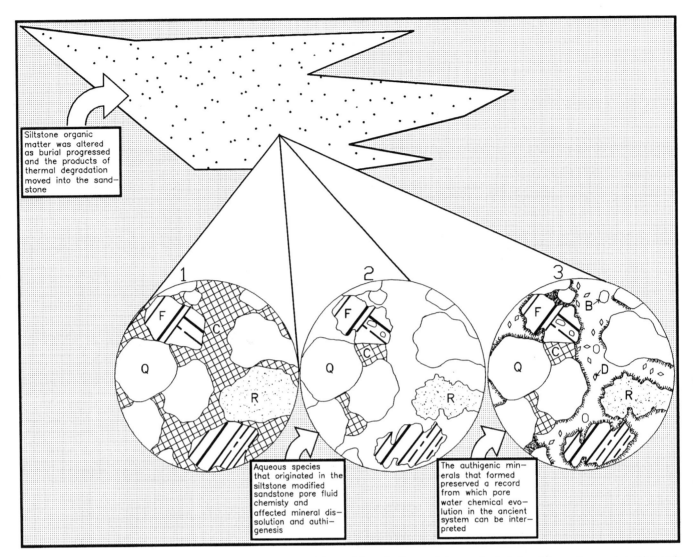

Siltstone organic
matter was altered
as burial progressed
and the products of
thermal degradation
moved into the sand-
stone

Aqueous species
that originated in the
siltstone modified
sandstone pore fluid
chemisty and
affected mineral dis-
solution and authi-
genesis

The authigenic min-
erals that formed
preserved a record
from which pore
water chemical evo-
lution in the ancient
system can be inter-
preted

FIG. 1.—Schematic diagram depicting the conceptual model for the Delaware Mountain Group and the perceived relations between the organic and inorganic diagenetic systems. Q = quartz grain, F = feldspar grain, R = rock fragment, C = calcite cement, Cl = clay minerals, D = dolomite, B = titanium oxide minerals and P = porosity.

affect physical and mineralogic properties. Distinctive suites of authigenic minerals that have resulted from rock/water interaction offer evidence from which diagenetic history and pore-water chemical evolution may be inferred.

Organic Matter Diagenesis

Sedimentary organic matter is ultimately derived from living organisms. Volumetrically, the most important producers of organic matter, and therefore the primary sources of organic matter preserved in sediments, are phytoplankton, zooplankton, the higher plants and bacteria. Remains and metabolic products of these organisms supply the starting material for the organic fraction of sediment; this material undergoes continual transformation in response to the changing biological and physiochemical conditions during sediment burial. Products of this transformation can have considerable impact on pore-water chemistry and diagenesis. Sedimentary organic matter under-

going diagenetic alteration by either biological or thermocatalytic mechanisms can affect pore-fluid chemistry by several means including: (1) buffering pH by serving as a source or—less commonly—a sink for hydrogen ions, (2) acting as a reducing agent and controlling Eh conditions, (3) altering the mobility of various inorganic species (such as aluminum or titanium) through formation of organometallic complexes and (4) controlling the carbon dioxide/carbonic acid system by serving as a source of alkalinity (HCO_{3^-}) or carbon dioxide (CO_2). Hence, organic diagenesis can affect important sandstone diagenetic processes such as precipitation or dissolution of carbonates, feldspars, and other important rock constituents, as well as formation and distribution of porosity.

The Nature and Occurrence of Organically-Derived Acids.—

The most commonly cited sources for fluids involved in creation of secondary porosity during diagenesis are: (1) meteoric water, (2) slightly acidic water released from smectite/illite conversion (McMahon, 1989) and (3) carboxylic- and carbonic acid-bearing fluids derived from thermal maturation of organic matter (Surdam et al., 1984; Crossey, 1985). In the Delaware Mountain Group, the first two sources can be discounted as important mechanisms in the generation of porosity. The presence of abundant, highly soluble halite and anhydrite cements and the lack of the distinctive light oxygen isotopic signature in formation water or authigenic phases precludes meteoric water influx as a valid mechanism. Clay transformation can be omitted as an important player in the diagenetic scheme because of the paucity of clay minerals or shale within the Delaware Basin. This mechanism requires high shale/sand ratios for large-scale generation of water and removal of cements—a condition that is not satisfied in the basin. The hypothesis of this study is that porosity was created through the remaining mechanism—dissolution of cements and detrital material by carboxylic and carbonic acids derived from thermal degradation of Delaware Mountain Group organic matter.

Thermal maturation of organic matter is a natural consequence of the burial of sediments. Maturation reactions result in the production of significant amounts of organic acids and CO_2 prior to and during oil generation. Concentrations of total organic acids measured in formation waters are as large as 10,000 ppm (parts per million) and average about 1,500 ppm. Monofunctional carboxylic acid anions are most common. Acetate, propionate, butyrate and valerate (in order of abundance) constituted 90% of dissolved organic species in studies by Willey et al. (1975), Carothers and Kharaka (1978) and Fisher (1987). Difunctional carboxylic acids, such as oxalic, maleic and malonic acids, have been identified in formation waters in lower concentrations as large as 3,000 ppm (Surdam et al., 1984; Kharaka, et al., 1985; Kharaka, et al., 1986). Experimental laboratory studies corroborate field determinations, showing that thermogenic degradation of kerogen—the refractory remnant of organic matter that survives early burial and is the predominant organic material remaining during late burial—evolves substantial quantities of carboxylic acids and CO_2 (Surdam et al., 1984; Crossey, 1985; Lundegard and Senftle, 1987). Formate, acetate, propionate and oxalate are the predominant organic acids generated through hydrous pyrolysis of kerogen as reported in these studies. At higher temperatures of 250 to 350°C, CO_2 appears to be the dominant end product of kerogen degradation (Lundegard and Senftle, 1987; Crossey, 1991). Many experimental studies have shown that these carboxylic and carbonic acids can create secondary porosity in sandstones by dissolution of carbonate cements and aluminosilicate minerals (Surdam et al., 1984; Crossey, 1985; Bevan and Savage, 1989). Examination of Delaware Mountain Group samples indicates that most of the porosity is secondary, brought about by removal of calcite. In a comparison of organic acid versus carbonic acid efficacy, Meshri (1986) examined stoichiometric and thermodynamic reactivities of these acids. Results of the study indicated that: (1) organic acids have greater stoichiometric reactivity; dissociation constant data show that proton donor capacities of organic acids are 6 to 350 times greater than for carbonic acid (Table 1); (2) organic acids have greater stoichiometric efficiency for dissolving carbonates. The organic acid anion is non-carbonate and does not contribute to dissolved carbonate upon dissociation, and some organic-acid calcium

TABLE 1.—DISSOCIATION CONSTANTS, CALCIUM COMPLEX EQUILIBRIUM CONSTANTS AND MAXIMUM REPORTED CONCENTRATIONS FOR CARBONIC ACID AND SOME COMMON ORGANIC ACIDS

Acid	pK[1]	pK[2]	Calcium Salt − log K (25°C)[1]	Calcium Salt − log K (100°C)[2]	Max. Reported Conc. (mg/kg)[3]
Carbonic	6.35	10.25	8.48*	9.27*	—
Formic	3.75		1.43	1.69	62.6
Acetic	4.76		1.24	1.46	10,000
Propionic	4.88		0.68	0.68	4,400
Oxalic	1.25	4.27	3.00	3.64	494
Malonic	2.85	5.71	2.50	3.39	2540

Measured log K values (1), calculated log K values (2) and maximum reported organic acid concentration values (3) compiled by Harrison and Thyne (1992). Calcite log K values from Morse and Mackenzie (1990).

salt solubilities are orders of magnitude greater than calcium carbonate solubility (Table 1). The difference in solubility is emphasized at the higher temperatures common to the late diagenetic environment; and (3) Gibbs energy of dissolution for calcite and feldspars shows better thermodynamic drive for organic acids than for carbonic acid; feldspar dissolution in the presence of acetic acid required 79 to 21 kcal/mole lower free energy than the free energies required in the presence of carbonic acid, and dissolution of calcite in acetic acid required 17 kcal/mole lower free energy than in carbonic acid.

Several studies have promoted the role of organic matter diagenesis in the creation of secondary porosity in sandstones, but these studies are often debated by proponents of the shale diagenesis model, which attributes porosity generation to the expulsion of clay-derived fluids during shale mineral transformation. The debate continues because the fluids involved in diagenesis typically originate in shales that are rich in organic matter and clay minerals, and resolving the contribution of each mechanism to the final, observed diagenetic outcome is difficult or impossible. Moncure et al. (1984) studied the Frio Formation of Oligocene age in the Texas Gulf Coast and concluded that secondary porosity development was driven by expulsion of organic maturation products from shale to sandstone. These conclusions were based on porosity distribution and elemental mass-balance considerations. Franks and Forester (1984) studied the Wilcox Group of Eocene age in the Gulf Coast and concluded that isotopically-derived temperature data indicated that the diagenetic period of calcite dissolution coincided with the time and temperature of thermal maturation of organic matter, and that CO_2 generated as a by-product of thermal alteration showed a strong correlation with zones of high secondary porosity. In contrast to the conclusions of these studies, Lundegard et al. (1984) discounted the importance of organically-derived acids in the Frio Formation. Their mass-balance calculations indicated that decarboxylation of organic matter could account for only 10 to 20% of observed secondary porosity. McMahon (1989) agreed with Lundegard et al. (1984) in his study of the Frio Formation. McMahon showed the potential of inorganic shale diagenesis to create porosity in the Frio Formation. Many other examples are available in the Gulf Coast area and elsewhere of studies that explain development of secondary porosity through coupled organic/sandstone diagenesis. Regardless of locale or investigator, these studies all seem to be unable to resolve the actual contribution of organic matter diagenesis to the development of secondary porosity with a high degree of certainty. Thus, organically-derived acids are theoretically and experimentally proven capable of creating secondary porosity in sandstones, and these acids are present in significant quantities in the subsurface, but detailed documentation of actual coupled organic/inorganic diagenetic systems is lacking.

GEOLOGIC BACKGROUND

Sandstones of the Guadalupian Delaware Mountain Group in the Permian Basin of West Texas are very fine-grained (mean grain-size <0.10 mm), subarkosic to arkosic sandstones interbedded with thick, organic-rich siltstones and relatively rare, thin limestones. Shales and detrital clays are notably rare. The Delaware Mountain Group includes, in descending order, the Bell Canyon, Cherry Canyon and Brushy Canyon Formations. Maximum thickness of the Delaware Mountain Group in the central basin is about 1,300 m, and the present burial depth is about 1,460 m (top of Bell Canyon).

The Tobosa Basin was the early Paleozoic precursor of the Permian Basin (Galley, 1958; Oriel et al., 1967; Hills, 1972). Slow deposition of carbonate and siliciclastic sediments took place in this broad, shallow basin from early to middle Paleozoic time. During Mississippian time, the Tobosa Basin was bisected by uplift of the Central Basin Platform. This north-south trending fault block created separate eastern and western basins—the Midland Basin to the east and Delaware Basin to the west—and established the structural configuration of the Delaware Basin, which persisted throughout Delaware Mountain Group deposition. The Delaware Basin is bound by the Central Basin platform to the east, the Ouachita-Marathon structural belt to the south, the Diablo platform to the west and the Pedernal uplift and Northwest slope to the north (Fig. 2). Throughout the basin's history, broad shelf areas surrounded the basin. The Delaware Basin was connected at the southern end to the Midland Basin by the Sheffield Channel and to the Marfa Basin by the Hovey Channel (Oriel et al., 1967; Hills, 1972).

During Delaware Mountain Group deposition, Guadalupian time, three distinct depositional environments existed in the Delaware Basin: (1) the shelf that contained the supratidal, intertidal, and lagoonal subenvironments; (2) the shelf margin that contained the reef and forereef slope subenvironments and (3) the basin environment (King, 1942, 1948; Silver and Todd, 1969; Meissner, 1972). These depositional environments have corresponding representative facies which are time equivalents of the basinal Delaware Mountain Group. Formations of the Artesia Group—the Tansill, Yates, Seven Rivers, Queen and Grayburg in descending order—are shallow subtidal to supratidal platform dolomite, evaporite and red sandstone equivalents of the Delaware Mountain Group. The shelf margin is represented by the Capitan "reef" and the Goat Seep Dolomite. The Delaware Mountain Group represents the basinal facies (Fig. 3) and predominantly is composed of siltstones and very fine-grained sandstones deposited in anoxic deep-water conditions. Water depths in the basin have been estimated at 300 to 600 m during Delaware Mountain Group deposition (King, 1934, 1942, 1948; Newell et al., 1953; Meissner, 1972).

MATERIALS AND METHODS

Cored intervals from 13 wells—eight from the Waha field area in Reeves and Pecos Counties, Texas and four from the Big Eddy Unit in Eddy County, New Mexico—were sampled and analyzed (Table 2). Oil and formation water samples were obtained from five of the wells for analysis. The methodology applied in the study incorporated use of the following techniques: (1) petrographic study to characterize mineralogy and relations of rock constituents, (2) electron probe microanalysis to determine chemical composition of selected grains and cements, (3) scanning electron microscopy coupled with energy dispersive spectroscopy to acquire morphological details and gross elemental composition of authigenic minerals, (4) X-ray diffraction to determine mineralogy of clay-mineral separates, (5) neutron activation analysis (NAA) to determine elemental composition of bulk samples, mineral separates and organic separates, (6) stable isotope mass spectrometry to determine carbon, oxygen, deuterium and sulfur isotopic compositions of authigenic minerals and organic matter and (7) total organic

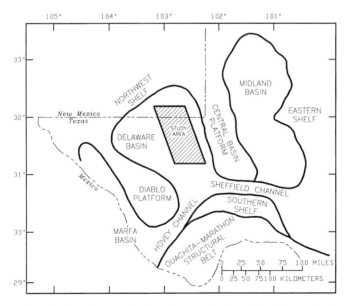

FIG. 2.—Location map showing the general setting of the Delaware basin and the study area.

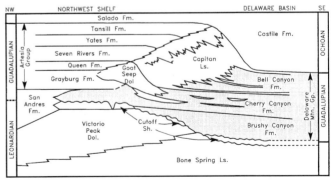

FIG. 3.—Diagrammatic cross section from the northwest shelf (NW) of the Delaware basin to the basin deep (SE) illustrating Permian stratigraphic relationships (modified from Hayes, 1964). Cross section not to scale.

TABLE 2.—INFORMATION ON CORED INTERVALS ANALYZED

Location	Well	Depth interval (ft)	Formation
Reeves Co., Tex.	Gulf, Barber A5	4,800–5,087	Bell Canyon
Reeves Co., Tex.	Gulf, Cleveland 1	4,797–5,010	Bell Canyon
Reeves Co., Tex.	Gulf, Cleveland 2	8,042–8,090	Brushy Canyon
Reeves Co., Tex.	Gulf, Jack Frost 1	4,948–5,493	Bell Canyon
Reeves Co., Tex.	Gulf, Hoefs 1	4,910–5,118	Bell Canyon
Reeves Co., Tex.	Gulf, Strain 2*	3,835–5,085	Bell Canyon
Reeves Co., Tex.	Gulf, Trees 2*	4,927–4,961	Bell Canyon
		5,845–6,460	Cherry Canyon
Pecos Co., Tex.	Sun, Hodge 2*	4,840–5,144	Bell Canyon
Eddy Co., N.M.	Bass, Poker Lk. 43	4,388–4,477	Bell Canyon
Eddy Co., N.M.	Bass, Big Eddy 45Y	3,659–4,986	Bell Canyon
Eddy Co., N.M.	Bass, Big Eddy 46	3,015–3,587	Bell Canyon
		6,100–6,137	Brushy Canyon
Eddy Co., N.M.	Bass, Big Eddy 48	4,665–4,781	Bell Canyon
		5,701–5,817	Cherry Canyon
		6,940–7,109	Brushy Canyon
Eddy Co., NM	Bass, Big Eddy 55	3,904–3,962	Bell Canyon

* indicates wells from which oil and water samples were taken.

carbon (TOC) analysis, Rock-Eval[1] pyrolysis, visual characterization and coupled gas chromatography/mass spectrometry (GC/MS) were employed to characterize organic matter abundance, composition and thermal maturity.

Biochemical marker analysis was carried out using facilities at Geochemical and Environmental Research Group, Texas A&M University. Visual organic matter characterization was performed by Geochem Labs in Houston, Texas. TOC and Rock-Eval analyses were performed using bulk rock samples by DGSI, The Woodlands, Texas. A summary of the important analytical methods applied in the study is set forth in the following section. Detailed analytical procedures and data are presented in Hays (1992) and Walling (1992).

Electron Microprobe Analysis

Calcium, magnesium, manganese, strontium, sodium, iron, sulfur and barium concentrations were obtained for calcite and dolomite in 13 selected thin-section samples using the Cameca SX50 electron microprobe at the Department of Geology and Geophysics, Texas A&M University. For all analyses, a beam potential of 15 kV was used; beam current was 25 nA, and beam width was 3 μm. Approximate detection limits for these elements were 400 ppm (Ca), 100 ppm (Mg), 200 ppm (Mn), 300 ppm (Sr), 70 ppm (Na), 200 ppm (Fe), 70 ppm (S) and 320 ppm (Ba).

Electron Microscopy

Scanning electron microscopy coupled with energy dispersive spectroscopy (EDS) was used to observe textural and morphological details and to acquire gross elemental compositions of authigenic minerals. Samples were attached to an aluminum stub with carbon paint and vapor-plated with gold-palladium alloy prior to analysis.

X-Ray Diffraction Analysis

The mineralogy and structure of clay mineral separates was determined using X-ray diffraction analysis (XRD). XRD pat-

terns were generated using a Rigaku automated diffractometer in the CuKα radiation mode. Oriented slides were prepared by pippeting slurries of ion-saturated sample onto glass; the slurries were then allowed to dry. Clay samples on the oriented slides were subjected to magnesium and potassium saturation prior to XRD analysis. Magnesium-saturated samples were treated with glycerol and analyzed at 25°C. Potassium-saturated samples were analyzed after heat treatment at 25°, 300° and 550°C increments to determine changes in basal reflections.

Neutron Activation Analysis

Elemental concentrations for all samples were determined by comparative NAA; a technique in which well characterized standard reference materials and unknowns were irradiated simultaneously. After irradiation, samples and standards were measured under identical conditions using the same detector. The technique of determining elemental concentrations versus standards as opposed to determining concentrations solely by measuring the number of gamma rays per unit time (the parametric method of NAA) eliminates uncertainties in the nuclear parameters (such as neutron flux, radionuclide reaction cross section and decay constant) and in detector efficiency. Two identical NBS standards and two types of quality control references (also NBS standards) were run for every six samples for both short-term pneumatic tube irradiations and long-term irradiations. Replicates of 75% of all samples were also prepared and analyzed. Samples and standards of 30 to 100 mg were weighed accurately into 0.3-ml polyethylene irradiation vials. Samples were first irradiated for 30 seconds using a pneumatic tube system, allowed to decay for 800 seconds and then analyzed for short half-life elements using a germanium detector. Samples were later irradiated for 14 hours for analysis of intermediate and long half-life elements. These were counted at 7, 11 and 28 day intervals. Irradiations were conducted at the research reactor of the Nuclear Science Center, Texas A&M University, while operating at a one megawatt level. Counting and concentration calculations were carried out using facilities of the Center for Chemical Characterization and Analysis facilities at Texas A&M University.

NAA analyses were conducted on bulk rock samples, silt and sand fractions, clay separates, bitumen extracts, kerogen separates and oil samples. Selected bulk rock samples were treated to remove organic matter and carbonates. Carbonate cements were removed in a pH 5 buffered solution of sodium acetate; organic material was removed by treatment in concentrated hydrogen peroxide. Sand, silt and clay fractions were obtained by sieving and centrifugation of dispersed samples. Bitumen (the component of kerogen extractable with organic solvents) was obtained from kerogen by Soxlet extraction of bulk rock siltstone samples. Powdered samples of about 30 g were extracted for 48 hours in a 70/30 solution of benzene and methanol as described by Lewan (1980). The extract was condensed to approximately 1 ml by rotoevaporation and placed in an NAA vial. Kerogen separates were obtained through a series of acid treatments and heavy liquid separation. A 30- to 50-g portion of crushed, extracted sample was combined with 500 ml of HCl (18%) at 50°C for 2 hours. The HCl was then siphoned off, and the sample was washed three times with 200 ml of deionized water, centrifuging and decanting the water after each wash. A

---

[1]The use of firm, brand, or trade names in this paper is for identification purposes only and does not constitute endorsement by the United States Geological Survey.

500-ml volume of concentrated HF (52%) was then added to the sample, and the mixture was placed in a 50°C water bath for 2 hours. The mixture was then left at room temperature for 15 hours. The spent acid was siphoned off, and the sample was washed in deionized water three times, centrifuging and decanting the water after each wash. A 150-ml volume of HCl (38%) was then added to the sample in a 200-ml centrifuge tube that was placed in a boiling water bath for 1 hour. The sample was then centrifuged; the acid was decanted, and the sample was washed in deionized water three times. After the final washing, the sample was mixed with a lithium metatungstate solution of 2.1 g/cm^3 density. This mixture was then centrifuged at 2,000 revolutions per minute for 20 minutes to yield a kerogen supernatant. The supernatant was collected and washed three times in a weak HCl solution (3%). Oil samples for NAA were collected and stored in plastic containers.

Stable Isotope Mass Spectrometry

Stable isotope mass spectrometry was used to characterize oxygen, carbon and sulfur isotopic compositions of authigenic minerals and organic components. Sulfur analyses were conducted at Coastal Laboratories, Austin, Texas; sulfur δ values are reported versus Central Diablo Troilite (CDT). For oxygen and carbon analyses, samples were analyzed on a Finnegan MAT 251 isotope ratio mass spectrometer with a triple collector system at the Department of Geology and Geophysics, Texas A&M University. Isotopic enrichments were determined relative to a laboratory reference gas which was calibrated to Pee Dee Belemnite (PDB) using the NBS 20 standard. Isotopic compositions were converted to PDB and corrected for ^{17}O using the method of Craig (1957). Analytical uncertainty was monitored by daily analysis of a laboratory reference standard and NBS 20. Maximum acceptable standard deviations for individual analyses were 0.04‰ for $\delta^{18}O$ and 0.03‰ for $\delta^{13}C$. For kerogen, carbon and oxygen isotopic composition were determined from extracted bulk rock samples and from pure kerogen separates. Bulk rock samples were acidified prior to combustion to remove carbonates. Kerogen separates were obtained as described for NAA. Bitumen samples were obtained by Soxlet extraction in methylene chloride. Bitumen extracts were split, and aromatic and aliphatic fractions were measured by high performance liquid chromatography (HPLC). Whole bitumen as well as aromatic and aliphatic fractions were analyzed. Oil samples were topped prior to analysis by placing them in a vacuum oven at 60°C overnight to remove light hydrocarbons up to approximately n-decane. Organic samples were sealed in quartz tubing and then combusted with cupric oxide and copper for 2 hours at 900°C and 2 hours at 600°C. Resultant CO_2 was transferred into the mass spectrometer and analyzed.

Due to the very fine-grained texture of Delaware Mountain Group rocks, clean separation of carbonate minerals was not possible, so dolomite and calcite cements were analyzed in bulk rock samples. Samples were selected on the basis of petrographic work in which the thin sections were stained for dolomite. Samples with high carbonate content (>4%) and either a high or low dolomite/calcite ratio were chosen for isotope analysis in order to better discriminate the isotope compositions of dolomite versus calcite. Carbonate samples were roasted under vacuum for 1 hour at 380°C to remove organics and other volatiles. The samples were then reacted with anhydrous phosphoric acid for 1 hour at 96°C to yield CO_2 gas. CO_2 generated was purified in a reduction furnace (McCrea, 1950; Mucciarone and Williams, 1990) prior to analysis. Mixed calcite/dolomite samples were leached by EDTA in the method of Videtich (1981); the dolomite remaining was then reacted with phosphoric acid.

Gas Chromatography/Mass Spectrometry

GC/MS analyses were carried out on bulk-sample siltstone extracts and oil samples from the sandstone reservoirs to determine the fractions and compositions of organic components present. 5β(H)-Cholane, deuterated phenanthrene, and deuterated chrysene were added as surrogate standards. Ground siltstone samples were extracted with hexane in a Soxlet apparatus for 12 hours. The extracts were concentrated by rotoevaporation, after which oil and extracts were similarly treated. Resins were removed by silica/alumina column chromatography. High performance liquid chromatography (Hewlett-Packard 1050 HPLC system with a Whatman Partisil preparative column) was employed to separate aliphatic and aromatic fractions. The aliphatic fraction was treated with molecular sieve beads to remove n-alkanes. An internal standard, deuterated pristane, was then added to each fraction to determine surrogate recovery. The aliphatic fraction was analyzed by GC/MS in the selected ion mode for triterpanes (m/z = 191), steranes (m/z = 217), monoaromatized steranes (m/z = 239, 253), and demethylated triterpanes (m/z = 177). The aromatic fraction was analyzed for triaromatized steranes (m/z = 231, 245), naphthalenes (m/z = 128, 142, 156, 170), and phenanthrenes (m/z = 178, 192, 206, 220). Peak assignments and calculations were performed using software developed by the Geochemical and Environmental Research Group; all computer peak assignments were manually checked.

PETROLOGY

Gross examination of Delaware Mountain Group cores shows that they are composed of gray to tan sandstone units 0.1 to 13 m in thickness interbedded with dark to light gray siltstone. The siltstone intervals are organic-rich and constitute as much as 80% of the Delaware Mountain Group section. Thin limestone beds are rare.

Detrital Mineralogy and Texture

Delaware Mountain Group sandstones observed in this study are very fine-grained subarkosic to arkosic sandstones (Figs. 4A, 4B, see p. 99). Mean sandstone detrital composition is 54% quartz, 38% feldspars, 4% lithic fragments, 1% detrital clay matrix and 3% other minerals (detrital mineral values are normalized to detrital constituent total, Table 3). Feldspars are largely plagioclase feldspars, but potassium feldspars are also present and account for about 10% of the total feldspar content. Most of the rock fragments (85% of the total) are finely crystalline volcanic rock fragments that contain plagioclase feldspar. The remainder of the lithics predominantly is composed of allochthonous limestone rock fragments apparently transported from the shelf margin of the Delaware Basin. The detrital composition of Delaware Mountain Group siltstones is similar

TABLE 3.—MEAN PETROGRAPHIC DATA (%) FOR REEVES AND PECOS COUNTY WELLS

Constituent	Hodge 2	Barber A5	Strain 2	Frost 1	Cleveland 1	Trees 2	Mean
Qz	51.4	51.7	57.0	49.1	54.1	54.7	53.2
PQz	0.3	0.2	0.2	0.3	0.3	0.5	0.3
Plag	37.8	35.7	30.6	32.5	35.6	31.8	34.7
Kfeld	1.9	5.7	3.7	4.2	3.6	4.2	3.5
VRF	3.6	3.5	2.8	2.2	2.7	3.2	3.2
SRF	0.0	0.0	0.1	0.2	0.1	1.8	0.4
Det mx	0.8	0.6	0.4	2.1	0.4	0.5	0.7
Musc	0.8	0.1	1.1	1.3	0.2	0.5	0.6
Tour	0.2	0.4	0.2	0.5	0.4	0.2	0.3
Zirc	0.5	0.4	0.6	0.7	0.5	0.5	0.5
Qz Og	0.9	0.9	0.7	0.3	1.4	0.9	0.9
Dolo	2.8	4.7	1.3	0.3	3.5	0.1	2.1
Halite	0.4	1.0	0.5	0.3	1.3	0.2	0.6
An/Gyp	2.8	0.7	4.0	2.6	2.3	0.7	2.2
Cal	9.3	2.9	6.5	5.1	4.2	6.3	6.5
Chlorite	9.5	13.3	16.3	15.0	15.6	12.5	13.0
TiO	0.3	0.4	0.1	0.1	0.4	1.7	0.6
Other	0.9	1.1	1.4	1.7	1.1	1.3	1.0

Abbreviations: Qz = quartz, PQz = polycrystalline quartz, Plag = plagioclase feldspar, Kfeld = potassium feldspar, VRF = volcanic rock fragments, SRF = sedimentary rock fragments, Det mx = detrital clay matrix, Musc = muscovite, Tour = tourmaline, Zirc = zircon, Qz Og = quartz overgrowth, Dolo = dolomite, An/Gyp = anhydrite/gypsum, Cal = calcite and TiO = titanium oxides.

to that of the sandstones. Mean siltstone detrital composition is 50% quartz, 38% feldspar, 2% lithic fragments, 5% detrital clay and 5% other (the "other" category being relatively high because of the difficulty of identifying some of the fine grained material).

Detrital clays notably are rare in Delaware Mountain Group sandstones and siltstones. Delaware Mountain Group sandstones contain a meager average of about 1% total detrital clay. The siltstones contain an average 5% of detrital clay. Detrital mica constitutes much of the total detrital clay content in the sandstones and the siltstones. True shales are very rare in the basin. Rock that appears to be shale to the casual observer generally is found to be dark, organic-rich siltstone with a negligible clay/mineral content (Fig. 4C, see p. 99).

Nature and Distribution of Authigenic Minerals

The major diagenetic constituents present in the Delaware Mountain Group sandstones are listed in Table 3 and include: (1) calcite, gypsum/anhydrite and halite cements, comprising the bulk of cements present; (2) chlorite and mixed-layered chlorite clays; (3) dolomite, potassium feldspar, titanium oxides and quartz, which comprise the important late diagenetic products and (4) other minor authigenic constituents including pyrite, illite, kaolinite and albite. Authigenic minerals average 26% of bulk rock volume. The suite of authigenic minerals present in the Delaware Mountain Group sandstones yields important information on pore-water chemistry evolution during burial.

The presence of a specific mineral phase indicates that chemical conditions were within the mineral's stability field at some time during burial. Conversely, dissolution or alteration of a mineral indicates that pore-water conditions fell outside the stability field of that mineral at some time. Spatial distributions and paragenetic relations of specific minerals allow determination of the relative timing of mineral formation or degradation

events in the diagenetic scheme, elucidating important changes in fluid chemistry in the rock.

The Early Cements.—

Calcite is the most abundant cement in Delaware Mountain Group sandstones and siltstones and primarily is present as a massive, poikilotopic cement which fills from 0 to 100% of intergranular volume (Fig. 4D, see p. 99). Mean calcite content for samples analyzed is 6%, ranging from a trace to 29% of the bulk rock. Microprobe data indicate a mean calcite composition of $Ca_{0.961} Mg_{0.013} Fe_{0.003} Mn_{0.023} CO_3$. Bulk-rock samples, which contained calcite (>90%) as the predominant carbonate cement, were analyzed for oxygen and carbon isotopes. Calcite $\delta^{18}O$ values are fairly consistent (Fig. 5), ranging from 0.3 to 1.8‰ PDB with the exception of one very low value at $-2.6‰$. Mean $\delta^{18}O$ is 0.5‰. Calcite $\delta^{13}C$ values are also fairly consistent ranging from -3.1 to 3.8‰ with the exception of one low value at $-3.1‰$. Mean $\delta^{13}C$ is 2.1‰. Petrographic data and comparison of calcite-rich zones containing massive, continuous calcite cement with calcite-depleted zones containing patchy, relict calcite cement showing dissolution features indicate that calcite was an early diagenetic feature that apparently was much more abundant earlier in burial history. Some pyrite appears to have formed early, coevally with calcite; pyrite crystals are commonly entombed in apparently undisturbed calcite.

Gypsum/anhydrite cement commonly occurs as a massive, pore-filling cement that fills from 0 to 100% of intergranular volume and appears less commonly as a patchy, irregularly distributed cement showing dissolution features (usually as gypsum). Gypsum/anhydrite cement makes up an average 2% of bulk rock volume. The cements appear to have been predominantly anhydrite prior to core recovery. Areas of anhydrite cement often have gypsum halos, which are probably indicative of post-recovery hydration.

Halite is a relatively common cement in the Delaware Mountain Group and generally is seen in thin sections as a massive, clean, isotropic mineral that fills from 0 to 100% of intergranular volume. Halite comprises a mean 0.7% of the bulk rock

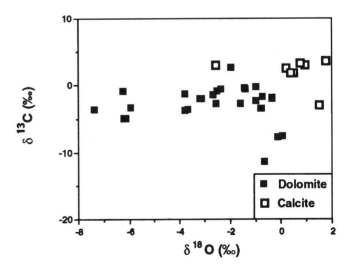

FIG. 5.—Diagram showing oxygen and carbon isotopic composition of dolomite and calcite cement samples.

and ranges up to 9% in some samples. Prior to coring and storage, in situ halite content probably was substantially greater than that exhibited in the cores at the time of this study. Halite dissolution features are common in core and thin section. Areas of extremely friable sandstone with no late stage authigenics in place show evidence of past halite cementation.

Quartz overgrowths are present as a minor early cement phase (Fig. 4B). Overgrowth formation appears to have preceded authigenic clay development.

The Authigenic Clays.—

Chlorite and mixed-layer chlorite/smectite minerals were identified by XRD techniques (Walling, 1992) and are a pervasive authigenic constituent in the Delaware Mountain Group, averaging 13% of bulk rock volume. Chlorite minerals are abundant in all samples except those in which the early carbonate, sulfate and halite cements remain intact. Optical and scanning electron microscopy show that chlorite is present, lining and filling pores (Figs. 4B, 6A) and replacing detrital grains.

Chlorite is present in all porous Delaware Mountain Group samples analyzed. Two textural chlorite end-members are manifest: (1) a finely crystalline chlorite in which individual clay platelets are poorly developed, difficult to distinguish in the chlorite mass, and show no chlorite "rosettes" (Fig. 6B); this morphology is commonly observed for interstratified chlorite/smectite (Barnhisel and Bertsch, 1989); and (2) a coarsely crystalline chlorite in which individual platelets are well formed, up to 0.1 mm across (Figs. 6A, 7A), and chlorite rosettes are common. Chlorites closer to the fine end-member are more abundant and constitute much of the pore-lining and pore-filling clay. The coarse chlorite appears to post-date the fine chlorite; some pores were lined with fine chlorite and subsequently filled with coarse chlorite (Fig. 7B). EDS shows that these coarse chlorites are often more iron-rich than the fine chlorite. The majority of chlorite appears to have formed prior to the development of authigenic dolomite, quartz and titanium oxides, but chlorite authigenesis did partially overlap onset of development of those later minerals (e.g., there are instances of chlorite formed over dolomite and authigenic titanium oxides). In these

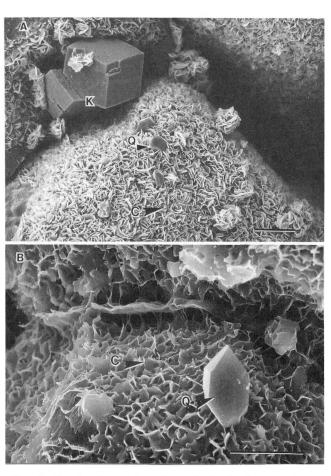

FIG. 6.—Scanning electron photomicrographs showing the nature of occurrence of authigenic chlorite (labeled C) within Delaware Mountain Group sandstones; scale bars represent 10 μm. (A) shows pore-lining chlorite with well developed platelets and chlorite rosettes; note the relationship of the chlorite with authigenic quartz (Q) and potassium feldspar (K). (B) shows poorly crystalline pore-lining and pore-filling chlorite; note the indistinct, anastomosing platelets, and the contrast in platelet crystal development between chlorites in A and B.

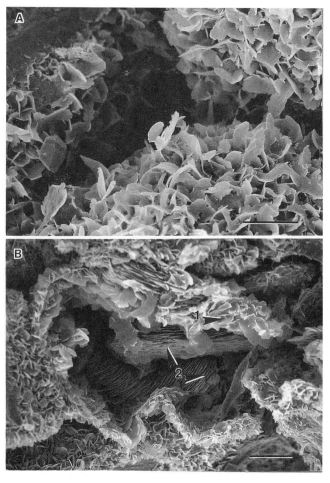

FIG. 7.—Scanning electron photomicrographs showing the nature of occurrence of authigenic chlorite within Delaware Mountain Group sandstones; scale bars represent 10 μm. (A) illustrates the morphology of coarse, well crystallized chlorite in a pore-lining habit. (B) highlights the paragenetic relation between (1) the fine, low-iron, pore-lining chlorite and (2) the coarse, high-iron chlorite which subsequently filled the pore.

instances of late chlorite authigenesis, the chlorite is usually of the coarser, well crystallized end-member.

The Late Stage Authigenics.—

Dolomite occurs in two distinct habits in the Delaware Mountain Group: as a massive, pore-filling cement that has formed through dolomitization of the original calcite cement and as discrete, euhedral dolomite rhombohedra that line and fill pores to varying degrees on top of earlier occurring chlorite (Figs. 4B, 8A). Dolomite accounts for an average 3% of bulk rock volume. Three dolomite compositional end-members were identified through electron microprobe analysis: (1) a "clean" dolomite of the average composition $Ca_{1.01} Mg_{0.98} Fe_{0.01} Mn_{0.01} (CO_3)_2$, (2) an iron-manganese enriched dolomite with an average composition of $Ca_{1.00} Mg_{0.88} Fe_{0.07} Mn_{0.06} (CO_3)_2$ and (3) a manganous dolomite of limited iron content with an average composition of $Ca_{1.00} Mg_{0.80} Fe_{0.04} Mn_{0.17} (CO_3)_2$. Figure 9 is a plot of manganese content (weight %) versus iron content (weight %) and illustrates compositional differences of the three end-members. Dolomites of end-members I and II plot as a homologous series

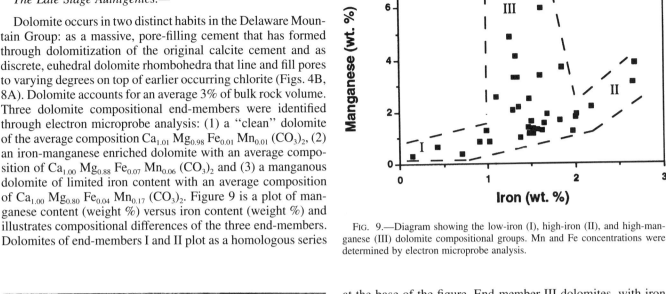

FIG. 9.—Diagram showing the low-iron (I), high-iron (II), and high-manganese (III) dolomite compositional groups. Mn and Fe concentrations were determined by electron microprobe analysis.

at the base of the figure. End-member III dolomites, with iron contents of 1 to 3%, plot as a distinct, manganese-enriched group falling above end members I and II. Compositional zoning observed within dolomite rhombs indicates that pore-water composition was variable with respect to iron and manganese during dolomite formation. No compositional distinction can be drawn between massive versus rhombohedral dolomite. The carbon isotopic composition of the dolomite is variable, ranging from −12 to 3‰ PDB (Fig. 5). Mean dolomite $\delta^{13}C$ is −2.8‰. The most positive $\delta^{13}C$ values (>3‰) approach the value expected if Permian sea water was the primary source of carbon (Given and Lohmann, 1986). This ^{13}C enriched carbonate was probably derived from detrital carbonate material and/or the early calcite cement. The majority of dolomite $\delta^{13}C$ values are isotopically depleted relative to calcite cement, falling between −10 and 0‰. Dolomite oxygen isotope ratios are also variable; $\delta^{18}O$ values range from −7.4 to 0.1‰ and average −2.6‰ (Fig. 5). This variation could be due to varying temperatures of dolomite formation, varying water source input, or greater rock-buffering effect in carbonate-poor bulk rock samples.

Paragenetic relations indicate that dolomite developed after the main episode of chlorite authigenesis and contemporaneously with feldspar, titanium oxide and quartz authigenesis. Dolomite rhombs have precipitated over chlorite and are interspersed with the coeval phases (Fig. 8A). Late stage quartz authigenesis resulted in formation of individual, euhedral crystals disseminated through pore spaces (Figs. 6A, 6B, 8A)

Titanium oxides are a unique and notable component of the authigenic mineralogy of the Delaware Mountain Group, comprising 0.5% of bulk rock volume. Anatase and brookite are identified on the basis of chemical composition, optical characteristics and crystal habit. Figure 8B shows the typical crystal habit of authigenic brookite.

PORE-WATER CHEMICAL EVOLUTION AND THE DIAGENETIC RECORD

Diagenetic events as determined from mineralogy and paragenetic relations of mineral phases in the Delaware Mountain

FIG. 8.—Scanning electron photomicrographs showing the nature of occurrence of the late stage authigenic minerals in the Delaware Mountain Group; scale bars represent 10 µm. (A) shows crystals of authigenic quartz (Q), dolomite (D), potassium feldspar (K), formed over pore-lining chlorite (C). (B) illustrates the crystal morphology of authigenic brookite (labeled B).

Group sandstones may be categorized into four main sequential stages: First, extensive cementation by calcite, halite and anhydrite occurred early in burial history. These cements partially arrested burial compaction and preserved a high percentage of intergranular volume. Second, large-scale dissolution of cements and some detrital material occurred during deep burial and created abundant secondary porosity. Third, pervasive chlorite authigenesis followed dissolution, partially occluding porosity and restricting fluid flow throughout the sequence. Fourth, authigenesis of dolomite, ferroan/manganous dolomite, potassium feldspar overgrowths, quartz, brookite and anatase began as chlorite formation ebbed. These constitute the important late authigenic phases. The four stages imply accompanying changes of a significant magnitude in pore-water chemistry through time, and the presence of specific late-stage authigenic products provides strong evidence for the impetus of pore-water chemical change. The early mineral phases were only passively involved in organic matter degradation and late burial diagenesis, thus the following discussion focuses on the third and fourth stages of diagenesis.

Chlorite Authigenesis

Chlorite and randomly interstratified chlorite/smectite are present in all porous samples of the Delaware Mountain Group. The parameters that control chlorite stability and development in the burial environment are pH, temperature and activities of component species (particularly magnesium and iron). High pH, high temperature and a high $[Fe^{2+}]$ $[Mg^{2+}]$ product favor chlorite stability (Kaiser, 1984). Chlorite and smectite develop by a reversible reaction in generalized form:

$$1.8 \text{ Calcium-montmorillonite} + 1.85 \text{ Mg}^{2+}$$
$$+ 2.3 \text{ Fe}^{2+} + 8.24 \text{ H}_2\text{O} = \text{Chlorite}$$
$$+ 0.29 \text{ Ca}^{2+} + 4.6 \text{ H}_4\text{SiO}_4 + 6.46 \text{ H}^+. \qquad (1)$$

Smectite or chlorite forms by recrystallization and ion diffusion (Ahn and Peacor, 1985) as pore-water chemistry changes from one stability field to another. Paragenetic relationships between the earlier, iron-poor, low chlorite/smectite clay and the late, iron-rich chlorite, as described in the petrology section, indicate that precipitation of mixed-layer chlorite/smectite probably post-dated most calcite and detrital grain dissolution. During the last stages of calcite and detrital grain dissolution, pH values would be expected to be low as the buffering capability of the rock decreased with dissolution of buffering minerals, and ionic species—such as silica, aluminum and iron—stripped from dissolved material would be more abundantly available in solution. Thus, geochemical conditions during the coincident, early stages of clay formation would have favored formation of clays with a low chlorite/smectite ratio. During the later stages of clay formation, the flux of acidic fluids decreased and metal cation activities increased, promoting chlorite precipitation and conversion of smectite to chlorite to increase chlorite/smectite ratios. This would cause the observed relation whereby poorly crystallized, low-iron chlorite/smectite lines pores and surrounds well crystallized, high-iron chlorite that fills the remainder of the pores.

Late Authigenesis

Authigenesis of dolomite, potassium feldspar and titanium oxide minerals occurred late during burial of the Delaware Mountain Group after most chlorite development was completed. Dolomite is an abundant late authigenic mineral that occurs in massive habit—probably formed through dolomitization of early calcite—and as discrete rhombohedral crystals. Dolomite formation indicates the presence of pore waters with high Mg^{2+}/Ca^{2+} ratios, high $HCO_3{}^-$ concentrations and low $SO_4{}^{2-}$ concentrations (Curtis and Coleman, 1986). Magnesium for dolomite formation (as well as for chlorite formation) was probably derived from brine that originated in the overlying Castile Formation evaporite sequence. High magnesium concentrations still prevail in the pore waters. Concentrations in the Reeves and Pecos Counties area average more than 5,000 ppm, about four times that of seawater.

Temperatures calculated from dolomite oxygen isotopic data are consistent with a late burial-setting interpretation of dolomite formation. Water $\delta^{18}O$ values (versus standard mean ocean water, SMOW) of $-0.7\%o$ and $-1.6\%o$, the "original" Permian seawater (Given and Lohmann, 1986) and current formation water $\delta^{18}O$ values, respectively, were used to determine formation temperatures in Land's (1985) dolomite revision of O'Neil et al.'s (1969) equation. Temperatures calculated with a $\delta^{18}O_{water}$ of $-0.7\%o$ range from 30° to 80°C and average approximately 45°C. Temperatures calculated with a $\delta^{18}O_{water}$ of $-1.6\%o$ range from 45° to 95°C and average about 60°C. The range of values is consistent with formation of dolomite during intermediate to late burial, and the maximum temperature calculated for dolomite formation (even using the less conservative $\delta^{18}O_{water} = -1.6\%o$) is in good agreement with the maximum temperature based on burial-setting interpretation for the rocks. However, the use of dolomite as a paleothermometer is handicapped by the assumption of a high effective water/rock ratio, the assumption of isotopic equilibrium and the poorly constrained value for water $\delta^{18}O$ during dolomite precipitation. Good evidence exists of multiple fluid sources being involved in diagenesis in these rocks (i.e., the Delaware Mountain Group and Castile Formation waters), and mixing of these fluids would result in variable water $\delta^{18}O$ values and a broad range of $\delta^{18}O$ values for dolomite forming in these pore waters (Fig. 5).

Dolomite minor element composition also has interesting implications for ambient pore-water chemistry at the time of its development. Cathodoluminescence and microprobe analysis show that the dolomite rhombs are compositionally zoned with respect to minor elements, indicating variable pore-water chemistry during crystal growth. The important minor elements in the dolomite, iron and manganese, define the three compositional end-members discussed in the dolomite petrology section, the (1) low-iron, low-manganese, (2) high-iron, moderate-manganese and (3) moderate-iron, high-manganese end-member compositions (Fig. 9). Although the original calcite cement present in the rocks of the Delaware Mountain Group may have supplied some of the iron and manganese to the dolomite, mass balance considerations indicate an additional source of these metals. Analyses of samples representing a unit volume of sandstone, 1 m^3 for example, indicate that the unit would contain an average 177,000 g of calcite (6.5 volume %) having a manganese content of 2,460 g and an iron content of 300 g based on mean calcite composition. The unit would also contain an average 71,300 g of dolomite (2.5 volume %) having 1,780 g of manganese and 1,060 g of iron based on mean dolomite composition. If all of the manganese and iron in the dolomite were assumed to be derived locally from calcite, then the rock had to have had a minimum original calcite content of 11.2% of bulk-rock volume with 4.7% of bulk-rock volume having been dissolved or dolomitized to yield enough manganese to explain the observed manganese content of the dolomite, and a minimum original calcite content of 23% of bulk-rock volume with 16.5% of bulk-rock volume having been dissolved to explain total iron in the dolomite. While the total amount of original calcite present in the Delaware Mountain Group could have equaled 23 volume %, and the amount dissolved or dolomitized could have equaled 16.5% of total rock volume, this is unlikely because manganese and iron partition coefficients for carbonates decrease with increasing temperature (Mucci and Morse, 1990). Under late burial conditions, it is unrealistic to expect dolomite to be efficient in scavenging manganese and iron, and the amount of calcite necessary to supply the required metals reaches unreasonable proportions. In addition, maximum iron and manganese concentrations in dolomite exceed maximum concentrations in calcite by a factor of 9.1 for iron and a factor of 2.6 for manganese, further indicating that an additional source of these metals was available.

Whereas iron may have several possible sources, the manganese in Delaware Mountain Group dolomite appears to be derived from organic matter. Manganese concentrations average 510 ppm in the organic matter extracts, 25,000 ppm in authigenic dolomite and 660 ppm in authigenic chlorite. Manganese in the detrital fraction averages only 140 ppm. The distribution of manganese indicates that organic matter may have been an important source. The metal was stored as organometallic complexes in organic matter during burial and released into pore water as the complexes degraded under thermal stress. Organic material may have been a source of iron as well, since iron is also present in substantial quantities in Delaware Mountain Group organic matter. These observations support a model that may explain the end-member compositions observed in the dolomite. Manganese organic complexes are less stable than iron complexes (Mantoura, 1981; Nissenbaum and Swaine, 1976; Surdam et al., 1989) and may have destabilized upon crossing a critical temperature threshold to yield large quantities of the complexed manganese during a short period of time as thermal maturation progressed. This would result in a pore-water concentration spike for manganese that was recorded by high-manganese end-member dolomite. The low-manganese, variable-iron dolomites were possibly precipitated while pore-water manganese concentrations were affected by more stable organic complexes.

Authigenic titanium oxides—brookite and anatase—occur as well developed crystals that post-date chlorite in the Delaware Mountain Group. The presence of abundant authigenic titanium oxides is an unmistakable indication of complexation of titanium by organic ligands. Titanium minerals are extremely insoluble under chemical and temperature conditions normal to weathering and burial diagenesis settings (Berrow et al., 1978; Ure and Berrow, 1982), and titanium has been shown to be mobilized only in the presence of organic acids and/or under conditions of extreme weathering (Correns, 1978; Ure and Berrow, 1982; Reed, 1990). These studies show that complexation of titanium by organic acids can increase concentrations of titanium in solution as high as 200 to 500 ppm. Titanium-bearing detrital minerals observed include ilmenite and rutile. These minerals often show obvious effects of chemical degradation (Fig. 10a) and served as a source of titanium for authigenic minerals. Local reprecipitation of titanium, as is common in weathering of titanium minerals under surface conditions (Milnes and Fitzpatrick, 1989; Bain, 1976), is not observed, indicating that a titanium mobilizing process of considerable efficacy was in action. Titanium oxide authigenesis evidently took place as titanium was first mobilized through complexation by organic acids generated during thermal degradation of Delaware Mountain Group organic matter and subsequently precipitated as brookite (Fig. 10b) and anatase after considerable movement (at least off of the scale of a thin section) as these organo-titanium complexes decayed.

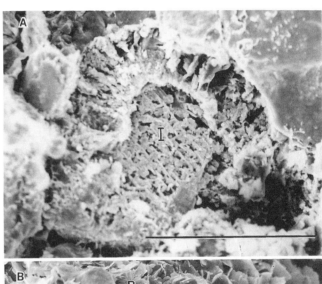

FIG. 10.—Scanning electron photomicrographs illustrating the nature of occurrence of titanium oxide minerals in the Delaware Mountain Group; scale bars represent 10 μm. (A) shows detrital ilmenite (I) that has experienced significant dissolution; note the lack of secondary titanium oxide minerals proximal to the detrital grain. (B) illustrates the abundant occurrence of authigenic titanium oxide minerals (brookite in this example, labeled B) within the sandstones that provides evidence of titanium mobilization through the action of organo-metallic complexes.

Thus, interpretation of Delaware Mountain Group petrology and diagenetic history yields strong circumstantial and intrinsic evidence implicating organic matter as a principal actor in controlling pore-fluid chemical evolution. Abundant organic matter appears to have affected diagenetic processes during late burial. However, interpretation of petrologic and geochemical data provides only the first argument for proving the case for the role of organic matter degradation in the diagenetic scheme. More complete understanding of the organic/inorganic synthesis requires corroboration through definitive characterization of organic geochemistry and analysis of natural tracers to be found in the organic and inorganic phases.

ORGANIC GEOCHEMISTRY

The sequence of diagenetic events experienced by Delaware Mountain Group rocks can be most reasonably interpreted as being driven by coupled organic/inorganic processes. Accrual of data on the organic matter in the siltstones is the necessary second step for completely evaluating the role of organic matter in controlling pore-water fluid chemical evolution and consequent diagenetic events in the sandstones; only if the organic matter of Delaware Mountain Group siltstones is of suitable type, of sufficient abundance and of an advanced state of thermal maturity can be a substantial release of carboxylic acids, CO_2 and organically complexed metals during burial have been a possibility.

Organic Matter Characterization

Total organic carbon (TOC) analysis shows that Delaware Mountain Group siltstones contain an average 3 wt% TOC with values ranging from 0.5 to 12%. The organic carbon content of the Delaware Mountain Group is well above the 0.5 to 1.0% minimum TOC generally acknowledged as the lower limit for a potential source rock (Tissot and Welte, 1978). In order for TOC data to be useful in determining actual organic matter abundance and in calculating acidic yields, a correction factor must be applied to compensate for other elements present in kerogen (i.e., hydrogen, oxygen, nitrogen and sulfur). Conversion factors reported by Tissot and Welte (1978) indicate that mean organic-matter content of the Delaware Mountain Group is approximately 4 wt%, ranging from about 0.7 to 17 wt%. Bitumen extracted from the siltstones by the Soxlet procedure averaged 0.8 wt%. By difference, kerogen averaged 3.2 wt%.

Rock-Eval pyrolysis identifies kerogen types based on the hydrogen and oxygen indices. The hydrogen and oxygen indices (HI and OI, respectively) are pyrolytic yields indicative of atomic hydrogen/carbon and oxygen/carbon ratios upon which the kerogen classification scheme is based (Tissot and Welte, 1978). Results were plotted in a van Krevelen-type diagram, plotting HI versus OI. Pure sample populations of the four kerogen types found in nature will plot along one of the four kerogen type lines on a van Krevelen diagram. Type I kerogen is typically derived from algae-rich lacustrine rocks and is the most hydrogen enriched type of kerogen. Type II kerogen is derived from marine phytoplankton. This type of organic matter is relatively hydrogen-rich and is prone to generation of significant quantities of organic acids and CO_2 during thermal maturation. Type III kerogen is derived from terrigenous plants. Type III is oxygen-rich and during thermal stress, generates the

greatest volumes of organic acids and CO_2 of the kerogen types. Type IV kerogen is oxidized kerogen in which hydrogen/carbon ratios have been reduced and oxygen/carbon ratios have been increased, resulting in a shifted placement on the van Krevelen diagram. Two distinct kerogen types were observed within the Delaware Mountain Group: Type II kerogen, which is more common in the deeper portions of the Delaware Basin in the Reeves and Pecos Counties area, and Type III kerogen, which is more common in the shallow northern reaches of the Delaware Basin in the Eddy County, New Mexico area (Fig. 11). In the Reeves and Pecos Counties area, approximately 85% of the siltstone samples contain predominantly Type II kerogen. Up slope in Eddy County, New Mexico, Type III kerogen is more important, comprising the majority of the kerogen in 75% of the samples. However, very high oxygen indices in some of the Type III samples may be evidence for partial oxidation of kerogen in the Eddy County area, biasing the Type II/Type III ratio. Van Krevelen-type diagrams generated from pyrolysis data indicate that no rock sample contains 100% of either kerogen type, but rather that all samples contain predominantly one type of kerogen with a minor amount of the other. Samples consistently plot off of the Type II and Type III best fit lines. The two kerogen types appear to have be deposited as discrete units. In Reeves and Pecos Counties, Type III kerogen-bearing siltstone units within thick intervals of Type II kerogen-bearing siltstone can be correlated from core to core on the basis of similar

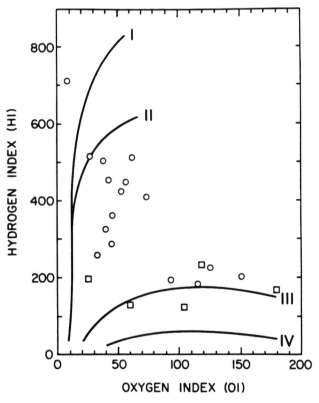

FIG. 11.—Rock-Eval pyrolysis hydrogen and oxygen indices plotted in a Van Krevelen-type format. The diagram shows the occurrence of Type II and Type III kerogens in siltstone samples. Pecos and Reeves County, Texas samples are indicated by circles; Eddy County, New Mexico samples are indicated by squares.

gamma ray, acoustical and resistivity response on geophysical logs.

Oil and Source Rock Thermal Maturity

Establishing the state of thermal maturity of Delaware Mountain Group siltstone organic matter is necessary to prove that these rocks were capable of generating organically-derived acids. The rocks could have released these acids in substantial quantities only if the threshold of incipient oil generation has been surpassed. Delaware Mountain Group maturity was assessed using Lopatin thermal modeling, Rock-Eval pyrolysis, optical indices and biomarker parameters to develop a credible, reconcilable appraisal of the degree of thermal maturation. These maturity indicators are referenced to vitrinite reflectance values (R_o) to provide a common frame of reference (Table 4).

Vitrinite Reflectance and Thermal Alteration Index.—

Reflected light analysis of kerogen separates show that vitrinite is present in minor amounts. Vitrinite is a type of humic organic matter often comprising some fraction of kerogen, and vitrinite reflectance is related to the degree of thermal stress to which the vitrinite has been subjected. Vitrinite reflectance (R_o) values range from a minimum 0.35 to a maximum of 0.87; average R_o is 0.54, indicating a moderately immature to mature ranking. The thermal alteration index (TAI), an index based on organic grain coloration in transmitted light, determined for Delaware Mountain Group kerogen samples ranges from 1.6 to 2.3. The mean TAI of 2.2 indicates moderate maturity.

Lopatin Modeling.—

Thermal modeling of the Delaware Basin using Lopatin's simple, empirical model (Waples, 1980) was conducted to assess the approximate thermal maturity of organic matter in terms of the timing of oil generation. The Lopatin method reports maturity on the time-temperature index (TTI) which can be compared to other maturation indices. A burial history curve based on the ages and present depths of the strata was constructed (Fig. 12) and superimposed on graphic subsurface isotherms through time. The intersections of the burial history curve with the isotherms define the time-temperature intervals used for calculating maturity. Input parameters used in the model are the are the time and temperature factors used to calculate the TTI. The TTI is an empirically-derived product of thermal stress and time required to bring kerogen into the oil-generative window (the period during which kerogen is ther-

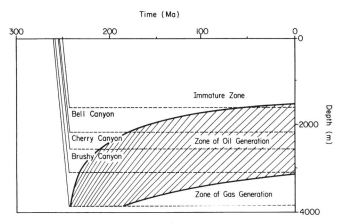

FIG. 12.—Diagram generated from Lopatin thermal maturation model calculations showing the top of the Delaware Mountain Group entering the oil generative window at an estimated 50 Ma.

mally cracked to form oil prior to the period of gas formation when the oil itself is thermally cracked). A constant thermal gradient of 25°C/km and upper and lower oil generative boundaries of 15 and 160 TTI were used as the input parameters. The TTI values used correspond to R_o values of 0.6 and 1.35 for Type II kerogens; these are the accepted approximate R_o values defining the beginning and end of oil generation. Application of the Lopatin model indicates that the upper part of the Delaware Mountain Group began generating oil at approximately 50 Ma in the middle Eocene (Fig. 12).

Rock-Eval Pyrolysis.—

Tmax is the temperature at which maximum pyrolytic yield occurs during pyrolysis of kerogen. Tmax is related to the degree of thermal stress that a sample has experienced—higher values indicating more advanced thermal maturity, lower values indicating less advanced thermal maturity. Rock-Eval Tmax values range from 417 to 440°C, averaging 431°C for samples analyzed (Table 4). This supports the Lopatin model for the Delaware Mountain Group, indicating that the organic matter has entered the oil generative window and is marginally to moderately mature, corresponding to a vitrinite reflectance (R_o) of 0.5 to 0.8 (Tissot, 1984).

Biomarkers.—

Biomarkers are geochemical fossils inherited in a relatively unaltered chemical state from the original organic material. Biomarkers can be useful as indicators of the thermal maturity of organic matter—in kerogen, bitumen and oil—and as a oil-source correlation tool. For this investigation, biomarkers in the terpane, sterane and aromatic sterane chemical groups which have been shown to be effective maturity and source indicators were studied in bitumen and oil. Bitumen, or bitumen extract, is the fraction of kerogen that can be extracted with organic solvents. Bitumen is the untransported, thermocatalytic product of whole kerogen.

As maturity indicators, biomarkers which are subject to isomerization and chemical transformation during thermal stress are useful up to the point at which chemical equilibrium is reached (for isomerization) or when the reactant has been

TABLE 4.—SOURCE ROCK MATURITY PARAMETER DATA SUMMARY

Index	Range	Mean	Ro Equivalent	N
Rock-Eval Tmax	417–440°C	431°C	0.5–0.8	23
Vitrinite reflectance	0.35–0.87	0.54	0.54	6
TAI	1.6–2.3	2.2	~0.70	6
MPI	0.43–0.83	0.6	~0.76	17
C31 Hopane R/R + S	0.57–0.67	0.58	>0.60	17
C29 Sterane R/R + S	0.24–0.57	0.34	<0.90	17
Triaromatic sterane C21/C21 + C28	0.31–0.50	0.38	~0.85	17

Abbreviations: Ro equivalent = approximate equivalent vitrinite reflectance range or value, Tmax = temperature of maximum pyrolytic hydrocarbon generation, N = number of samples, MPI = methylphenanthrene index, TAI = thermal alteration index.

completely transformed into the daughter product (for chemical transformation). For different biomarkers, chemical equilibrium or complete transformation is reached at different levels of maturity, making specific biomarkers more valuable at various stages of maturity. Biomarker maturity indicators can be separated into two groups based on these chemical changes: (1) biomarkers resulting from a change in the number of carbon atoms in the molecule and (2) biomarkers resulting from isomeric changes at certain chiral centers in the carbon chain. Maturity is evaluated by deriving a molecular ratio of precursor to daughter product. The methylphenanthrene index (MPI) is of the first type; this method determines phenanthrene and its four monomethyl homologues (1-, 9-, 2-, 3-methylphenanthrenes) in the triaromatic fraction of sediment extracts (Curiale et al., 1989). The ratio of 2- and 3- methylphenanthrenes (produced during thermal stress) to the originally occurring 1- and 9-methylphenanthrenes is the methylphenanthrene index, which empirically follows a maturity trend. The triaromatic sterane transformation fraction, another maturity index, is also of the first type. C20 and C21 triaromatic steranes are created by cleavage of carbon-carbon bonds in the C27 and C28 triaromatic sterane precursors. Extent of the C27 to C20 and C28 to C21 transformations are reported by the C20/(C20 + C27) and C21/(C21 + C28) ratios.

The most widely used epimerization maturity indicators are those involving hydrogen atom configuration at the C-22 position in the hopanes and the C-20 position in the steranes. The initial biological form of these molecules has the R (rectus or right-handed) stereochemical configuration; thermally-induced isomerization creates increasing abundances of the S (sinister or left-handed) configuration until an equilibrium mixture is achieved.

The biomarker maturity data are consistent with the Lopatin and Rock-Eval Tmax assessments. The mean methylphenanthrene index (MPI) for Delaware Mountain Group source rocks is 0.60, corresponding to an R_o of 0.76 (Table 4). Delaware Mountain Group oils appear slightly more mature with an MPI of 0.77. The extent of C31 17α(H)-hopane isomerization at C-22 indicates that the source rocks are at near-equilibrium values for the R and S isomers and that they are well within the oil generative window. Bitumen extracts exhibit C31 R/(R+S) ratios of 0.57 to 0.67 with a 0.58 mean. C32, C33, C34 and C35 17α(H)-hopane S/(R+S) ratios are very similar with average values of 0.57, 0.63, 0.64 and 0.60 respectively. Oil S/(R+S) values for the 17α(H)-hopanes are similar to the rock extracts: 0.66 for C31, 0.61 for C32, 0.71 for C33, 0.64 for C34 and 0.51 for the C35 oil (Table 4). C29 sterane 20S/(20R+20S) ratios for Delaware Mountain Group source-rock bitumen extracts range from 0.24 to 0.57, averaging 0.34. This is below the sterane equilibrium value of 0.52 and indicates an R_o maturation level of less than 0.9 (Curiale et al, 1989; Sofer et al., 1986). C29 sterane S/(R+S) values for Delaware Mountain Group oils are relatively constant with a mean of 0.55. This higher apparent maturation level is at or above the accepted equilibrium value for sterane epimerization (0.52 to 0.55) and is most likely due to biomarker fractionation during migration. 20S steranes migrate faster than 20R steranes. Thus, oils are typically enriched relative to the associated source bitumen (Zhusheng et al., 1988). This conclusion is supported by further migration parameter data which will be discussed later. Triaro-

matic steranes were monitored for the C20 to C27 and C21 to C28 transformations. Oil and source rock exhibit very similar values. The transformation fractions C20/(C20 + C27) and C21/(C21 + C28) for the source-rock extracts are equivalent, averaging 0.38 for both. Mean transformation fractions for the oils are 0.39 for C20/(C20 + C27) and 0.45 for C21/(C21 + C28). These values correspond to an R_o of approximately 0.85 (Mackenzie, 1984; Sofer et al., 1986).

The various parameters investigated to establish the degree of thermal maturity of the organic matter give a consistent appraisal. The voluminous siltstones of the Delaware Mountain Group contain organic matter that is marginally to moderately mature, and this organic material appears to be well into the oil generative phase.

Oil/Source Correlation

Correlation of Delaware Mountain Group siltstone organic matter and oils reservoired in Delaware Mountain Group sandstones provides evidence that siltstone organic matter did yield the fluids associated with thermal stress into the sandstones. A multiparameter approach to correlation was applied, comparing results of carbon and sulfur stable isotope analyses and comprehensive biomarker analysis. Oils reservoired in the sandstones were compared with bitumen samples from the siltstones, the suspected source rock.

Stable Isotopes.—

The carbon isotopic composition of kerogen is dependent on the original isotopic composition of the source organic matter and on isotope fractionation during kerogen formation. Isotopic distribution in biologic compounds is related to chemical structure (Galimov, 1973). For example, aliphatic chain carbon and methoxyl group carbon are depleted in ^{13}C relative to carboxyl, phenol, carbonyl and amine group carbon. As a result, lipids are ^{13}C depleted while proteins and carbohydrates are ^{13}C enriched. During the polycondensation, insolubilization and polymerization processes of kerogen formation (Tissot and Welte, 1978), selective elimination of carboxylic and other functional groups results in kerogen that is slightly depleted relative to the starting material. With the onset of oil generation, thermal cracking of hydrocarbon bonds releases hydrocarbon molecules from the source-kerogen structure that are depleted in $\delta^{13}C$ by 1 to 4‰. Bitumens typically have an intermediate $\delta^{13}C$ value (Tissot and Welte, 1978) between oil and associated source kerogen.

Carbon isotopic analysis of Delaware Mountain Group kerogen separates, bitumen extracts and Delaware Mountain Group reservoired oils show a slight, progressive decrease in $\delta^{13}C$ moving from kerogen to extract to oil (Fig. 13), indicating a good correlation and supporting the idea that kerogen in the siltstones was the source for oils reservoired in the sandstones. The mean $\delta^{13}C$ for kerogen is $-27.3‰$. Mean extract $\delta^{13}C$ is $-28.5‰$, and the mean $\delta^{13}C$ for the oils is $-29.2‰$. Delaware Mountain Group kerogen is enriched an average 1.9‰ relative to the oil, giving a good match between oil and suspected source.

Most of the sulfur present in kerogen and oil is in organic combination. The majority of the sulfur is ultimately derived from seawater sulfate, moving via open-system diffusion into

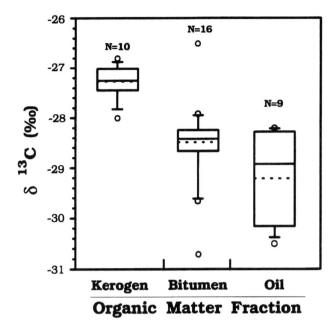

FIG. 13.—Box plot of $\delta^{13}C$ data of organic matter fractions studied showing excellent correlation between Delaware Mountain Group kerogen, kerogen extracts (bitumen), and oils. A slight decrease in mean $\delta^{13}C$ from kerogen to extract to oils is apparent and indicates correlation of these phases. N indicates the number of samples. The 25th and 75th percentiles are defined by the boxes; the solid lines across the interior of each box delineate median values; the dashed lines indicate mean values. The vertical bars delineate the 10th and 90th percentiles, and all outlier data are plotted as open circles.

shallow-sediment interstitial water. There the sulfur is reduced by microbial action and incorporated into organic matter during very early burial (Orr, 1974). The predominant sulfur compounds found in kerogen are diagenetic products bearing little resemblance to the original biological sulfur compounds (Orr, 1978). Thus, early sulfate cements and sulfur found in kerogen and oil ultimately have the same sulfur source, seawater. In the case of the Delaware Mountain Group, the system appears not to have been a completely open system throughout the time of sulfate cement formation and incorporation of sulfur into the organic matter in the siltstones because the pore-water sulfate isotopic composition was modified somewhat from the accepted, original seawater values. The mean $\delta^{34}S$ for sulfate cements sampled from sulfate-filled syndepositional fractures and fenestral anhydrite structures in Delaware Mountain Group siltstone is $+14.5\%o$. This value is enriched in ^{34}S by about $4\%o$ as compared to the expected value for Permian cements (the late Permian minima being about $+10\%o$) and is indicative of partial closing of the pore-water system. Pore-water sulfate was enriched in ^{34}S as sulfate reduction progressed during early burial. This interpretation is corroborated by trace metal and stable isotope data accrued on calcite and pyrite (Hays, 1992).

Fractionation of pore-water sulfur during microbial reduction and incorporation into organic matter causes a ^{34}S depletion of approximately $15\%o$ in the organic matter relative to coeval sulfate cements (Thode and Monster, 1965). During intermediate burial, sulfur isotopic composition appears to be quite stable as the original organic matter undergoes transformation to kerogen. The majority of sulfur in oils originates from organic sulfur

compounds in source kerogen (Thode and Monster, 1965; Faure, 1986). Therefore, sulfur isotopic composition of any oil is closely tied to source organic matter, as well as to coeval sulfate cements, and sulfur isotope values in oil and source-rock sulfate cements can serve as a source correlation parameter. Observed oil $\delta^{34}S$ values are usually identical to or enriched by less than $2\%o$ relative to source kerogen (Orr, 1985) and are depleted by approximately $15\%o$ as compared to source-rock sulfate cements (Thode and Monster, 1965). Delaware Mountain Group oils have a mean $\delta^{34}S$ of $+0.7\%o$ and are depleted by $13.8\%o$ with respect to the source-rock sulfate; this is in the range expected for an oil and its associated source rock. Thus, Delaware Mountain Group sulfur isotope data give a good correlation between sulfate of the suspected rock and the oil.

Terpane Biomarker Compounds.—

Correlation using biomarker data was achieved through (1) visual inspection and comparison of oil and extract mass chromatograms for the various biomarker compounds, (2) comparison of biomarker concentration data derived by integration of mass chromatogram peak height and referenced to an internal standard (α cholestane) and (3) comparison of certain key biomarkers present in oils and extracts.

Diterpanes and triterpanes were determined by examination of m/z 177 and 191 mass chromatograms and parent ion mass chromatograms. Diterpanes from C20 to C30 (Fig. 14) and triterpanes from C27 to C35 (Fig. 14) were identified in all samples. Visual comparison of oil and source-rock extract mass chromatograms shows excellent correlation between terpane distributions. Total terpane content is nearly equivalent between extracts and oils. Extracts average 36.4% total terpanes; oils average 38.0%. Hopane C30/C29 and C30/C28 ratios are valuable source correlation parameters and show good agreement between source extracts and oils. Extracts average 1.47 for C30/C29 and 0.05 for C30/C28. Oils average 1.48 for C30/C29 and 0.1 for C30/C28. The presence of the C30 triterpane 'oleanane' in Delaware Mountain Group oils and source-rock extracts is another good correlation parameter. Oleanane is a relatively rare biomarker which has been used as an indicator of terrigenous organic matter input from Late Cretaceous and Tertiary age sources (Ekweozor and Udo, 1988). The presence of oleanane in the Delaware Mountain Group is unique in that it has not been previously identified in rocks older than late Cretaceous. The oleanane can be assumed to be indigenous to these rocks because no thermally mature Tertiary or late Cretaceous section exists near the basin. Oleanane is much more abundant in extracts of terrigenous Type III kerogen in the Delaware Mountain Group, having a mean concentration of 35.0 ppm. It is also present to a lesser degree in Type II kerogen extracts, averaging 7.9 ppm, indicating that Delaware Mountain Group organic matter is not homogeneous; some terrigenous organic matter is present even in kerogen classified as Type II by Rock-Eval pyrolysis. The oils contain an average 1.6 ppm oleanane, indicating that the more abundant, oil-prone Type II kerogen generated most Delaware Mountain Group oil rather than the oleanane-rich Type III kerogen. The correlation between oil and source-rock terpane content is very good; however, one difference that exists between oils and extracts is seen in the abundance of hopanes relative to tricyclic terpanes. Terpane content in source-rock extracts shows a mean 12.2% diterpanes and 69.4% ho-

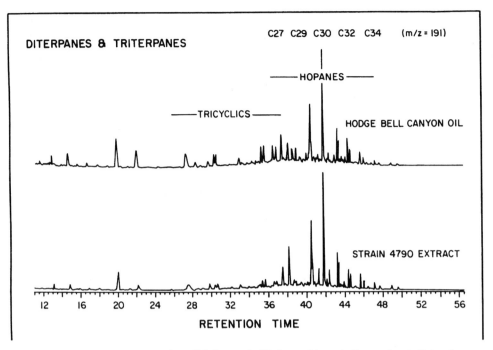

Fig. 14.—Terpane mass chromatograms (m/z = 191) for a typical Delaware Mountain Group oil and siltstone kerogen extract.

panes. Terpane content in the oils averages 45.0% diterpanes and 42.1% hopanes. Like the C20 sterane S/(R + S) anomaly discussed earlier, the variation in tricyclic terpane and hopane content between oil and source rock is a result of biomarker fractionation during migration. Tricyclic terpanes have been shown to migrate more quickly than pentacyclic terpanes in laboratory and field studies (Seifert et al., 1979; Zhusheng et al., 1988; Pu et al., 1989); thus, oils should be enriched in tricyclic terpanes relative to their source rocks precisely as is observed in Delaware Mountain Group oils and extracts.

Sterane Biomarker Compounds.—

The steranes were determined by mass chromatography, using m/z 217, 218, 231, 400 and 414 chromatograms and molecular ion plots. The oils and extracts contain high concentrations of steranes; regular steranes (C21, C22 and C27–C30) comprise a mean 50.4% of total biomarker compounds in the oils and 48.0% of the source-rock extracts. Visual comparison of mass chromatograms shows excellent correlation between Delaware Mountain Group oil and source-rock extract patterns and reveals several important correlation parameters. Figure 15 shows the occurrence and similar distributions of low molecular weight C21 and C22 steranes, diasteranes and C27 through C30 regular steranes, which contribute to a viable correlation between Delaware Mountain Group oils and source rock. Differences exist between the relative proportions of diasteranes versus regular steranes and between $\alpha\alpha\alpha$ versus $\alpha\beta\beta$ regular steranes in oils and extracts. These differences can be explained by the different processes to which biomarkers in oils and extracts are subjected. Diasteranes are abundant, comprising an average 47.3% of the total steranes in the oils and 39.5% of steranes in the extracts. Diasteranes are formed by acid-catalyzed backbone rearrangement of regular steranes in clay-rich rocks and differ from regular steranes only in the location of

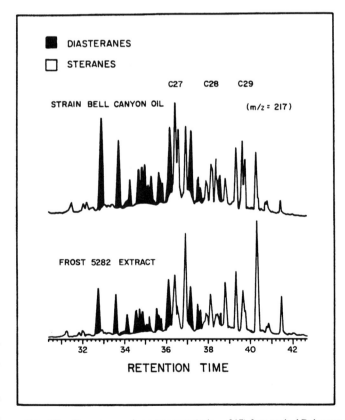

Fig. 15.—Sterane mass chromatograms (m/z = 217) for a typical Delaware Mountain Group oil and siltstone kerogen extract. Diasterane peaks are shaded for reference.

two methyl substituents (Clayton, 1989). High diasterane content is diagnostic of extracts and oils derived from clay mineral-bearing siliciclastic rocks and is a good correlation parameter. Higher concentrations of diasteranes occur in oils than in their respective source rocks because diasteranes migrate more rapidly than regular steranes (Seifert and Moldowan, 1978). This appears to be the case in the Delaware Mountain Group and is in agreement with other molecular indicators of migration (i.e., C20 sterane R/S ratios, 5α(H), 14α(H), 17α(H)/5α(H), 14β(H), 17β(H) sterane ratios and tricyclic terpane/hopane ratios). Delaware Mountain Group oils and extracts contain similar abundances of C30 steranes, averaging 2.5 and 3.4% respectively. The presence of C30 steranes is indicative of extracts and oils derived from a marine source. C27–C29 sterane distributions also correlate well between oils and extracts. A ternary diagram of C27, C28 and C29 sterane abundances shows a very tight grouping of oil and extract samples (Fig. 16). The oils contain about 5% less C29 steranes than the average source-rock extract. This may be explained by a greater contribution of marine Type II kerogen to oil generation; organic matter derived from a marine source generally contains lower concentrations of C29 steranes relative to C27 steranes. The extent of isomerization at C-14 and C-17 as measured by the αα/ββ sterane ratio is greater in the oils than in the source-rock extracts. The difference in the αα/ββ sterane ratio between oil and extract is most likely a migrational effect. ββ steranes migrate faster than αα steranes and may be concentrated in oils relative to the source rock (Zhusheng et al., 1988).

Aromatic Sterane Biomarker Compounds.—

Further support of the Delaware Mountain Group oil source rock correlation is seen in the aromatic sterane distributions and in the presence of β-carotane. Both monoaromatic (C-ring) and triaromatic (ABC-ring) steroid hydrocarbons are present in significant quantities in oil and in siltstone extracts. The aromatic

steroid /(steroid + aromatic steroid) ratio averages 24.61% and 20.17% for extracts and oils respectively.

Comparison of mono- and triaromatic sterane mass chromatographs (m/z 253 and m/z 231) shows that the profiles are practically identical (Fig. 17). β-carotane is a fairly rare biomarker which has been variously interpreted as an indicator of terrestrially-derived organic matter (Moldowan et al., 1985) or of organic matter deposited in hypersaline environments (Mello et al., 1988a, b; Jiang and Fowler, 1986). β-carotane is present in Delaware Mountain Group oils and extracts. Interestingly, β-carotane is more abundant in the terrestrially-derived Type III kerogen extracts than in either the Type II extracts or oils.

Organic geochemical study of Delaware Mountain Group siltstones and modern fluids shows that the organic matter is of suitable type, adequate abundance and a sufficient state of thermal maturity such that, during late stage diagenesis, pore-water chemistry within the sandstones must have been strongly influenced by organic matter maturation products. In addition, correlation of oil and source rock in the Delaware Mountain Group shows that Delaware Mountain Group siltstone organic matter did yield the fluids associated with thermal stress and that these fluids did migrate into the sandstones. The final test is to look for evidence of organic influence on pore-water chemical evolution along flow pathways in the permeable sandstones. Fluids moving from the organic-rich, encapsulating siltstones into the sandstones where Delaware Mountain Group oil is reservoired should have left distinctive isotopic and trace element signatures in the authigenic minerals that formed during later stages of diagenesis if in fact there was a trenchant effect on pore-water chemistry and diagenetic process.

TRACER INVESTIGATION OF COUPLED ORGANIC/INORGANIC DIAGENESIS

Organic geochemical data and geochemical indicators of organic control over fluids involved in late diagenesis, such as the occurrence of authigenic titanium oxide minerals and minor element chemistry of chlorite and dolomite, build a strong case for the importance of coupled organic-inorganic diagenesis in the Delaware Mountain Group. Yet, perhaps the case is not incontrovertible with these arguments alone. The third argument, a tracer study, provides the third leg by which the case can stand alone. By characterizing the isotopic and trace element composition of Delaware Mountain Group siltstone organic matter, it is possible to identify useful natural tracers; these tracers can be used to establish a genetic link between the organic matter and authigenic products in the sandstones and show that the products of organic matter degradation did exercise some control over pore-fluid chemical evolution in the sandstones. No ideal tracer exists in this system; however, several tracers are available that do give evidence of a source/sink relationship for products of siltstone organic matter degradation and late-stage authigenic minerals in the sandstones.

Carbon Isotopes

Carbon isotopes of dolomite are the most patent indicator of organic involvement in pore-fluid chemistry. Carbon isotope fractionation is relatively independent of temperature, but considerable fractionation of carbon occurs between the organic and inorganic carbon reservoirs. As a result, a large variance exists between the isotopic composition of carbon derived from

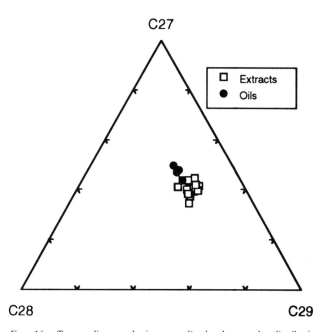

FIG. 16.—Ternary diagram plotting normalized carbon number distributions of 5α, 14α, 17α regular steranes for oils and extracts.

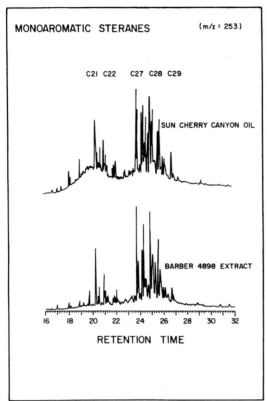
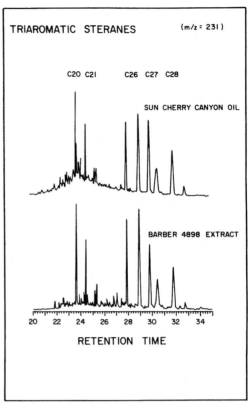

Fig. 17.—Aromatic sterane mass chromatograms (m/z = 253, 231) for a typical Delaware Mountain Group oil and siltstone kerogen extract.

a carbonate source that formed at equilibrium with seawater (about +4‰ for Permian carbonate) and carbon derived from organic matter (< −25‰ for Delaware Mountain Group siltstone organic matter). The $\delta^{13}C$ values of authigenic dolomite will reflect the carbon isotopic composition of the carbon incorporated into the carbonate structure whether from an organic source, an inorganic source or some combination. Dolomite $\delta^{13}C$ values in Delaware Mountain Group sandstones vary from −12 to 3‰ PDB and have a mean value of −2.8‰ (Fig. 5). The most positive $\delta^{13}C$ values (>0‰) indicate a minimal organic contribution to dolomite carbonate carbon. However, the majority of dolomite $\delta^{13}C$ values are much lower, falling between 0 and −10‰. These indicate a definite organic contribution to dolomite carbon. Using $\delta^{13}C$ values of −25‰ and +4‰ respectively for the organic- and marine-derived carbon, the organic matter contribution to dolomite is calculated as an average 25% for all samples analyzed. The maximum organic contribution is approximately 50%. The maximum organic contribution calculated here is in good agreement with the theoretical maxima determined by Meshri (1986) for a pore-water system bearing organic acids in which pH is buffered by a pre-existing marine carbonate. Meshri showed that decarboxylation of organic acids would result in organically derived acidity being represented by the carbonic acid system. Dissociation of the CO_2 would bring about dissolution of pre-existing carbonate with a heavy, marine carbon isotopic composition, and carbonate species in solution would originate from both ^{13}C-depleted organic carbon and ^{13}C-enriched marine carbonate; such a signal is clearly evident in Delaware Mountain Group dolomite.

Thus, the dolomite—which was a late burial authigenic product—shows an unmistakable organic signature.

Trace Metals

The Nature of Trace Metals in Organic Matter.—

Much of the metal in kerogens is incorporated in porphyrin complexes (Tissot and Welte, 1978; Lewan, 1980). Porphyrins are tetrapyrrollic rings inherited from chlorophyll. The tetrapyrrole framework consists of a ring of four pyrrole molecules connected at the a position carbons by four methine (CH) molecules (Fig. 18). Within the center of the structure, two of the pyrroles hold imine hydrogen atoms. Under slightly acidic conditions, these hydrogen atoms are easily replaced by bivalent cations (e.g., VO^{+2}, Ni^{+2}, Mg^{+2}, Mn^{+2}, Fe^{+2}, Co^{+2}, Zn^{+2} and Cu^{+2}). Cations of small radii are favored (Lewan, 1980). Metals bond with opposing tetrapyrrole nitrogen atoms; these are tenacious bonds which have high thermal stability (Morandi and Jensen, 1964) and which resist exchange reactions between complexed metal and cations in the surroundings (Dean and Girdler, 1960). Considerable metal quantities can be released only through destruction of the complexes as thermal degradation progresses during burial. Experimental studies show that destruction of tetrapyrrole rings begins at moderate levels of thermal maturation with increasing breakdown occurring with increased thermal stress.

Humate complexes are organometallic complexes in which oxygen atoms serve as ligands, usually in carboxyl or phenolic functional groups (MacCarthy and Mark, 1976; Surdam, et al.,

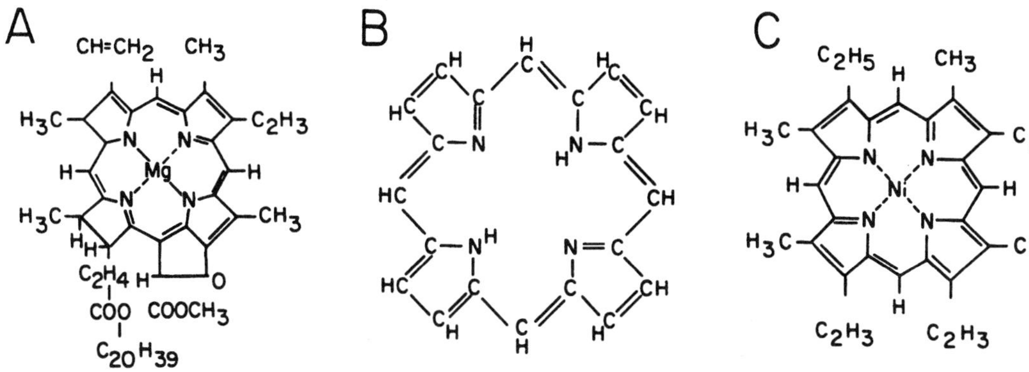

FIG. 18.—Examples of (A) chlorophyll—a biologic precursor to tetrapyrrole complexes, (B) the basic framework of tetrapyrrole complexes, and (C) insular porphyrin—a tetrapyrrole common at moderate levels of thermal stress (modified from Lewan,1980).

1989). Humate complexes form strong bonds with metal cations at neutral to basic pH conditions but are unstable at low pH (Lewan, 1980). Humate complexation of metals is thought to be important during early burial but only up to the early stages of thermal maturation, where conditions become increasingly reducing and acidic and humate demetallation takes place (Tissot and Welte, 1978; Lewan, 1980). Because of the differences in stability, porphyrins are thought to be the more important complexer in the deep-burial diagenetic environment.

The sources of metals occurring in kerogen and oil are metals originally incorporated by the living organism and metals derived from interstitial water. In the major types of organic matter present in sediments, endemic metal concentrations are controlled by concentrations in the growth medium (Cannon, 1963; Riley and Roth, 1971). In a comprehensive appraisal of nickel and vanadium contents of organic matter and kerogen, Lewan (1980) noted that ligninic kerogen (Type III) extracts usually did not require an enrichment process to explain nickel and vanadium content when compared with the precursor organic matter, while original concentrations in precursor material of liptinitic kerogen (Type I and Type II) were insufficient to account for metal concentrations and did require an additional source. To account for the differences observed, formation of organometallic complexes after sedimentation must also be important. In an open sediment system, ion diffusion between a volume of sediment, the water column and surrounding sediments can supply metals to organic matter. As metals in pore water react to form organometallic complexes, additional cations diffuse into the sediment volume to replace extracted metals. This model has been widely applied to explain concentration of metals in organic-rich sediments, such as in formation of sedimentary iron sulfides (Berner, 1969) and manganese oxides (Bonatti et al., 1972) and organic incorporation of nickel and vanadium (Lewan, 1980).

Organic matter—kerogen, bitumen and oils—have been shown to hold a distinctive trace metal signature during thermal degradation and generation of oil in numerous studies (Taguchi, 1975; Tissot and Welte, 1978; Thompson et al., 1990). Trace metals in organic phases have been used to classify and fingerprint oils and correlate different organic matter phases (i.e., correlation of kerogen and bitumen in source rocks to oils in reservoir rocks). These studies have most commonly used trace metal proportionality as a comparative parameter.

With the thermal stress associated with burial, organic matter undergoes a rearrangement toward a more thermally stable form; heteroatomic bonds are broken, and volatiles (e.g., water, carbon dioxide, carboxyl and carbonyl groups), are released—all a part of the oil generation process. Trace metals are also released with thermal degradation of organometallic complexes. Pore waters enriched with these metals make them readily available for incorporation into authigenic minerals.

Trace Metals as Geochemical Indicators.—

Trace metals in clays have proven useful in studies of many systems as indicators of geochemical environment and as tracers delineating flow paths, among other uses (e.g., Riese and Brookins, 1984; Hansley, 1986; Coveny et al., 1987). However, most studies make rather limited use of trace metal data as a tool, using this tool for generalized assessments rather than for making specific determinations. More rigorous and exacting use of tracers in clay and in organic phases should also be possible, specifically for definitively linking organic and inorganic diagenetic systems.

The use of metals as tracers is problematic in an ancient geological system. Investigation is limited to trace-metal-bearing phases that can be easily separated and analyzed by techniques with usable limits of detection. Because of these constraints, dolomite was analyzed for trace elements only by electron microprobe analysis; chlorite—which is easily isolated without alteration—was analyzed by neutron activation analysis after assessment of sample purity using XRD. Organic-matter bitumen extracts of organic-rich siltstones were also analyzed by NAA. Bitumen was chosen for study because the organic solvent extraction technique used for bitumen has been shown to have no significant effect on trace metal content and gives a good appraisal of organic matter trace-metal composition, whereas the acid digestion technique for acquiring kerogen separates has a considerable effect (Lewan, 1980; Hays, 1992).

Choosing informative tracers is another problem. Any tracer of value must have been involved in (1) the demetallation reactions that first caused release of metals from the parent organic matter during thermal maturation and (2) the precipitation reactions that incorporated metals within the important authigenic minerals in detectable quantities. The only detectable tracer of value for dolomite is manganese. The structure of chlorite lends itself to more extensive substitutions, and several metals present in chlorite have been identified as valuable tracers in this study: manganese, zinc, nickel, cobalt, vanadium and chromium. Unfortunately, most of the metals detected in chlorite are present to some degree in every fraction of the system under consideration (i.e., siltstone organic matter, siltstone and sandstone detrital minerals and sandstone authigenic minerals). Therefore, simply using a metal's presence or absence as a criterion would yield ambiguous results since there are multiple possible sources. The exceptions were those cases where the metal is absent or relatively rare in the detrital fraction while present in significant quantities in the organic and authigenic fractions, such as for manganese and vanadium. In these cases, a mass balance argument can be applied to establish a relation. Another problem to be addressed for these non-conservative tracers is their distribution between dissimilar phases—organic matter and chlorite—during release and incorporation processes. Various metals are bound within organic matter by bonds of differing strength and are released differentially as the bonds are ruptured by thermal stress. Measures for the stability of organometallic bonds include Gibbs free energy for metallation, demetallation activation energy and ligand field stabilization energy, LFSE (Lewan, 1980). Unfortunately, a dearth of thermodynamic data leaves LFSE as the only indicator of organometallic bond strength for all of the metals of concern. LFSE is related to the distribution of electrons in different energy level orbitals within a bound cation and increases with increasing stability of the bound cation. Cations with similar LFSE values should behave comparably during thermal destruction of organometallic complexes and be released at similar times during thermal maturation and in similar concentration ratios. In the sink which is under consideration for these metals—chlorite, dissimilar cations incorporated into the chlorite structure will have different partition coefficients, causing an inconsistent representation of pore-fluid chemistry during chlorite formation if partitioning is not considered. Both the strength of metal cation bonding with organic matter and the affinity of metal cations for incorporation into specific sites within the chlorite structure are strongly dependent upon ionic radius and charge. To minimize the effects of nonuniform release of various metals from organic matter and the effects of partitioning into chlorite (parameters that are not well constrained) and still be able to detect any tracer signal relating or differentiating the three fractions, ratios of trace metals were used to correlate organic matter (the bitumen fraction) to chlorite. Using ratios of metals that have affinities for similar structural positions in chlorite and have similar bond stabilities in organic matter complexes as tracer parameters allows uses of metals that may be present in all fractions and minimizes the effects of partitioning of metal out of organic matter and into chlorite during diagenesis. Ionic radii, charge and LFSE values for the six metals used as tracers are listed in Table 5. LFSE data used in this study are from Lewan (1980); radii data are from Zhadanov (1965).

The six metals used as tracers were relatively abundant in the organic matter and the chlorite fractions and tended to be less abundant in the detrital fraction (Table 6). The metals were grouped into ratio pairs based on (1) similarities in incorporation into chlorite as determined by charge and ionic radii of the cations commonly included in the chlorite structure and (2) on similarities in organic bond strengths as related to ligand field strength. The ratio pairs used were $Mn/Mn + Zn$, $Ni/Ni + Co$

TABLE 5.—RADII AND LFSE* OF TRACER CATIONS

Cation	Radius (Å)	LFSE*
Mn + 2	0.80	0
Zn + 2	0.83	0
Ni + 2	0.74	1.20
Co + 2	0.78	0.80
V + 3	0.67	—
Cr + 3	0.64	—
VO + 2	0.63	1.20
V + 2	0.72	—
Cr + 2	0.83	0.60

*LFSE = ligand field stabilization energy; values are calculated for octahedral coordination. Radii data are from Zhadanov (1965); LFSE data are from Lewan (1980).

TABLE 6.—AVERAGE TRACER ABUNDANCES AND RATIOS

Elements & Ratios	Fraction Concentrations (ppm) and Ratio Values*		
	Bitumen (N = 25)	Chlorite (N = 16)	Detrital (N = 16)
Mn	512	661	143
Zn	370	246	22
Mn/Mn + Zn	0.67	0.73	0.88
Ni	736	113	9
Co	22	14	4
Ni/Ni + Co	0.96	0.86	0.17
V	255	166	31
Cr	28	160	59
V/V + Cr	0.62	0.52	0.35

*Ratio values are the mean of individual metal analysis ratios. N indicates the number of samples.

and $V/V + Cr$; mean values of the metal ratios calculated for individual samples are reported in Table 6. The average $Mn/Mn + Zn$ ratios for bitumen organic matter and chlorite are 0.67 and 0.73, respectively, a difference of 0.06. The detrital fraction $Mn/Mn + Zn$ ratio is 0.88, a difference of 0.15 as compared with chlorite. The average $Ni/Ni + Co$ ratios for bitumen organic matter and chlorite are 0.96 and 0.86, respectively, a difference of 0.10. The detrital fraction $Ni/Ni + Co$ ratio is 0.17, a difference of 0.69 when compared with chlorite. The average $V/V + Cr$ ratios for bitumen organic matter and chlorite are 0.62 and 0.52, respectively, a difference of 0.10. The detrital fraction $V/V + Cr$ ratio is 0.35, a difference of 0.17 when compared with chlorite. These tracer indices show a good correlation between organic matter bitumen extracted from the siltstones and authigenic chlorite separated from the sandstones. Ratio values determined for the bitumen organic matter and chlorite are similar, while ratios determined for the detrital fraction differ considerably.

When normalized $Mn/Mn + Zn$, $Ni/Ni + Co$ and $V/V + Cr$ ratios are plotted on a ternary diagram, the correlation between bitumen organic matter and chlorite tracer content becomes strikingly apparent (Fig. 19). The bitumen organic matter population exhibits variable $V/V + Cr$ and $Mn/Mn + Zn$ ratios and fairly consistent $Ni/Ni + Co$ ratios. The chlorite population has two subpopulations: a population falling along the $Ni/Ni + Co$ zero line—having $Ni/Ni + Co$ ratios at or very near zero, and a population falling in a zone near the center of the plot—having subequivalent metal ratios. The detrital population shows fairly consistent, moderate to low $V/V + Cr$ ratio; variable, relatively high $Mn/Mn + Zn$ ratios and variable, low $Ni/$

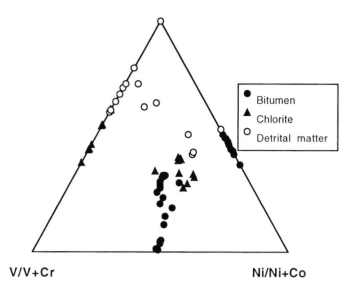

FIG. 19.—Ternary diagram plotting normalized ratios of the important metal tracer pairs for bitumen organic matter, authigenic chlorite, and detrital mineral fractions.

Ni + Co ratios with a significant fraction of the population having Ni/Ni + Co ratios at or near zero. The bitumen organic matter population and central chlorite population show considerable overlap while most of the detrital population is distinctly separated from the other two populations, with the exception of two detrital samples that have relatively high Ni/Ni + Co ratios. Bitumen organic matter and chlorite tend to have lower Mn/Mn + Zn and higher Ni/Ni + Co and V/V + Cr ratios than the detrital fraction. Chlorite ratio values generally are intermediate to the values of the bitumen and detrital mineral fractions, which are the possible sources for the metals; this is probably indicative of some input from each source. The zero-line chlorite population ratio values are similar to detrital population values and probably indicate the predominance of detrital material as the source of the metals in those clay samples. However, the central chlorite population comprises the majority, and the ratio values are much closer to those of the bitumen organic matter—evidence that organic matter was exercising the dominant control on the concentration of these metals in solution during the time of formation of most chlorite and evidence of a genetic relation between organic matter and chlorite.

Thus, tracer investigation of the Delaware Mountain Group shows a strong organic signature preserved in the late stage authigenic products (e.g., chlorite and dolomite). Evidence gained from natural tracers—carbon isotopes and trace metals—indicates that pore-fluid chemistry in the sandstones was controlled to a considerable extent by organic matter alteration in the organic-rich siltstones that encapsulate the sandstones. When this evidence is combined with petrologic, geochemical and organic geochemical evidence, it allows construction of a powerful conceptual model and builds a case for organic matter playing a critical role in diagenetic processes that is essentially unequivocal. As thermal maturation of organic matter progressed, thermal cracking and decarboxylation of organic matter and destruction of organometallic complexes produced oil and introduced isotopically light carbon and metals into pore

waters; all of which migrated into the sandstones encapsulated by the organic-rich siltstone source. Experimental work suggests that these processes occur coevally with the production of carboxylic acids and CO_2—processes apparently responsible for the dissolution event in Delaware Mountain Group sandstones. So, pore fluids bearing the products of organic matter alteration controlled mineral dissolution as well as mineral authigenesis, and the authigenic minerals forming preserved an isotopic and chemical record of the ambient pore-fluid chemistry.

SUMMARY AND CONCLUSIONS

(1) The Delaware Mountain Group is composed of very fine-grained, subarkosic sandstone and siltstone. Sandstones are interbedded with thick, organic-rich siltstones that comprise up to 80% of the sequence. Shales and detrital clays notably are rare within the group. Therefore, shale diagenesis can be discounted as inconsequential in affecting diagenetic processes in the sandstones. Alteration of organic matter during burial played the predominant role in controlling diagenesis.

(2) Diagenetic events, as determined from mineralogy and paragenetic relations of mineral phases in the Delaware Mountain Group sandstones, may be categorized into four main sequential stages:
   A. Authigenic cements (predominantly calcite) developed during early burial, partially arrested burial compaction and preserved a high percentage of intergranular volume.
   B. Large-scale dissolution of the cements and some detrital material occurred during deep burial and created abundant secondary porosity.
   C. Pervasive chlorite authigenesis followed dissolution, partially occluding porosity and restricting fluid flow throughout the sequence.
   D. Authigenesis of dolomite, ferroan/manganous dolomite, potassium feldspar overgrowths, quartz, brookite and anatase began as chlorite formation ebbed.

(3) Porosity in the sandstones largely is secondary in origin, created by dissolution of calcite cement. Effective porosity has been decreased by late-stage authigenic minerals, particularly by chlorite and dolomite.

(4) Data accrued during this investigation built a three-thrusted argument springing from application of petrology and inorganic geochemistry, organic geochemistry and geochemical tracers to the problem and allow construction of a conceptual model of organic involvement in the evolution of pore-water chemistry during late burial of this sequence of rocks. Aqueous species (e.g., carboxylic acids, carbonic acids and metal cations) derived from degradation of organic matter in Delaware Mountain Group siltstones modified pore-fluid pH and Eh, cation activities and mineral solubilities in the sandstones, controlling dissolution and authigenesis processes during late burial. The importance of the role that organic matter played in the diagenetic scheme is strongly evident:
   A. Delaware Mountain Group organic matter is sufficiently abundant, of suitable type and of a suitably advanced state of thermal maturity such that it could be expected

to have generated carboxylic and carbonic acids in volumes needed to create observed secondary porosity and control pore-water chemical evolution to some degree.

B. The organic matter has yielded the fluids associated with thermal stress, and these fluids moved from the siltstones into the sandstones. Correlation of siltstone organic matter with oil in Delaware Mountain Group sandstone reservoirs shows that the siltstones were the source of much of the oil. Experimental work by other researchers has shown that large volumes of carboxylic acids, carbonic acid and water are released from source-rock kerogen prior to and during the stage of oil generation; because Delaware Mountain Group kerogen has generated oil, it has certainly yielded these other products of thermal stress (i.e., carboxylic acids, carbonic acid and water).

C. Carbon isotopes give clear evidence of the presence of organically-derived aqueous species in pore water. Low $\delta^{13}C$ values of late-stage authigenic products—dolomite— are evidence of an organic contribution to pore-water carbonate carbon.

D. The presence of abundant authigenic titanium oxides is another valuable indication of the involvement of organically-derived compounds in late diagenetic processes. The formation of authigenic titanium oxides indicates that titanium mobility was elevated by formation of organometallic titanium complexes—the only natural mechanism for enhancing titanium solubility. Organic ligands derived from degradation of Delaware Mountain Group organic matter must have been present in pore water during late burial to account for the mobilization of this extremely insoluble metal.

E. Natural inorganic tracers in the organic matter (manganese, zinc, nickel, cobalt, vanadium and chromium) establish a source/sink relation between the organic matter in the siltstones and authigenic minerals in the permeable sandstones. The relation is manifest in the trace element content of late-stage authigenic products that formed in pore fluids whose chemistry was controlled by organic matter degradation. Mass balance considerations show that organic matter was the most likely source for some of the metals incorporated into late-stage authigenic chlorite and dolomite; this is especially true for manganese. Where a mass balance approach cannot be effectively applied, comparison of ratios of similar metal cations shows that these ratios are preserved when comparing organic matter and chlorite, but are significantly different from the other possible source—the detrital fraction. The various lines of tracer evidence indicate that fluids carrying the products of organic matter degradation moved from the organic-rich siltstones into the sandstones where authigenic products formed and preserved a record of ambient pore-water chemistry.

ACKNOWLEDGMENTS

This project was supported by the State Lands Energy Recovery Optimization Project of the Texas Governor's Office. The authors also express their gratitude to Dr. Ethan Grossman, Texas A&M Department of Geology and Geophysics, whose help, guidance and stable isotope lab were critical to collection of stable isotopic data; Dr. Dennis James for his assistance with NAA and use of the Texas A&M Center for Chemical Characterization facilities; and Dr. Charles Kennicutt and Dr. Tom MacDonald at the Texas A&M Geochemical and Environmental Research Group for their help and use of GERG facilities, without which much of the geochemical work included here would have been impossible. We also thank the Texas A&M Nuclear Science Center for use of the research reactor.

REFERENCES

AHN, J. H. AND PEACOR, D. R., 1985, TEM study of diagenetic chlorite in Gulf Coast argillaceous sediments: Clays and Clay Minerals, v. 33, p. 228–236.

BAIN, D. C., 1976, A titanium rich soil clay: Journal of Soil Science, v. 27, p. 68–70.

BARNHISEL, R. I. AND BERTSCH, P. M., 1989, Chlorites and hydroxy interlayered vermiculite and smectite, in Dixon, J. B. and Weed, S. B., eds., Minerals in Soil Environments: Madison, Soil Science Society of America, 1244 p.

BERNER, R. A., 1969, Migration of iron and sulfur within anaerobic sediments during early diagenesis: American Journal of Science, v. 267, p. 19–42.

BERROW, M. L., WILSON, M. J., AND REAVES, G. A., 1978, Origin of extractable titanium and vanadium in the A horizons of Scottish podzols: Geoderma, v. 21, p. 89–103.

BEVAN, J. AND SAVAGE, D., 1989, The effect of organic acids on the dissolution of K-feldspar under conditions relevant to burial diagenesis: Mineralogical Magazine, v. 53, p. 415–425.

BOLES, J. R. AND FRANKS, S. G., 1979, Clay diagenesis in Wilcox Sandstones of southwest Texas: Implications of smectite diagenesis on sandstone cementation: Journal of Sedimentary Petrology, v. 49, p. 55–70.

BONATTI, E., KRAEMER, T., AND RYDELL, H., 1972, Classification and genesis of submarine iron- manganese deposits, in Horn, A., ed., Ferromanganese Deposits on the Ocean Floor: New York, Arden House, p. 149–166.

BURST, J. F., 1969, Diagenesis of Gulf Coast clayey sediments and its possible relation to petroleum migration: American Association of Petroleum Geologists Bulletin, v. 53, p. 73–93.

CANNON, H. L., 1963, The biochemistry of vanadium: Soil Science, v. 96, p. 196–204.

CAROTHERS, W. W. AND KHARAKA, Y. K., 1978, Aliphatic acid anions in oilfield waters-implications for the origin of natural gas: American Association of Petroleum Geologists Bulletin, v. 62, p. 2241–2453.

CLAYTON, J. L., 1989, Geochemical evidence for Paleozoic oil in Lower Cretaceous O Sandstone, northern Denver basin: American Association of Petroleum Geologists Bulletin, v. 73, p. 977–988.

CORRENS, C. W., 1978, Titanium, in Wedepohl, K. H., ed., Handbook of Geochemistry: Berlin, Springer-Verlag, p. 220.

COVENY, R. M., JR., LEVENTHAL, J. S., GLASCOCK, M. D., AND HATCH, J. R., 1987, Origins of metals and organic matter in the Mecca Quarry Shale Member and stratigraphically equivalent beds across the Midwest: Economic Geology, v. 82, p. 915–933.

CRAIG, H., 1957, Isotopic standards for carbon and oxygen correction factors for mass spectrometric analysis of carbon dioxide: Geochimica et Cosmochimica Acta, v. 12, p. 133–149.

CROSSEY, L. J., 1985, The origin and role of water soluble organic compounds in clastic diagenetic systems: Unpublished Ph.D. Dissertation, University of Wyoming, Laramie, 115 p.

CROSSEY, L. J., 1991, Thermal degradation of aqueous oxalate species: Geochimica et Cosmochimica Acta, v. 55, p. 1515–1527.

CURIALE, J. A., LARTER, S. R., SWEENEY, R. E., AND BROMLEY, B. W., 1989, Molecular thermal maturity indicators in oil and gas source rocks, in Naeser, D. and McCulloh, T. H., eds., Thermal History of Sedimentary Basins: New York, Springer-Verlag, p. 53–72.

CURTIS, C. D. AND COLEMAN, M. L., 1986, Controls on the precipitation of early diagenetic calcite, dolomite, and siderite concretions in complex depositional sequences, in Gautier, D., ed., Roles of Organic Matter in Sediment Diagenesis: Tulsa, Society of Economic Paleontologists and Mineralogists Special Publication 38, p. 23–34.

DEAN, R. A. AND GIRDLER, R. B., 1960, Reaction of metal etioporphyrins on dissolution in sulfuric acid: Chemical Industry, January, p. 100–101.

DENHAM, M. E., 1992, Minor non-carbonate authigenic components as indicators of late-stage diagenetic processes in the Smackover Limestones: Unpublished Ph.D. Dissertation, Texas A&M University, College Station, 112 p.

EKWEOZOR, C. M. AND UDO, O. T., 1988, The oleananes: Origin, maturation and limits of occurrence in Southern Nigeria sedimentary basins: Organic Geochemistry, v. 13, p. 131–140.

FAURE, G., 1986, Principles of Isotope Geology, 2nd ed.: New York, Wiley and Sons, 464 p.

FISHER, J. B., 1987, Distribution and occurrence of aliphatic acid anions in deep subsurface waters: Geochimica et Cosmochimica Acta, v. 51, p. 2459–2468.

FRANKS, S. G. AND FORESTER, R. W., 1984, Relationships among secondary porosity, pore-fluid chemistry and carbon dioxide, Texas Gulf Coast, in McDonald, D. A. and Surdam, R. C., eds., Clastic Diagenesis: Tulsa, American Association of Petroleum Geologists Memoir 37, p. 63–79.

GALIMOV, E. M., 1973, Carbon isotopes in oil and gas geology (original in Russian): Moscow, Nedra, English translation-Washington, NASA Publication TT F-682, 22 p.

GALLEY, J. E., 1958, Oil and geology in the Permian basin of Texas and New Mexico, in Weeks, L. G., ed., Habitat of Oil: Tulsa, American Association of Petroleum Geologists, p. 395–446.

GALLOWAY, W. E., 1984, Hydrogeologic regimes of sandstone diagenesis, in McDonald, D. A. and Surdam, R. C., eds., Clastic Diagenesis: Tulsa, American Association of Petroleum Geologists, p. 3–13.

GIVEN, R. K. AND LOHMANN, K. C., 1986, Isotopic evidence for the early meteoric diagenesis of the reef facies, Permian reef complex of West Texas and New Mexico: Journal of Sedimentary Petrology, v. 56, p. 183–193.

HANOR, J. S., 1987, Variations in the chemical composition of oil-field brines with depth in northern Louisiana and southern Arkansas: Implications for mechanisms and rates of mass transport and diagenetic reaction: Gulf Coast Association of Geological Societies Transactions, v. 34, p. 55–61.

HANSLEY, P.L., 1986, Relationship of detrital, nonopaque heavy minerals to diagenesis and provenance of the Morrison Formation, Southwestern San Juan Basin, New Mexico, in Turner-Peterson, C. E., Santos, E. S., and Fishman, N. S., eds., A Basin Analysis Case Study: The Morrison Formation Grants Uranium Region New Mexico: Tulsa, American Association of Petroleum Geologists, p. 257–276.

HARRISON, W. J. AND THYNE, G. D., 1992, Predictions of diagenetic reactions in the presence of organic acids: Geochimica et Cosmochimica Acta, v. 56, p. 565–586.

HAYES, P. T., 1964, Geology of the Guadalupe Mountains, New Mexico: Washington, D.C., United States Geological Survey Professional Paper 446, 69 p.

HAYS, P. D., 1992, The role of organic matter alteration in sediment diagenesis—The Delaware Mountain Group of West Texas and southeast New Mexico: Unpublished Ph.D. Dissertation, Texas A&M University, College Station, 177 p.

HILLS, J. M., 1972, Late Paleozoic sedimentation in West Texas Permian Basin: American Association of Petroleum Geologists Bulletin, v. 26, p. 217–255.

HOWER, J. E., ESLINGER, E. V., HOWER, M. E., AND PERRY, E. A., 1976, Mechanism of burial metamorphism of argillaceous sediment: Mineralogical and chemical evidence: Geological Society of America Bulletin, v. 87, p. 725–737.

JAMES, N. P. AND CHOQUETTE, P. W., 1984, Limestones—the meteoric diagenetic environment: Geoscience Canada, v. 11, p. 161–194.

JIANG, Z. S. AND FOWLER, M. G., 1986, Carotenoid derived alkanes in oils from northwestern China: Organic Geochemistry, v. 10, p. 831–839.

KAISER, W. R., 1984, Predicting reservoir quality and diagenetic history in the Frio Formation (Oligocene) of Texas, in McDonald, D. A. and Surdam, R. C., eds., Clastic Diagenesis: Tulsa, American Association of Petroleum Geologists Memoir 37, p. 195–215.

KHARAKA, Y. K., HULL R. W., AND CAROTHERS, W. W., 1985, Water-rock interactions in sedimentary basins, in Relationship of Organic Matter and Mineral Diagenesis: Tulsa, Society of Economic Paleontologists and Mineralogists Short Course Notes 17, p. 79–176.

KHARAKA, Y. K., LAW, L. M., CAROTHERS, W. W., AND GOERLITZ D. R., 1986, Role of organic species dissolved in formation waters from sedimentary basins in mineral diagenesis, in Gautier, D., ed., Roles of Organic Matter in Sediment Diagenesis: Tulsa, Society of Economic Paleontologists and Mineralogists Special Publication 38, p. 111–122.

KING, P. B., 1934, Permian stratigraphy of trans-Pecos Texas: Geological Society of America Bulletin, v. 45, p. 697–798.

KING, P. B., 1942, Permian of West Texas and southeastern New Mexico: American Association of Petroleum Geologists Bulletin, v. 26, p. 535–763.

KING, P. B., 1948, Geology of the southern Guadalupe Mountains, Texas: Washington, D.C., United States Geological Survey Professional Paper 215, 183 p.

LAND, L. S., 1985, The origin of massive dolomite: Journal of Geological Education, v. 33 p. 112–117.

LEWAN, M. D, 1980, Geochemistry of vanadium and nickel in organic matter of sedimentary rocks: Unpublished Ph.D. Dissertation, University of Cincinnati, Cincinnati, 353 p.

LONGSTAFFE, F. J., 1984, The role of meteoric water in diagenesis of shallow sandstones: Stable isotope studies of the Milk River Aquifer and Gas Pool, southeastern Alberta, in McDonald, D. A. and Surdam, R. C., eds., Clastic Diagenesis: Tulsa, American Association of Petroleum Geologists Memoir 37, p. 81–98.

LUNDEGARD, P. D., LAND, L. S., AND GALLOWAY, W. E., 1984, Problem of secondary porosity: Frio Formation (Oligocene), Texas Gulf Coast: Geology, v. 12, p. 399–402.

LUNDEGARD, P. D. AND SENFTLE, J. T., 1987, Hydrous pyrolysis: a tool for the study of organic acid synthesis: Applied Geochemistry, v. 2, p. 605–612.

MACCARTHY, P. AND MARK, H. B., 1976, Perspectives in humic acid research Part II: Principles of humic acid chemistry: Cincinnati, University of Cincinnati Graduate Research Journal, v. 2, p. 63–74.

MACKENZIE, A. S., 1984, Applications of biological markers in petroleum geochemistry, in Brooks, J. and Welte, D., eds., Advances in Petroleum Geochemistry: New York, Academic Press, p. 115–214.

MANTOURA, R. F. C., 1981, Organo-metallic interactions in natural waters, in Duursma, E. K. and Dawson, R., eds., Marine Organic Chemistry: New York, Elsevier, p. 179–224.

McCREA, J. M., 1950, On the isotopic chemistry of carbonates and a paleotemperature scale: Journal of Chemical Physics, v. 18, p. 849–857.

McMAHON, D. A., 1989, Secondary porosity in sandstones and shale diagenesis, Oligocene, South Texas: Unpublished Ph.D. Dissertation, Texas A&M University, College Station, 177 p.

MEISSNER, F. F., 1972, Cyclic sedimentation in Middle Permian strata of the Permian Basin, West Texas and New Mexico, in Elam, J. and Chuber, S., eds., Cyclic Sedimentation the Permian Basin, 2nd ed.: Midland, West Texas Geological Society, p. 203–232.

MELLO, M. R., GAGLIANONE, P. C., BRASSELL, S. C., AND MAXWELL, J. R., 1988a, Geochemical and biological marker assessment of depositional environments using Brazilian oils: Marine and Petroleum Geology, v. 5, p. 205–223.

MELLO, M. R., TELNAES, N., GAGLIANONE, P. C., CHICARELLI, M. I., BRASSELL, S. C., AND MAXWELL, J. R., 1988b, Organic geochemical characterization of depositional palaeoenvironments of source rocks and oils in Brazilian marginal basins: Organic Geochemistry, v. 13, p. 31–45.

MESHRI, I. D., 1986, On the reactivity of carbonic and organic acids and generation of secondary porosity, in Gautier, D., ed., Roles of Organic Matter in Sediment Diagenesis: Tulsa, Society of Economic Paleontologists and Mineralogists Special Publication 38, p. 123–128.

MILNES, A. R. AND FITZPATRICK, R. W., 1989, Titanium and zirconium minerals, in Dixon, J. B. and Weed, S. B., eds., Minerals in Soil Environments: Madison, Soil Science Society of America, p. 1131–1206

MOLDOWAN, J. M., SEIFERT, W. K, AND GALLEGOS, E. J., 1985, Relationship between petroleum composition and depositional environment of petroleum source rocks: American Association of Petroleum Geologists Bulletin, v. 69, p. 1255–1268.

MONCURE, G. K., LAHANN, R. W., AND SIEBERT, R. M., 1984, Origin of secondary porosity and cement distribution in a sandstone/shale sequence from the Frio Formation (Oligocene), in McDonald, D. A. and Surdam, R. C., eds., Clastic Diagenesis: Tulsa, American Association of Petroleum Geologists Memoir 37, p. 151–161.

MORSE, J. W. AND MACKENZIE, F. T., 1990, Geochemistry of Sedimentary Carbonates: New York, Elsevier, 707 p.

MORANDI, J. R. AND JENSEN, H. B., 1964, Porphyrin skeleton survives retorting: Chemical Engineering News, v. 42, p. 48.

MUCCI, A. AND MORSE, J. W., 1990, Chemistry of low temperature abiotic calcites: Experimental studies on coprecipitation, stability, and fractionation: Reviews in Aquatic Sciences, v. 3, p. 217–254.

MUCCIARONE, D. A. AND WILLIAMS, D. F., 1990, Stable isotope analyses of carbonate complicated by nitrogen oxide contamination: A Delaware Basin example: Journal of Sedimentary Petrology, v. 60, p. 608–614.

NEWELL, N. D., RIGBY, J. K., FISHER, A. G., WHITEMAN, A. J., HICKOX, J. E., AND BRADLEY, J. S., 1953, The Permian Reef Complex of the Guadalupe Mountains Region, Texas and New Mexico: San Francisco, Freeman and Co., 236 p.

NISSENBAUM, A. AND SWAINE, D. J., 1976, Organic matter—metal interactions in recent sediments: The role of humic substances: Geochimica et Cosmochimica Acta, v. 40, p. 809–816.

O'NEIL, J. R., CLAYTON, R. N., AND MAYEDA, T. K., 1969, Oxygen isotope fractionation in divalent metal carbonates: Journal of Chemical Physics, v. 51, p. 5547–5558.

ORIEL, S. S., MYERS, D. A., AND CROSBY, E. J., 1967, The West Texas Permian basin region, *in* Paleotectonic Investigations of the Permian System in the United States: Washington, D.C., United States Geological Survey Professional Paper 515-C, p. 18–60.

ORR, W. L., 1974, Biogeochemistry of sulfur, *in* Wedepohl, K. H., ed., Handbook of Geochemistry: Berlin, Springer, 343 p.

ORR, W. L., 1978, Sulfur in heavy oils, oil sands, and oil shales, *in* Strausz, O. P. and Lown, E. M., eds., Oil Sand and Oil Shale Chemistry: Berlin, Verlag, p. 223–243.

ORR, W. L., 1985, Kerogen/asphaltene/sulfur relationships in sulfur rich Monterey oils: Organic Geochemistry, v. 10, p. 499–516.

PU, F., PHILP, R. P., ZHENXI, L., AND GUANGGUO, Y., 1989, Geochemical characteristics of aromatic hydrocarbons of crude oils and source rocks from different sedimentary environments: Organic Geochemistry, v. 16, p. 427–435.

REED, C. L., 1990, The role of oxalic acid on the dissolution of granitic sand: An experimental investigation in a hydrothermal flow-through system: Unpublished M.S. Thesis, Texas A&M University, College Station, 53 p.

RIESE, W. C. AND BROOKINS, D. G., 1984, The Mount Taylor uranium deposit, San Mateo, New Mexico, USA: Uranium, v. 1, p. 189–209.

RILEY, J. P. AND ROTH, I., 1971, The distribution of trace elements in some species of phytoplankton grown in culture: Journal of Marine Biological Assessment, v. 51, p. 63–72.

SEIFERT, W. K. AND MOLDOWAN, J. M., 1978, Applications of steranes, terpanes, and monoaromatics to the maturation, migration and source of crude oils: Geochimica et Cosmochimica Acta, v. 42, p. 77–95.

SEIFERT, W. K., MOLDOWAN, J. M., AND JONES, R. W., 1979, Application of biological marker chemistry to petroleum exploration: Reprints of the 10th World Petroleum Congress, p. 425–440.

SILVER, B. A. AND TODD, R. G., 1969, Permian cyclic strata, northern Midland and Delaware Basins, West Texas and southeastern New Mexico: American Association of Petroleum Geologists Bulletin, v. 53, p. 2223–2251.

SOFER, Z., ZUMBERGE, J. E., AND LAY, V., 1986, Stable carbon isotopes and biomarkers as tools in understanding genetic relationship, maturation, biodegradation, and migration of crude oils in the Northern Peruvian Oriente (Maranon) Basin: Organic Geochemistry, v. 10, p. 377–389.

SURDAM, R. C., BOESE, S. W., AND CROSSEY, L. W., 1984, The chemistry of secondary porosity, *in* McDonald, D. A. and Surdam, R. C., eds., Clastic Diagenesis: Tulsa, American Association of Petroleum Geologists, p. 127–134.

SURDAM, R. C., DUNN, T. L., HEASLER, H. P., AND MACGOWAN, D. B., 1989, Porosity evolution in sandstone/shale systems, *in* Hutcheon, I. E., ed., Short Course in Burial Diagenesis: Montreal, Mineralogical Society of Canada, p. 61–134.

TAGUCHI, K., 1975, Geochemical relationships between Japanese Tertiary oils and their source rocks: 9th World Petroleum Congress Proceedings, v. 2, p. 193–194.

THODE, H. G. AND MONSTER, J., 1965, Sulfur isotope geochemistry of petroleum, evaporites, and ancient seas, *in* Young, A. and Galley, J. E., eds., Fluids in Subsurface Environments: Tulsa, American Association of Petroleum Geologists Memoir 4, p. 367–377.

TISSOT, B. P., 1984, Recent advances in petroleum geochemistry applied to hydrocarbon exploration: American Association of Petroleum Geologists Bulletin, v. 68, p. 545–563.

TISSOT, B. P., AND WELTE, D. H., 1978, Petroleum Formation and Occurrence: New York, Springer-Verlag, 538 p.

THOMPSON, K. F. M., KENNICUTT II, M. C., AND BROOKS, J. M., 1990, Classification of offshore Gulf of Mexico oils and condensates: American Association of Petroleum Geologists Bulletin, v. 74, p. 187–198.

URE, A. M. AND BERROW, M. L., 1982, The elemental constituents of soils: Environmental Chemistry, v. 2, Royal Society of Chemistry, p. 94–204.

VIDETICH, P. E., 1981, A method for analyzing dolomite for stable isotopic composition: Journal of Sedimentary Petrology, v. 51, p. 661–662.

WALLING, S. D., 1992, Authigenic clay minerals in sandstones of the Delaware Mountain Group: Bell Canyon and Cherry Canyon Formations, Waha field, West Texas: Unpublished M.S. Thesis, Texas A&M University, College Station, 63 p.

WAPLES, D. W., 1980, Time and temperature in petroleum formation: application of Lopatin's method to petroleum exploration: American Association of Petroleum Geologists Bulletin, v. 64, p. 916–926.

WILLEY, L. M., KHARAKA, Y. K., PRESSER, T. S., RAPP, J. B., AND BARNES, I., 1975, Short-chain aliphatic acids in oilfield waters of Kettleman North Dome oil field, California: Geochimica et Cosmochimica Acta, v. 39, p. 1707–1710.

ZHADANOV, G. S., 1965, Crystal Physics: New York, Academic Press, 261 p.

ZHUSHENG, J., PHILP, R. P., AND LEWIS, C. A., 1988, Fractionation of biological markers in crude oils during migration and the effects on correlation and maturation parameters: Organic Geochemistry, v. 13, p. 561–571.

ORIGIN AND TIMING OF CARBONATE CEMENTS IN THE ST. PETER SANDSTONE, ILLINOIS BASIN: EVIDENCE FOR A GENETIC LINK TO MISSISSIPPI VALLEY-TYPE MINERALIZATION

JANET K. PITMAN
U.S. Geological Survey, 939 Denver Federal Center, Lakewood, CO 80225, USA

AND

CHRISTOPH SPÖTL
Institut für Geologie und Paläontologie, Universität Innsbruck, Innrain 52, 6020 Innsbruck, Austria

ABSTRACT: The Ordovician St. Peter Sandstone in the Illinois Basin has undergone complex diagenetic modification involving (1) early K-feldspar, dolomicrospar, and illite precipitation, (2) late quartz, planar and baroque dolospar, anhydrite, calcite, and illite cementation and (3) carbonate-cement and K-feldspar dissolution. In southern Illinois, burial reconstruction in combination with silicate mineral age dates indicate that late-diagenetic cementation in the St. Peter Sandstone occurred during deep burial (~3,300 m) in Late Pennsylvanian and Early Permian time when major ore-forming (MVT) events were taking place in the region. Maturation kinetics suggest that precipitation temperatures at this depth were ~140°C, assuming the major heat source was from the basement. The range of burial temperatures predicted for the St. Peter Sandstone (~65°–140°C) compares closely with the temperatures of hydrothermal ore-forming fluids (~80°–180°C) suggesting the fluids involved in diagenesis (i.e., dolomitization and quartz and anhydrite precipitation) may have been part of the same (paleo)hydrologic system that caused MVT mineralization. In the shallow northern part of the basin, fluid inclusion data indicate that dolospar precipitated at higher temperatures (110°–115°C) than would be expected in these otherwise low temperature (<50°C) rocks. Provided these values are reliable, the St. Peter Sandstone was affected by a heat source that was not burial related. Hydrothermal fluids associated with the Upper Mississippi Valley District could account for these temperatures. Fluid inclusion and isotopic data indicate that the fluids involved in burial cementation throughout the basin were saline and comparable in composition to the brines responsible for MVT mineralization (~20 wt% NaCl equivalent). In the absence of igneous activity, warm, topographically driven fluids (i.e., low temperature (<200°C) brines) moving updip from the southern tectonic margin of the basin can explain much of the dolospar and the associated mineral cements in the St. Peter Sandstone. ^{18}O-depleted dolospar concentrated along the La Salle anticline in east-central Illinois suggest that this structural feature was a major conduit for the movement of these hot fluids through the basin. The ultimate source of the fluids may have been the Ouachita fold belt or the Reelfoot rift. There also is some evidence that fluids were expelled from the Arkoma and Black Warrior Basins.

INTRODUCTION

The Ordovician St. Peter Sandstone in the Illinois basin (Fig. 1) has undergone diagenetic modification that included extensive early- to late-stage dolomitization, calcite precipitation, feldspathization, quartz crystallization, early to late-stage illitization and porosity reduction and enhancement. Although there have been some general studies on the composition, diagenesis and reservoir quality of the St. Peter Sandstone (Odom et al., 1979; Hoholick, 1980), little detailed research on the mineral paragenesis, composition of individual diagenetic phases, age and origin of cements, and vertical and lateral variation of cements has been done. The purpose of this study was to (i) conduct a detailed petrographic and geochemical characterization of the St. Peter Sandstone on a basin-wide scale with an emphasis on dolomitization, (ii) determine the nature and timing of different dolomite types and the regional importance of early versus late-stage dolomite, (iii) assess the extent to which the St. Peter Sandstone acted as a paleo-aquifer for regional fluid flow, and specifically, whether it served as a principle conduit for the mineralizing fluids that produced the lead-zinc-fluorine mineral deposits of the Mississippi Valley-type (MVT) adjacent to the basin and (iv) develop a model of dolomitization that is integrated on a regional hydrologic and tectonic scale. The chemical and isotopic characteristics of authigenic carbonate minerals in the St. Peter Sandstone in conjunction with the St. Peter's burial and thermal history can be used to decipher potential mechanisms of diagenesis and times during which diagenesis occurred. The age and origin of these mineral alterations, in turn, place constraints on the nature and timing, and the source of paleofluids that moved through the basin. This is especially important with respect to dolomitization because of its potential genetic link to MVT mineralization in the basin. Understanding the age and origin of mineral alterations and their relation to fluid-flow also provides insight into porosity reduction and enhancement.

The St. Peter Sandstone underlies the Upper Mississippi Valley (UMV) district at the northern end of the Illinois basin and may have been a major conduit system for the basinal fluids involved in MVT mineralization. Two-dimensional, finite-difference, hydrologic models developed by Garven and Freeze (1984a, b), Bethke (1985, 1986), and Bethke et al. (1988) simulated the movement and direction of fluid flow in the Illinois Basin. According to these models, significant heat can be transported in the subsurface of interior cratonic basins if hot fluids are channelled along a regional conduit system. The models call for low temperatures of waters in recharge areas and high temperatures in discharge areas along tectonic margins of foreland basins. On the basis of paleohydrologic modeling, it has been suggested that the UMV district north of the Illinois basin (Fig. 1) formed from low-temperature (<200°C) hydrothermal brines during a period of gravity-driven flow, initiated by the uplift of the Pascola arch on the southern margin of the basin (Bethke, 1986). The results of our study suggest that integrated mineralogical and geochemical data coupled with inferences from structural styles can be used to constrain the reaction processes governing cementation and the movement and rates of fluid transport within the (paleo)hydrologic system in the basin.

GEOLOGIC SETTING

The Illinois Basin in the midcontinent region (Fig. 1) initially was a broad, open, cratonic embayment that formed in conjunction with the New Madrid rift complex during breakup of a supercontinent in the late Precambrian (Burke and Dewey, 1973; Ervin and McGinnis, 1975). Activity within the rift controlled subsidence, sediment accumulation rates, depositional environments and formation of younger geologic structures

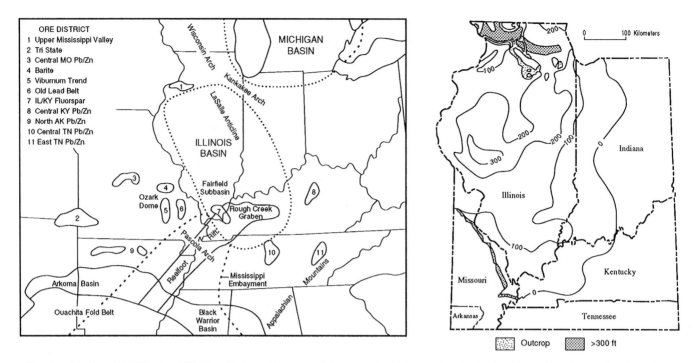

FIG. 1.—Location of the Illinois and Michigan Basins, major structural features, and principal ore-forming districts in the midcontinent region. Isopach map (modified from Kolata, 1991) shows regional thickness variations in the St. Peter Sandstone in the Illinois Basin.

(Braile et al., 1986). From Late Cambrian through Middle Ordovician time, thermal subsidence was the dominant mechanism controlling the development of the proto-Illinois Basin (Kolata and Nelson, 1991). More than 7,000 m of sediment accumulated in the depocenter, situated near the Rough Creek graben and overlying the rift complex with lesser amounts of sediment deposited elsewhere in the basin. During the remainder of Paleozoic time, basin subsidence was tectonically controlled. Initially, tectonic subsidence was relatively slow but later, compression along continental margins associated with the Ouachita and Alleghenian orogenies caused the basin to subside more rapidly (Heidlauf et al., 1986). The Illinois basin experienced multiple periods of tectonic deformation in response to the Taconic, Acadian and Ouachita orogenies. In Pennsylvanian and Permian time, during the Ouachita orogeny, compressional stresses near the Rough Creek graben led to extensive folding, faulting and igneous intrusion along the southern margin of the basin (Nelson, 1991). MVT mineralization south of the basin also was related to Ouachita tectonism in the Permian (Pan et al., 1990; Snee and Hayes, 1992; Chesley et al., 1994). The Illinois Basin attained its present structural configuration following the uplift of the Pascola arch during the early Mesozoic Era (Kolata and Nelson, 1991). The arch is now covered by approximately 900 m of Late Cretaceous and Early Tertiary siliciclastics of the Mississippi embayment (Kolata and Nelson, 1991).

The St. Peter Sandstone is continuous across Illinois and western Indiana and can be traced into Ohio and western Kentucky. In the Illinois Basin, the St. Peter Sandstone varies from 9–90 m in thickness, thins southward across the basin and is buried at depths varying from less than 1,000 m in the northern part to more than 2,000 m in the Fairfield subbasin to the south

(Figs. 1, 2). In southern Illinois, the St. Peter unconformably overlies the Everton Dolomite and underlies the Joachim Dolomite and Dutchtown Limestone (Fig. 2). In northern Illinois, the St. Peter Sandstone unconformably rests on the Prairie du Chien Group. Within the Illinois Basin proper, the St. Peter Sandstone comprises three members: the basal Kress Member, a conglomerate that is locally present; the Tonti Member, a fine-grained, quartzarenite comprising the main body of the formation, and the Starved Rock Member, a medium-grained sandstone that is confined to a belt trending northeast-southwest across the basin. Depositional environments represented by these members include marine shelf, offshore bar, beach and eolian (Shaw, oral commun., 1993).

ANALYTICAL TECHNIQUES

Core samples from 65 wells throughout the Illinois Basin were utilized in the study. Thin sections of representative sandstone samples that were blue-dye epoxy impregnated were stained to facilitate mineral identification. Staining with Alizarin red-S and potassium ferricyanide permitted Fe-free carbonate minerals to be distinguished optically from Fe-bearing carbonate minerals; sodium cobaltinitrate stain aided in the identification of K-feldspar. Samples were analyzed using a combination of the following techniques: optical petrography, scanning electron microscopy (SEM), cathodoluminescence (CL) microscopy, electron microprobe analysis, fluid inclusion microthermometry and stable isotope analysis. CL microscopy was employed to study carbonate cements in all carbonate-bearing samples. We used a cold-cathode Technosyn 8200 Mk II model operating at ~20 kV, 500–600 mA and ~7 Pa pressure. Using an energy-dispersive X-ray analytical system, SEM was

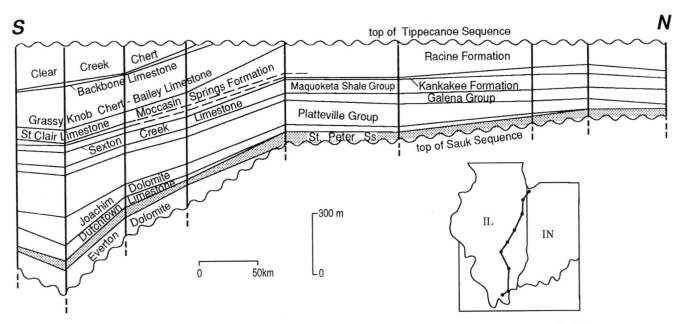

Fig. 2.—North-south cross section of lower Middle Ordovician (Whiterockian) through Lower Devonian (Ulsterian) rocks in Illinois (Tippecanoe sequence of Kolata, 1991).

routinely employed to identify mineral phases in back-scat-tered-electron mode. In addition, growth relationships of indi-vidual cements were studied on freshly broken sandstones in secondary-electron mode. The chemical composition of indi-vidual carbonate cement generations, located using CL photo-micrographs, was determined by electron microprobe analysis (JEOL Superprobe 733). Operating conditions were 15-kV-ac-celerating voltage, 50-nA-beam current, and a 10-μm-beam di-ameter. Based on counting statistics, the detection limits are estimated to be 0.02 wt% CaO, 0.02 wt% MgO, 0.05 wt% FeO, 0.05 wt% MnO and 0.03 wt% SrO (all measured by wave-length-dispersion). Fluid inclusion microthermometric mea-surements were performed on doubly polished wafers using a U.S.G.S.-type gas-flow-heating/freezing system calibrated against known melting point standards (analyst: T. J. Reynolds, Fluid Inc., Englewood, CO). Homogenization temperatures (T_h) were not pressure corrected. Carbonate-bearing samples were analyzed isotopically at the U. S. Geological Surveys Isotope Geochemical Laboratory (analyst: Augusta Warden). Carbon and oxygen isotope ratios were obtained by a timed-dissolution procedure based on different reaction rates for chemically dis-tinct carbonate phases (Walters et al., 1972). To prevent con-tamination by CO_2 from organic matter during acid digestion, kerogen was removed from the samples. Upon reaction with phosphoric acid, CO_2 gas evolved in the first hour was attributed to calcite and CO_2 gas evolved after several hours was assigned to dolomite. All isotope results are reported as the per mil (‰) difference relative to the Peedee belemnite (PDB) standard us-ing the delta (δ) notation. Data reproducibility is precise to ±0.2‰.

RESULTS

Detrital Mineralogy

The St. Peter Sandstone is predominantly a quartzarenite composed of monocrystalline, subrounded to well-rounded,

fine- to medium-grained quartz, with minor amounts (<5%) of feldspar, lithic fragments, heavy minerals and matrix clay (Dap-ples, 1955; Odom et al., 1979; Hoholick et al., 1984; this study). In the northern part of the Illinois Basin, sandstones locally are enriched in variable amounts (≤5%) of subangular to sub-rounded, silt- to fine-sand-size detrital feldspar and are subar-kosic, whereas sandstones in the southern part of the basin are generally quartzarenites containing only accessory feldspar (<1–2%). In east-central Illinois, K-feldspar is the dominant feldspar type and consists predominantly of detrital, sub-rounded, silt to fine-sand size grains with clear, discontinuous, euhedral K-feldspar overgrowths. Throughout the basin, lithic fragments are mainly fine-crystalline chert, polycrystalline quartz and granite. Matrix is present in variable amounts and commonly shows the effects of replacement by authigenic K-feldspar or clay minerals.

Sandstone Diagenesis

The St. Peter Sandstone in the Illinois Basin shows a variety of diagenetic cements and textures, most notably quartz, car-bonate and sulfate. Mechanical compaction, chemical compac-tion, including stylolitization and pressure solution, and disso-lution of authigenic minerals are features also common in the St. Peter Sandstone. The general paragenetic sequence of au-thigenic minerals and their distribution pattern are shown in Figures 3 and 4, respectively. The diagenetic assemblage in the central and southern portions of the basin includes late-stage illite, quartz, multiple generations of dolomite, calcite and an-hydrite; minor K-feldspar, chlorite and pyrite and traces of sphalerite and fluorite are present locally. The diagenetic min-eral assemblage in the northern part of the basin consists of early illite, K-feldspar, early dolomite and quartz as well as minor hematite and kaolinite. The basin-wide apportionment of cements is similar to that of Hoholick et al. (1984), although

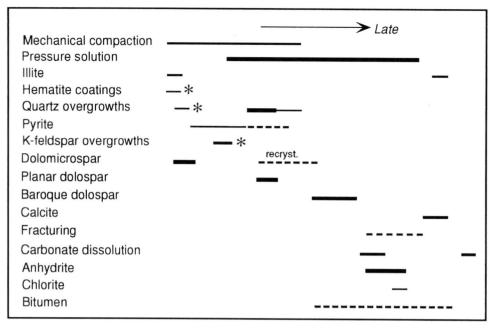

FIG. 3.—Mineral paragenesis in the St. Peter Sandstone. Diagenetic phases marked by an asterisk are largely confined to the northern part of the basin; relative thickness of bar reflects abundance and significance of diagenetic event.

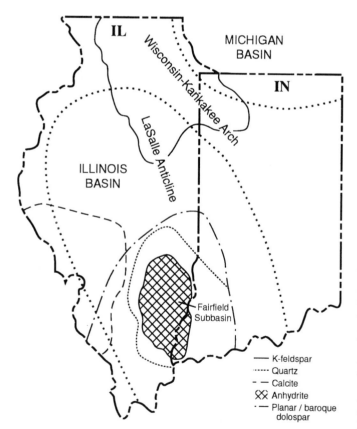

FIG. 4.—Areas of maximum cement abundance. Note that authigenic K-feldspar predominates in the northern part of the basin, in contrast to authigenic quartz, planar and baroque dolospar, calcite and anhydrite which are dominant in the southern portion of the basin.

there are some differences in the distribution of calcite, authigenic K-feldspar and quartz (see Fig. 4).

The degree of pressure solution in sandstones is highly variable even on a millimeter-scale and is clearly related to the presence of illitic material (detrital and authigenic) and framework grain size (Stackelberg, 1987; this study), whereby illitic grain coatings and finer grain sizes promote chemical compaction. K-feldspar leaching and repeated episodes of carbonate dissolution also occurred, although their exact timing is not well constrained petrographically. The abundance and distribution of porosity in sandstones is a function of geographic location within the basin and depth of burial (see also Hoholick et al., 1984). In the shallow northern part of the basin near the outcrop, the sandstones are excellent reservoir rocks (porosity exceeds 20%) with a large, well-developed network of primary and secondary pores. Dissolution of authigenic K-feldspar overgrowths occurs locally in some samples. In the deep southern part of the basin, there has been substantial porosity reduction (porosity varies from 0–10%) due to mechanical compaction and mineral cementation. Where there is evidence for dolomite and/or calcite dissolution, porosity is enhanced. Solution features such as discontinuous (corrosive) contacts between successive carbonate phases and relict cement in pores commonly are present and suggest that carbonate dissolution coincided with burial dolomitization and calcite precipitation.

On the basis of optical properties and crystal size, four major varieties of authigenic carbonate—dolomicrospar, planar dolospar, baroque dolospar and (poikilotopic) calcite—can be distinguished in the St. Peter Sandstone (Fig. 5, see p. 100). All four carbonate phases generally do not coexist in a single sandstone sample nor do they always occur in a single well. Figure 6 and Table 1 provide an overview of the chemical composition

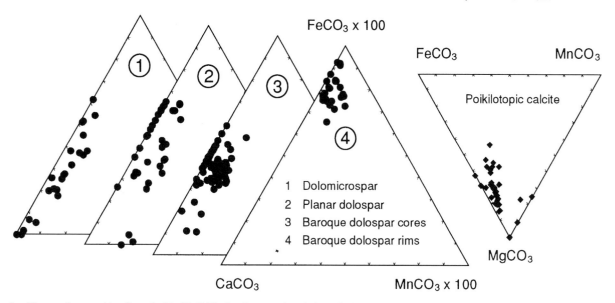

FIG. 6.—Elemental composition (in mole %) of individual carbonate-mineral phases in the St. Peter Sandstone. Note overall similarity in composition between all dolomite cements except the ferroan baroque dolomite rims.

TABLE 1.—SUMMARY OF ELEMENTAL AND STABLE ISOTOPIC COMPOSITIONS OF CARBONATE CEMENTS IN THE ST. PETER SANDSTONE

Cement Type	CaCO$_3$ (mole %)	MgCO$_3$ (mole %)	FeCO$_3$ (mole %)	MnCO$_3$ (mole %)	N	Σ	δ^{13}C (‰, PDB)	δ^{18}O (‰, PDB)	N
Dolomicrospar	51.6 (49.8–56.0)	48.1 (43.7–50.0)	0.3 (<DL-0.8)	0.03 (<DL-0.09)	4	23	−5.3 (−9.2 to −2.2)	−3.7 (−6.8 to −1.4)	45
Planar dolospar	50.2 (49.0–52.0)	49.2 (47.5–50.3)	0.5 (<DL-1.4)	0.06 (<DL-0.2)	4	35	−3.4 (−6.5 to −0.9)	−5.2 (−9.2 to −1.4)	24
Baroque dolomite core	50.6 (49.3–51.6)	49.0 (47.9–50.1)	0.4 (0.1–1.2)	0.05 (<DL-0.2)	8	64	−4.7	−6.2	7
Baroque dolomite rim	50.8 (49.3–52.6)	46.5 (44.8–48.2)	2.5 (1.8–3.6)	0.2 (<DL-0.5)	5	24	(−7.5 to −3.4)	(−7.5 to −4.6)	
Calcite	98.7 (97.7–99.8)	1.0 (0.2–1.9)	0.3 (<DL-0.6)	0.09 (<DL-0.3)	5	40	−4.6 (−6.6 to −3.1)	−9.1 (−10.6 to −6.5)	19

Elemental compositions determined by electron microprobe analysis. Stable isotope ratios of baroque dolospar represent mixtures of cores and rims. Values in parentheses give total range; N, number of samples analyzed; Σ, total number of individual spots probed in all samples; DL, detection limit. Analytical data shown on Table 2.

of the cements and summarize the isotopic variations in these minerals.

Dolomicrospar.—

Fine-crystalline (4–25 μm) dolomicrospar is an abundant (2–10%, locally 20%) texturally early cement in several wells in the northern part of the basin. Generally, it occurs as thin laminae within the sandstone or it occupies intergranular pores. Dolomicrospar is composed of planar dispersed crystals and, less commonly, nonplanar interlocking crystals of nonferroan dolomite. Typically the mineral shows moderately bright, orange-red CL. Luminescent zoning was not observed. Less common than dolomicrospar is dolomicrite (crystal size ≤4μm). Some dolomicrospar samples reveal distinct microfacies features, including dark-micritic, rounded components and patches, and peloidal and mottled textures. Other samples show rare elongate features that strongly resemble root casts known from calcretes and related carbonates (e.g., Esteban and Klappa, 1983; Wright, 1990). Silt-size detrital K-feldspar crystals are common in dolomicrospar-cemented samples from wells in the northern part of the basin. Typically K-feldspar grains are sub-

rounded and free of authigenic feldspar overgrowths which attests to their detrital origin. Traces of iron sulfide, principally pyrite, are associated with dolomicrospar. Pyrite is distributed as individual cubes (5–30 μm) or as clusters of euhedral crystals mostly on dolomicrospar grains, inferring that pyrite precipitation followed dolomite formation. Electron microprobe analyses of dolomicrospar crystals show low iron concentrations (Table 1), consistent with the CL observations. Four out of five analyzed samples are near-stoichiometric dolomite with ≤52 mole % CaCO$_3$; one sample yielded significantly higher excess Ca values (~55 mole %, Table 1, Fig. 6). Manganese values are close to the microprobe detection limit (see Methods), while all strontium measurements were below detection limit.

Planar Dolospar.—

A large fraction of the dolomite (>10%) in the central and southern portions of the basin (Fig. 4) and a minor component of dolomite (<5%) in the northern part of the basin are coarse-crystalline, pore-filling dolospar, varying in crystal size from 20–300 μm. On the basis of straight crystal boundaries and even extinction in polarized light, this dolomite classifies as planar-e (following Sibley and Gregg, 1987), hereafter referred to as planar dolospar. Planar dolospar-cemented sandstones are characterized by fairly high intergranular volumes (2–10%, locally 15–20%) and only minor chemical compaction of quartz framework grains. Individual dolospar crystals are characterized by abundant solid inclusions, giving rise to a cloudy appearance in transmitted light. In the central part of the basin, planar dolospar contains rare detrital K-feldspar grains and authigenic pyrite. Some dolomicrospar samples also show larger, euhedral planar dolospar crystals "floating" in a groundmass of dolomicrospar, suggesting incipient growth of planar dolospar at the expense of dolomicrospar. Other planar dolospar is clearly neoformed, as indicated by euhedral, inclusion-poor crystal faces protruding into open pores. In sandstones where planar dolospar is abundant, individual crystals commonly form an interlocking mosaic of equant crystals with little or no intercrystalline porosity. Bright orange-red CL, indistinguishable from the luminescent color emitted by dolomicrospar, suggests that planar dolospar also is non-ferroan. The absence of luminescent zoning argues that the Fe/Mn ratio in planar dolospar crystals is fairly constant. Electron microprobe traverses across individual crystals corroborate the lack of compositional zoning suggested from CL observations. Microprobe analysis shows low Fe and Mn concentrations that are similar to the Fe and Mn concentrations measured in dolomicrospar (Table 1, Fig. 6). $CaCO_3$ values of planar dolospar are typically ~50 mole % (i.e., stoichiometric dolomite). In wells from the southern part of the basin, planar dolospar crystals often are overgrown by ferroan dolomite (see below) and in some samples, planar dolospar coexists with authigenic quartz, calcite and anhydrite. The relative timing of quartz and planar dolospar precipitation is difficult to determine, but generally quartz formed earlier. However, that quartz precipitation continued locally after planar dolospar formation cannot be ruled out.

Baroque Dolospar.—

Some dolospar (<3%), predominantly in sandstones from the southern portion of the Illinois Basin (Fairfield subbasin, Fig. 4), has optical properties characteristic of baroque (or saddle) dolospar (a subcategory of nonplanar dolomite according to Sibley and Gregg, 1987). This dolomite is coarse-crystalline, has slightly to moderately curved crystal boundaries, and displays broad sweeping extinction. CL microscopy reveals that baroque dolomite crystals are composed of two generations of dolomite, an earlier bright orange-red luminescent core and a thin, mostly dull brown luminescent rim. The luminescent core encompasses at least two thirds of the volume of the entire crystal and is equivalent to planar dolospar. Chemical analysis shows that the core area has a homogeneous non-ferroan composition similar to that of planar dolospar (Fig. 6, Table 1). The outer, dull luminescent generation is chemically distinct and contains 1.8–3.6 mole % $FeCO_3$ (i.e., ferroan dolomite, Table 1). No significant difference in mole % $CaCO_3$ was found between cores and rims of baroque dolomite crystals (Table 1). The development of sweeping extinction seems to be unrelated to the formation of ferroan overgrowths, because incipient sweeping extinction was also observed in some larger planar dolospar crystals. No correlation exists between the develop-

ment of sweeping extinction and the stoichiometry of the dolomite in these rocks (cf. Barber et al., 1985).

Calcite.—

Calcite occurs in minor amounts (<5% by volume average) in the St. Peter Sandstone and is most abundant in the southwestern part of the basin (Fig. 4). Calcite is poikilotopic, nonferroan to slight ferroan and exhibits bright orange CL. It occurs both as pore-filling cement and, less commonly, as a replacement mineral. Locally, in the southern and central parts of the basin, calcite contains traces of micron-size fluorite and sphalerite. Electron microprobe analysis of five samples show Mg concentrations averaging 1.0 mole % $MgCO_3$. Fe and Mn values are low, typically 0.1–0.5 mole % $FeCO_3$ and ≤0.1 mole % $MnCO_3$ (Table 1, Fig. 6). According to petrographic analysis, calcite postdates quartz cement as well as baroque dolospar and other authigenic mineral phases, thus it appears to be one of the latest cements to have formed in the St. Peter Sandstone. Some calcite shows etched boundaries suggesting that it underwent dissolution; thus, calcite previously may have been more widespread before it was subsequently dissolved.

Noncarbonate Cements.—

Other cements in the St. Peter Sandstone include minor authigenic quartz, K-feldspar, anhydrite and illite. Barite, hematite, chlorite and kaolinite are volumetrically insignificant. The petrographic characteristics of some of these minerals are summarized briefly. For the paragenesis of these noncarbonate cements and the area of the basin where these minerals commonly occur, the reader is referred to Figures 3 and 4, respectively.

Authigenic quartz is the most abundant (<5%) noncarbonate cement in the St. Peter Sandstone and occurs as syntaxial overgrowths on detrital quartz grains. Although by far the majority of the samples examined lack authigenic quartz or contain only incipient and incomplete overgrowths, a few contain abundant quartz cement (up to 20%). Quartz cement is best developed in deeply buried sandstones (>2,000-m burial depth) in the southern part of the basin (Fig. 4). Where quartz overgrowths are present, they generally predate planar dolospar and are engulfed by baroque dolospar. If the sandstones exhibit porosity enhancement, overgrowths are commonly embayed or, in some samples, contain voids that are evidence for dissolution of other mineral cements. Quartz cement often is immediately adjacent to millimeter- to centimeter-thick layers of sandstone showing extensive pressure solution. This spatial relationship suggests that at least some of the silica in authigenic quartz was derived from chemical compaction of framework grains. Preliminary fluid inclusion data for one well in the southern part of the basin yielded homogenization temperatures of 90°–110°C for primary fluid inclusions in quartz overgrowths; salinity estimates based on ice melting temperatures suggest ~15 wt% NaCl equivalent.

Authigenic K-feldspar is a common, though volumetrically minor constituent (<5%) in many sandstones, particularly in wells from the northern part of the basin and along the LaSalle anticline in east-central Illinois (Fig. 4). In these areas, fine-grained detrital K-feldspar grains, characterized by a bright-blue CL color, are overgrown by clear, euhedral, non-luminescent K-feldspar rims. Authigenic K-feldspar also occurs as sub- to anhedral aggregates replacing illitic matrix. These aggregate crystals appear to lack detrital cores (cf. Odom et al., 1979). K-feldspar overgrowths predate planar dolospar and may be as early as dolomicrospar. In sandstones with secondary porosity, authigenic K-feldspar overgrowths often display a skeletal morphology inferring dissolution, whereas the detrital K-feldspar cores appear unaltered. In these sandstones, K-feldspar cement also is distributed as patches, suggesting that its present occurrence might be dissolution-controlled.

Minor illitic clay minerals (a few percent, locally as high as 10%) are present in many samples. An early, webby-flaky generation can be distinguished from a later-stage, fibrous-filamentous illite. SEM studies show that early-diagenetic illite predates quartz overgrowths and mechanical compaction, while late-diagenetic illite fills secondary pores caused by carbonate (calcite?) dissolution. We did not determine the mineralogical composition of these two generations, but their micromorphology as seen under the SEM suggests that they are probably illite/smectite mixed-layer clays.

Anhydrite occurs sparsely (<5%) in wells from the Fairfield subbasin (Fig. 4). Its mode of occurrence ranges from small scattered patches to large euhedral crystals that preferentially replace mineral cements, including planar dolospar, baroque dolospar, and quartz. In some samples, anhydrite also replaces calcite although the paragenesis of anhydrite and calcite is unclear.

Stable Isotope Geochemistry and Fluid-Inclusion Analysis of Carbonate Cements

Representative samples of dolomite and calcite in the St. Peter Sandstone were analyzed for their carbon and oxygen-isotope compositions (Tables 1, 2; Fig. 7). Dolomicrospar samples were collected mostly from wells in central and northern Illinois where this cement is widespread. Samples of planar and baroque dolospar are predominantly from wells in central and southern Illinois although a few samples of planar dolospar are from the southern part of the Michigan Basin. Calcite samples generally were taken from the southwest portion of the Illinois Basin where it is most abundant.

The $\delta^{13}C$ and $\delta^{18}O$ ratios of dolomicrospar are highly variable, particularly in carbon. $\delta^{13}C$ ratios are uniformly negative and range from -2.2 to $-9.2‰$ PDB; $\delta^{18}O$ ratios are moderately depleted in ^{18}O and range from -1.4 to $-6.8‰$ PDB (Tables 1, 2; Fig. 7). Some of the more negative oxygen isotope compositions reflect a component of planar dolospar replacing dolomicrospar. Note that the $\delta^{18}O$ isotope ratios of dolomicrospar samples are consistently lighter than those of carbonate precipitated from Ordovician seawater (e.g., Lohmann and Walker, 1989; Fig. 7). The $\delta^{13}C$ and $\delta^{18}O$ isotope ratios of planar dolospar are extremely variable and largely overlap the values for dolomicrospar and baroque dolospar. Values of planar dolospar vary widely ranging from -6.5 to $-0.9‰$ PDB ($\delta^{13}C$) and -9.2 to $-1.4‰$ PDB ($\delta^{18}O$; Tables 1, 2; Fig. 7). The carbon and oxygen isotope ratios of baroque dolospar range from -7.5 to $-3.4‰$ PDB and from -7.5 to $-4.6‰$ PDB, respectively (Tables 1, 2; Fig. 7). It is noteworthy that the $\delta^{18}O$ ratios of baroque dolospar in sandstones are comparable to those of ore-related dolomite in mineralized carbonate rocks (-7 to $-2‰$ PDB; Hall and Friedman, 1969; Gregg and Sibley, 1984; Garvin et al., 1987; Lee, 1994) suggesting that dolo-

mite cementation and ore formation may have involved fluids of similar oxygen-isotopic composition and/or temperature. The carbon and oxygen isotopes of calcite vary from -3.1 to $-6.6‰$ PDB and -6.5 to $-10.6‰$ PDB, respectively (Tables 1, 2). Figure 7 illustrates that the $\delta^{18}O$ ratios of calcite tend to be lighter than the $\delta^{18}O$ ratios of burial dolospar, whereas the $\delta^{13}C$ ratios of calcite and dolomite are within the same range.

The spatial distribution of mean $\delta^{18}O$ isotopic ratios in burial carbonate cements shows some interesting relationships (Fig. 8). In northern Illinois near the UMV district, the lightest $\delta^{18}O$ ratios of dolomicrospar and planar, and baroque dolospar are in samples from shallow-buried strata in the Michigan Basin and on the Kankakee arch. Farther south, in the east-central part of the basin where the St. Peter Sandstone is moderately to deeply-buried, the $\delta^{18}O$ ratios of these dolospars display consistently lighter values northward along the north-south trending La Salle anticline. Away from the anticline, the $\delta^{18}O$ ratios of burial dolospar are enriched in ^{18}O by 2 to 3‰ PDB. In southern Illinois, the $\delta^{18}O$ ratios of planar and baroque dolospar show significant depletion in deeply buried sandstones close to the Illinois-Kentucky fluorspar district.

A few samples from a well on the north flank of the basin contain primary, two-phase fluid inclusions in baroque dolospar. In one sample, inclusions located at the boundary between planar dolospar and baroque dolospar overgrowths are very small (<3 μm) and have consistent liquid to vapor ratios. Homogenization temperatures of these inclusions range from 110°–115°C; inclusions in other dolospar samples have significantly lower temperatures (~65°C). Final ice melting temperatures of all inclusions are consistently less than $-20°C$, suggesting highly saline, pore-water compositions >20 wt% NaCl equivalent.

DISCUSSION

Based on the timing of individual cementation events relative to compaction of the St. Peter, the diagenetic history can be broadly divided into a shallow-burial, early diagenetic stage and a deep-burial, late diagenetic stage. The origin and timing of the mineral phases and the nature of the fluids from which they precipitated are discussed below.

Carbonate Cement Diagenesis

The four stages of carbonate diagenesis documented in the St. Peter Sandstone appear to be consistent throughout the basin. Dolomicrospar is the earliest carbonate cement in the samples examined. The high intergranular volumes in dolomicrospar-cemented sandstones suggest formation occurred at shallow burial depths possibly penecontemporaneous with sedimentation. Whether this fine-grained dolomite formed from a $CaCO_3$ precursor or represents an early diagenetic dolomite, cannot be determined on the basis of our data. Most dolomicrospar samples exhibit clear-cut evidence of recrystallization, indicating that this early carbonate phase underwent extensive burial diagenetic alteration. A few samples from the southwestern part of the basin consist of almost pure dolomite layers interbedded with quartzose sandstones. These dolomites reveal relic features that resemble structures known from modern and ancient calcretes, and dolocretes (see above). Microfacies features diagnostic of shallow marine carbonate deposition, how-

JANET K. PITMAN AND CHRISTOPH SPÖTL

TABLE 2.—STABLE ISOTOPE COMPOSITION OF CARBONATE CEMENTS IN THE ST. PETER SANDSTONE

ID	State	Well Name	Depth (ft)	δ¹³C (‰, PDB)	δ¹⁸O (‰, PDB)
		DOLOMICROSPAR			
C 58	Illinois	Stanolind Oil & Gas Co., Leiner #1	4068.0	−4.86	−2.33
C 58	Illinois	Stanolind Oil & Gas Co., Leiner #1	4073.0	−3.26	−5.46
C 58	Illinois	Stanolind Oil & Gas Co., Leiner #1	4076.8	−3.41	−2.34
C 58	Illinois	Stanolind Oil & Gas Co., Leiner #1	4077.1	−4.55	−4.00
C 1090	Illinois	Eason Oil Co., Edna C. Thomas #1	1865–67	−6.93	−3.74
C 1196	Illinois	Herndon Drilling Co., W. J. Fecht #1	1450.0	−2.91	−4.94
C 1196	Illinois	Herndon Drilling Co., W. J. Fecht #1	1452.0	−2.33	−4.51
C 2239	Illinois	W. C. McBride, Inc., A. V. Hunleth #1	3943.0	−4.63	−1.79
C 2239	Illinois	W. C. McBride, Inc., A. V. Hunleth #1	3945.0	−5.84	−4.04
C 2239	Illinois	W. C. McBride, Inc., A. V. Hunleth #1	3946.0	−5.28	−1.82
C 2740	Illinois	Superior Oil Co., H. C. Ford et al., #C-17	7270.6	−4.32	−4.68
C 2740	Illinois	Superior Oil Co., H. C. Ford et al., #C-17	7298.1	−3.21	−5.89
C 2740	Illinois	Superior Oil Co., H. C. Ford et al., #C-17	7304.3	−2.44	−4.59
C 2740	Illinois	Superior Oil Co., H. C. Ford et al., #C-17	7332.4	−4.44	−5.40
C 2800	Illinois	Mississippi River Fuel Corp., Theobald #A-15	971.0	−7.37	−2.81
C 3200	Illinois	Sun Oil Co., Maurice E. Vale	1943.0	−3.79	−1.67
C 3500	Illinois	Pike County Gas Assoc., Conkright #2	1012.0	−4.60	−3.07
C 3533	Illinois	Magnolia Oil Co., M. T. Rodda #1	5242.0	−7.63	−4.73
C 3533	Illinois	Magnolia Oil Co., M. T. Rodda #1	5250.0	−6.62	−6.81
C 3533	Illinois	Magnolia Oil Co., M. T. Rodda #1	5254.0	−6.61	−5.90
C 3533	Illinois	Magnolia Oil Co., M. T. Rodda #1	5255.0	−6.59	−6.07
C 3533	Illinois	Magnolia Oil Co., M. T. Rodda #1	5259.0	−6.80	−3.73
C 3533	Illinois	Magnolia Oil Co., M. T. Rodda #1	5260.0	−6.70	−5.65
C 3533	Illinois	Magnolia Oil Co., M. T. Rodda #1	5261.0	−6.96	−3.49
C 3533	Illinois	Magnolia Oil Co., M. T. Rodda #1	5266.0	−6.34	−4.02
C 3598	Illinois	Pure Oil Co., C. T. Montgomery #18-B	7534.0	−7.94	−3.36
C 3598	Illinois	Pure Oil Co., C. T. Montgomery #18-B	7551.0	−8.19	−4.23
C 3598	Illinois	Pure Oil Co., C. T. Montgomery #18-B	7563.0	−8.82	−3.61
C 3598	Illinois	Pure Oil Co., C. T. Montgomery #18-B	7576.0	−9.17	−3.95
C 3750	Illinois	Maryland Services Co., E. F. Kercheis #1-S	2846.0	−7.30	−3.36
C 4213	Illinois	R. A. Davis, South #1	1248.6	−2.21	−3.48
C 4831	Illinois	Texaco Inc., R. S. Johnson #1	5237.0	−6.11	−2.12
C 4831	Illinois	Texaco Inc., R. S. Johnson #1	5237.10	−5.84	−2.78
C 186	Indiana	H. Hyslope & W. Simpson, L. Bishop #1	1703.0	−3.65	−3.23
C 186	Indiana	H. Hyslope & W. Simpson, L. Bishop #1	1795.0	−4.04	−3.46
C 332	Indiana	Indiana Gas & Ewater Co. Inc., Carl P. Ross #1	1452.7	−2.52	−2.38
C 437	Indiana	Northern Indian Public Service Co., O. E. Morphet #1	1402.0	−2.98	−5.89
C 495	Indiana	General Electric Co., WD-2	7893.3	−5.81	−3.79
C 640	Indiana	Buttercup Energy, Rachel Pensinger #1	3189.6	−4.32	−3.50
TP-W1	Missouri	Phillips Petroleum, Winge #TP-W1	469.2-.5	−2.70	−3.16
1701–1	Missouri	Getty Mining Co., Mineral Test 1701–1	120.2-.4	−5.42	−2.69
1701–1	Missouri	Getty Mining Co., Mineral Test 1701–1	120.2-.4	−5.43	−2.18
1701–1	Missouri	Getty Mining Co., Mineral Test 1701–1	120.2-.4	−6.14	−2.02
1701–1	Missouri	Getty Mining Co., Mineral Test 1701–1	146.8	−5.87	−3.55
1701–1	Missouri	Getty Mining Co., Mineral Test 1701–1	154.6	−7.72	−1.42
		PLANAR DOLOSPAR			
C 1090	Illinois	Eason Oil Co., Edna C. Thomas #1	1940–42	−3.74	−9.22
C 2740	Illinois	Superior Oil Co., H. C. Ford et al., #C-17	7281.0	−3.80	−4.88
C 2740	Illinois	Superior Oil Co., H. C. Ford et al., #C-17	7331.3	−4.14	−4.59
C 2740	Illinois	Superior Oil Co., H. C. Ford et al., #C-17	7361.0	−5.83	−6.06
C 2740	Illinois	Superior Oil Co., H. C. Ford et al., #C-17	7616.0	−4.95	−6.22
C 2740	Illinois	Superior Oil Co., H. C. Ford et al., #C-17	7623.0	−4.70	−7.13
C 2740	Illinois	Superior Oil Co., H. C. Ford et al., #C-17	7624.0	−4.17	−7.85
C 3533	Illinois	Magnolia Oil Co., M. T. Rodda #1	5253.0	−6.49	−6.63
C 3598	Illinois	Pure Oil Co., C. T. Montgomery #18-B	7526.9	−5.39	−8.19
C 3766	Illinois	Central Light Co., Mary Wooding #1	1422.0	−0.92	−2.71
C 3766	Illinois	Central Light Co., Mary Wooding #1	1423.0	−1.07	−2.54
C 4636	Illinois	Northern Illinois Gas Co., Feinhold #1	1202.0	−2.04	−4.43
C 7214	Illinois	Chicago Sanitary District, Deep Tunnel Test DH-9	990.0	−3.60	−4.51
C 7214	Illinois	Chicago Sanitary District, Deep Tunnel Test DH-9	991.0	−2.21	−3.03
C 7214	Illinois	Chicago Sanitary District, Deep Tunnel Test DH-9	993.0	−3.00	−3.99
C 7607	Illinois	New Jersey Zinc, Koefer H-2	949.0	−1.16	−1.37
C 7607	Illinois	New Jersey Zinc, Koefer H-2	950.0	−2.66	−2.42
C 7607	Illinois	New Jersey Zinc, Koefer H-2	950.2	−2.04	−3.47
C 9475	Illinois	Metropolitan Sanitary District, Chicago DH 71–9	853.9-.11	−4.10	−7.48
C 9475	Ilinois	Metropolitan Sanitary District, Chicago DH 71–9	854.2-.3	−3.36	−6.56
C 10180	Illinois	Northern Illinois Gas Co., Pyne #1	1560.0	−2.71	−4.34
C 435	Indiana	Northern Indiana Public Service Co., Boezman #1	1437.4	−1.93	−4.80
C 435	Indiana	Northern Indiana Public Service Co., Boezman #1	1442.0	−2.39	−4.36
C 495	Indiana	General Electric Co., WD-2	7872.4	−5.14	−8.91

TABLE 2.—*Continued*

ID	State	Well Name	Depth (ft)	δ¹³C (‰, PDB)	δ¹⁸O (‰, PDB)
		BAROQUE DOLOSPAR			
C 3598	Illinois	Pure Oil Co., C. T. Montgomery #18-B	7435.0	−7.45	−4.63
C 4785	Illinois	Texaco Inc., Cuppy #1	7616.0	−5.09	−5.93
C 4785	Illinois	Texaco Inc., Cuppy #1	7623.0	−4.00	−7.51
C 4785	Illinois	Texaco Inc., Cuppy #1	7624.0	−3.42	−6.45
C 4785	Illinois	Texaco Inc., Cuppy #1	7626.0	−4.30	−6.64
C 4785	Illinois	Texaco Inc., Cuppy #1	7627.0	−4.13	−7.36
C 4785	Illinois	Texaco Inc., Cuppy #1	7630.0	−4.30	−5.21
		CALCITE			
C 28	Illinois	Miller, B.W., Sample #1	2666.9	−3.82	−7.80
C 58	Illinois	Stanolind Oil & Gas Co., Leiner #1	4076.0	−3.89	−9.88
C 2800	Illinois	Mississippi River Fuel Corp., Theobald #A-15	905.0	−4.87	−10.14
C 2800	Illinois	Mississippi River Fuel Corp., Theobald #A-15	918.0	−3.38	−6.52
C 2800	Illinois	Mississippi River Fuel Corp., Theobald #A-15	931.0	−3.08	−6.74
C 2800	Illinois	Mississippi River Fuel Corp., Theobald #A-15	940.0	−4.96	−9.29
C 2800	Illinois	Mississippi River Fuel Corp., Theobald #A-15	948.0	−4.89	−9.50
C 2800	Illinois	Mississippi River Fuel Corp., Theobald #A-15	957.0	−4.01	−9.32
C 3200	Illinois	Sun Oil Co., Maurice E. Vale	1853.0	−3.15	−8.01
C 13443	Illinois	Illinois Power Co., Truitt #2	3158.1	−5.20	−9.09
C 13443	Illinois	Illinois Power Co., Truitt #2	3159.0	−4.10	−8.78
C 13443	Illinois	Illinois Power Co., Truitt #2	3160.0	−4.00	−9.63
C 13443	Illinois	Illinois Power Co., Truitt #2	3160.0	−4.24	−9.58
C 13443	Illinois	Illinois Power Co., Truitt #2	3160.1	−5.37	−8.84
C 13443	Illinois	Illinois Power Co., Truitt #2	3161.0	−6.23	−10.50
C 13443	Illinois	Illinois Power Co., Truitt #2	3162.0	−6.58	−9.51
C 13443	Illinois	Illinois Power Co., Truitt #2	3162.1	−6.52	−9.64
C 13443	Illinois	Illinois Power Co., Truitt #2	3164.0	−6.21	−9.84
C 332	Indiana	Indiana Gas & Water Co., Inc., Carl P. Ross #1	1471.1	−3.07	−10.61

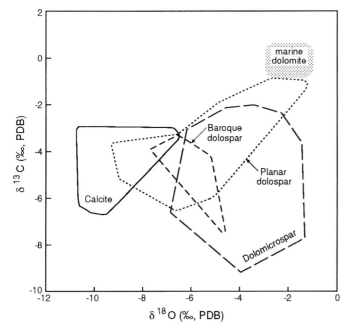

FIG. 7.—Stable isotopic composition of various carbonate cements in the St. Peter Sandstone. Note large overlap between the three petrographic types of dolomite cement. The approximate composition of dolomite in isotopic equilibrium with Ordovician seawater is taken from Lohmann and Walker (1989).

more attention in the future, because pedogenic carbonates older than Late Ordovician are hitherto unknown (Retallack, 1992).

Despite their wide scatter, the carbon isotopic compositions of dolomicrospar from individual wells show consistent values. In some samples, the carbon composition partially overlaps with the presumed marine field (see Fig. 7). The more positive δ¹³C ratios of dolomicrospar are interpreted to be controlled largely by the paleoenvironmental conditions during initial carbonate precipitation. The more negative δ¹⁸O ratios likely reflect variable reequilibration during burial, consistent with petrographic observations showing recrystallization. Diagenetic alteration of the carbon isotopic composition cannot be ruled out, but is regarded as insignificant in the light of water/rock interaction models (cf. Banner et al., 1988) that show δ¹³C to be rock-buffered in carbonate-bearing systems. In the absence of additional geochemical indicators such as Sr isotope data, we favor a predominantly non-marine initial water composition at the time of St. Peter Sandstone deposition for these samples.

Planar dolospar postdates mechanical and minor chemical compaction of the sandstones. Two petrographic observations suggest that a significant fraction of planar dolospar is a replacement of dolomicrospar: (i) high intergranular volumes occupied by planar dolospar are characteristic of uncompacted sand and (ii) many dolospar crystals are cloudy and inclusion-rich. These observations suggest that at least some of the planar dolospar is a replacement of an earlier cement, perhaps diagenetic dolomicrospar that had stabilized the sand fabric and thus had prevented compaction. Similar CL properties, elemental compositions, and stable isotope ratios of planar dolospar and dolomicrospar also suggest they are genetically related. It is therefore proposed that the apparent recrystallization (dolomit-

ever, are absent in these samples. Based on these preliminary observations, at least some discrete dolomite beds within the St. Peter Sandstone appear to be continental (pedogenic) rather than marine-carbonate formations. These occurrences deserve

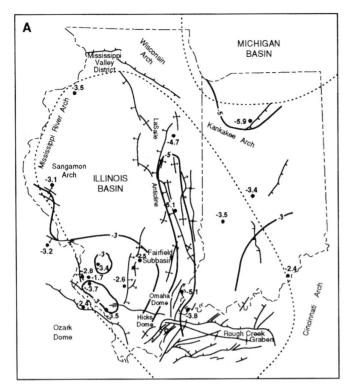

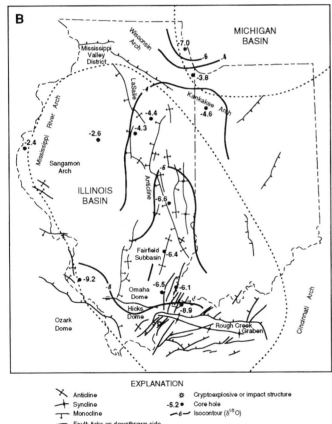

EXPLANATION

✕ Anticline   ✲ Cryptoexplosive or impact structure
✚ Syncline   -5.2 ● Core hole
┬ Monocline   ╌5╌ Isocontour (δ¹⁸O)
┯ Fault; ticks on downthrown side

FIG. 8.—Regional variations in the mean oxygen-isotope compositions of
dolomite in the St. Peter Sandstone. (A) δ¹⁸O ratios in dolomicrospar. (B) δ¹⁸O
ratios in planar and baroque dolospar. Note similarity in isotopic pattern of
early- and late-diagenetic dolospar on a basin-wide scale. Major structural features taken from Nelson (1991).

ization?) of dolomicrospar and/or its precursor occurred during
the same diagenetic event that resulted in planar dolospar
growth at the expense of finer-grained dolomicrospar matrix.
This process obviously reached completion only locally, as indicated by the coexistence of dolomicrospar (and rare dolomicrite) and coarse planar dolospar in some samples.

Precipitation of ferroan baroque dolomite overgrowths on
planar dolospar was a localized diagenetic process generally
restricted to the southern (i.e., deeper) part of the basin. Baroque
dolospar (and most planar dolospar) formed after quartz precipitation but before the development of late-stage anhydrite.
The isotopic composition of baroque dolospar represents
mixtures of planar dolospar cores and ferroan baroque dolospar
rims. Therefore, it is not surprising that the isotopic ratios of
baroque and planar dolospar show significant compositional
overlap (Fig. 7). The higher iron content of the baroque dolomite rims probably reflects changes in the iron supply and/or
the redox chemistry of the fluids. The close spatial association
of planar and baroque dolospar together with their overlapping
$\delta^{13}C$ and $\delta^{18}O$ isotopic compositions suggest that these dolomite
cements formed from fluids with similar isotopic compositions.
The $\delta^{13}C$ ratios of most planar and baroque dolospar (-1 to
$-8\%_0$ PDB) are substantially lighter than the estimated $\delta^{13}C$
ratio of Ordovician marine carbonates ($+1$ to $-1\%_0$ PDB; Fig.
7) and thus, are interpreted to reflect incorporation of a component of ^{12}C-enriched bicarbonate produced by decarboxylation of organic matter. Correspondingly light $\delta^{18}O$ values of
these dolomites (-5 to $-9\%_0$ PDB) are consistent with higher
temperatures during dolomitization.

Calcite is clearly the latest carbonate to form in the St. Peter
Sandstone. Carbon isotope ratios (-3 to $-7\%_0$ PDB) are
within the range of the dolomite values, suggesting rock-buffering. Calcite $\delta^{18}O$ ratios (-7 to $-11\%_0$ PDB) are consistently
lighter than the dolospar values (Fig. 7), which can be explained
by isotopic fractionation effects.

Constraints on Timing of Diagenetic Events

Three authigenic minerals in the St. Peter Sandstone, K-feldspar, illite and calcite are of particular significance with respect
to the chronology of the carbonate cements, because absolute
age dates are available on the timing of their precipitation. Authigenic K-feldspar preserved in the northern part of the basin
is a texturally early cement that predated quartz cementation
and major sediment compaction thus, its age could provide a
useful maximum age for the onset of carbonate cementation.
Although few age dates are currently available on authigenic
minerals in the St. Peter Sandstone within the Illinois Basin
proper (Hay et al., 1988), a fairly large number of K-feldspar
ages have been reported for the St. Peter Sandstone and K-
bentonites adjacent to the basin. The results of studies on the
St. Peter Sandstone in the vicinity of the Wisconsin arch (Marshall et al., 1986) and the Michigan Basin (Barnes et al., 1992b),
and on altered K-bentonite tuffs from the Cincinnati arch (Elliott and Aronson, 1987; 1993) and Mississippi Valley area
(Hay et al., 1988; Lee and Aronson, 1991) indicate that formation of K-feldspar throughout the midcontinent occurred
largely during Early Devonian time (~400 Ma). Older ages
have been reported (Krueger and Woodard, 1972; Marshall et
al., 1986) but they may reflect sample contamination (cf. Hay

et al., 1988). In summary, available data suggest that the age of K-feldspar authigenesis in the St. Peter Sandstone is some 60 my younger than the age of deposition. Although these results are at least partially consistent with the petrographic data, suggesting feldspar precipitation predated major burial cements (Fig. 3), they are not consistent with early diagenetic formation of this feldspar (cf. Odom et al., 1979; this study). The age dates suggest that little diagenetic alteration occurred in the St. Peter Sandstone during the Late Ordovician and Silurian, until the unit was buried to a depth of ≤700 m in the central part of the basin.

Late-stage illite in the St. Peter Sandstone postdated burial dolospar cementation and carbonate dissolution. The age of this illite is important because it places constraints on the minimum age of burial-carbonate cementation and the timing of dissolution. Available age data on authigenic illite (and illite/smectite) are difficult to interpret because of the wide spread in the ages which, according to this study, reflect multiple episodes of clay mineral diagenesis. Data from the Illinois Basin proper are lacking and currently available data were reported only from locations adjacent to the basin. Conventional K/Ar isotopic age dates for the Mississippi Valley area show a wide spectrum of ages that define two distinct episodes of illitization in Middle Ordovician bentonites (Hay et al., 1988); an older one in the Devonian and Mississippian at ~360 Ma and a younger one of Permian age at ~265 Ma. Lee and Aronson (1991) also found evidence for an older, ~400 Ma (Devonian) illitization event that was coeval with K-feldspar diagenesis and a young (Middle Triassic) episode of diagenetic illitization with K/Ar ages between 230–215 Ma. East of the Illinois Basin (along the Cincinnati Arch) and south of the basin, illitization in K-bentonites apparently occurred during Permian time, at ~277 to 251 Ma (Elliott and Aronson, 1993). The broad range of ages for authigenic illite in Ordovician bentonites adjacent to the Illinois Basin are only partially consistent with the illite age dates reported for Ordovician sandstones in the central Michigan Basin. The paragenetic sequence of alterations in the St. Peter Sandstone in the Michigan Basin (Odom et al., 1979; Barnes et al., 1992a) is fairly similar to that of the St. Peter in the Illinois Basin, suggesting that rocks of the same age experienced a similar diagenetic history in the two basins. In the Michigan Basin, illite is a late-stage cement that postdated baroque dolospar and anhydrite (Barnes et al., 1992a). However, K/Ar model ages indicate that this late illite precipitated during Late Devonian and Mississippian time (367–327 Ma, mean 346 ± 11 Ma; Barnes et al., 1992b), suggesting that dolospar and anhydrite must have formed fairly early during burial.

The age of late-diagenetic calcite cement in the St. Peter Sandstone has not been determined but, the age of vein-fill calcite in superjacent Ordovician carbonate rocks has been dated. U-Pb systematics indicate that calcite in the UMV district formed in Middle Jurassic time (~162 Ma), approximately 100 my after MVT mineralization (Brannon et al., 1993). At present, no genetic link has been established between the calcite cementing and vein filling events. However, trace amounts of fluorite and sphalerite found in calcite cement indicate that precipitation occurred in the presence of metalliferous (ore-forming) fluids which suggests that a Jurassic age for calcite cementation is unlikely. The timing of calcite emplacement in sandstones thus is regarded to be Permian.

Paleothermal Conditions during Diagenesis

Little direct information is available on the paleothermal conditions that prevailed during St. Peter diagenesis and, in particular, during dolomite precipitation. Combined petrographic, geochemical and preliminary fluid-inclusion data are consistent with formation of planar and baroque dolospar and associated mineral cements in a burial regime governed by conductive heat flow. However, hydrothermal ore-forming fluids focused northward through the basin could have served as an additional source of heat during burial diagenesis and, therefore, must be considered when evaluating the thermal history and origin of authigenic minerals in the St. Peter Sandstone.

Vitrinite reflectance (R_o) data, conodont alteration indices (CAI) and burial and thermal-history modeling were used to constrain the maximum thermal conditions that Ordovician strata in the southern part of the Illinois Basin experienced during burial. Vitrinite reflectance profiles that extend through Paleozoic strata in the Fairfield subbasin (Barrows, unpubl. data, 1985) typically display low R_o values which result in very steep depth-R_o gradients (see Fig. 9 for example). Following Barrows and Cluff (1984), anomalously low R_o values such as those displayed in Figure 9 reflect suppression of organic matter of the type commonly reported for rocks dominated by type II kerogen. Organic-matter suppression effects are important to recognize because temperature estimates determined from suppressed vitrinite reflectance data will be systematically lower than the actual temperatures attained by the rocks (Lo, 1993). According to the representative (uncorrected) R_o profile for the Fairfield subbasin (Fig. 9), the St. Peter Sandstone, although devoid of organic matter, has been subjected to temperatures that correspond to a vitrinite reflectance of ~0.75% R_o. Using the calibration of Barker and Pawlewicz (1994), an R_o value of 0.75% corresponds to a temperature range of ~112°–115°C, depending on the heating rate. Because of the suppressed nature of the reflectance data, values of ~112°–115°C should be regarded as a minimum estimate of the maximum temperature reached by the rocks. No generally accepted method exists for correcting suppressed R_o values, thus the deviation of this value from the "real" maximum temperature cannot be assessed.

To further constrain the thermal history of the deep part of the Illinois basin, the interval analyzed for vitrinite reflectance was modeled kinetically using BASINMOD™, integrating stratigraphic and lithologic data from Bond et al. (1971a, b) and Shaver (1985, Fig 9). A total of 1,200 m of erosion and a transient heat flow of 570 to 100 Ma = 38 mW/m² and 100 Ma to present = 55 mW/m² were used in the model because these conditions produced a profile that closely matched the measured curve in younger Pennsylvanian age coals where the vitrinite values are most reliable. Except in Pennsylvanian strata where measured and hypothetical values are similar, the kinetic-maturation model consistently predicted R_o values that were higher than the measured values for rocks of Permian age and older in the Fairfield subbasin (see Fig. 9). On the basis of the kinetic curve, the St. Peter Sandstone attained a thermal maturity corresponding to an R_o value of ~1.1%, which, according to Barker and Pawlewicz (1994), indicates a burial-heating temperature of ~140°C. A "corrected" temperature of ~140°C is regarded as a more realistic estimate of the maximum burial temperature in this area than the "uncorrected" estimate of ~112°–115°C (see above).

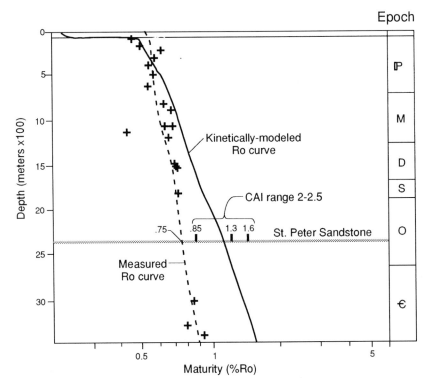

Fig. 9.—Measured vitrinite-reflectance (R_o) curve (dashed line; Barrows, 1985, unpubl. data) versus a kinetically-modeled R_o curve (solid line) for the Fairfield subbasin, southern Illinois. Range of R_o values based on CAI data for Ordovician carbonate rocks in central and southern Illinois are shown for comparison. Differences between the curves reflect organic-matter suppression effects (see text for discussion).

Using Lopatin analysis, Cluff and Byrnes (1991) predicted thermal maturity levels for different depths in the Illinois Basin that were consistent with measured R_o values in Pennsylvanian coals. For the Maquoketa Shale in the Fairfield subbasin, R_o values between 0.80 and 0.95% were predicted. Since the base of the Maquoketa Shale is approximately 300 m upsection from the St. Peter Sandstone, R_o values for the St. Peter, although not explicitly reported by the authors, are probably on the order of ~0.90 to 1.0%. These reflectance values correspond to a burial temperature range of ~127–140°C, using the calibration curve of Barker and Pawlewicz (1994) and thus are in fair agreement with the BASINMOD™-derived estimates (see above).

CAI data, although less precise, are consistent with the previous findings. Throughout central and southern Illinois, CAIs in Ordovician rocks range from 2 to 2.5 (mean CAI = 2; Norby, 1993, written commun.), which corresponds to R_o values of 0.85 to about 1.6% (mean R_o = 0.85–1.3%; Harris, 1979; Fig 9). Based on the CAI-temperature scale of Nowlan and Barnes (1987), these vitrinite values correspond to a temperature range of ~60°–170°C (mean = ~60°–140°C). Because CAI is a function of both time and temperature, conodonts with a CAI ≥2 indicate exposure to temperatures >100°C for more than 100,000 years (Harris, 1979).

Thermal history modeling based on reaction kinetics in conjunction with burial reconstruction provides an additional means of evaluating the temperature conditions in the southern part of the Illinois Basin (see discussion of kinetic maturation above). Figure 10 is a burial-history curve of the St. Peter Sand-

stone derived by BASINMOD™ for the deep Fairfield subbasin. The effects of sediment compaction and differences in thermal conductivity between adjacent rock units were accounted for in the model. Superimposed on the burial curve is the generalized mineral paragenesis and the isotopic ages of authigenic K-feldspar (~400 Ma; Marshall et al., 1986; Barnes et al., 1992b) and the youngest reported diagenetic illite (230–215 Ma; Lee and Aronson, 1991). The age of MVT mineralization (~270 Ma) is shown on the curve for comparison. Paleoisotherms, calculated using the kinetic model of Sweeney and Burnham (1989), were used to extrapolate the burial temperatures that existed during sandstone diagenesis. The kinetically-derived temperatures are based on the premise that the major heat source was from the basement and do not take into account the possible influence of a heat source from hydrothermal fluids (Bethke, 1986; Garven and others, 1993). On the basis of the silicate mineral age dates, the burial-thermal model predicts post-Silurian (Devonian to Middle Triassic) temperatures of ~65°–140°C for mineral authigenesis in the southern part of the basin (Fig. 10). The estimated temperature of K-feldspar precipitation, 65°C, is regarded as a reasonable estimate for the minimum temperature of sandstone diagenesis. Based on homogenization temperatures of fluid inclusions in quartz overgrowths, quartz crystallized at temperatures as high or higher than 110°C. This implies that dolospar (and anhydrite) postdating quartz precipitated at temperatures exceeding 110°C. As previously discussed, late-diagenetic illite in the St. Peter Sandstone postdates most other mineral phases. If this illite formed at ~265 Ma (Early Permian; Hay et al., 1988) or as late as 230–215 Ma

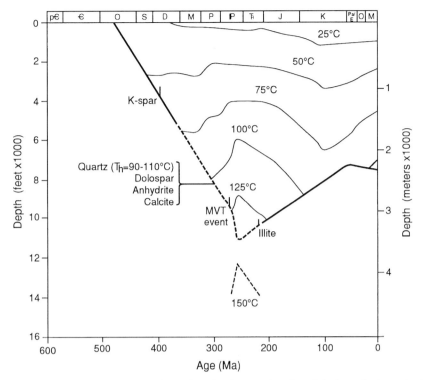

FIG. 10.—Burial-history curve of the St. Peter Sandstone showing timing of carbonate-mineral precipitation and other major diagenetic events (dashed line) in the Fairfield subbasin, southern Illinois. 1,200 m of erosion and variable heat flow (570 to 100 Ma = 38 mW/m²; 100 Ma to present = 55 mW/m²) were assumed in the model. Isotopic ages of authigenic K-feldspar and illite (see text for references) along with homogenization temperatures of secondary quartz constrained timing of authigenic mineral phases; age of lead-zinc-fluorine mineralization (MVT event) is shown for comparison. Paleoisotherms were calculated using the kinetic-maturation model of Sweeney and Burnham (1989) and do not take into account the influence of a potential hydrothermal heat source. Note that most carbonate diagenesis took place when rocks were near maximum burial (see text for discussion).

(Middle Triassic; Lee and Aronson, 1991), then late-diagenetic illite and mineral phases predating late illite (i.e., quartz, dolospar, anhydrite and calcite) precipitated when the St. Peter Sandstone was at or close to it's maximum burial and temperature (~3,300 m, ~140°C), assuming diagenesis occurred in a burial heat-flow regime. According to the burial-history model, the St. Peter reached it's maximum depths and temperatures in Late Pennsylvanian and Early Permian time when major ore-forming events were taking place in the region. It is noteworthy that the estimated burial temperatures predicted for the St. Peter Sandstone in the southern part of the Illinois Basin (~65°–140°C) are comparable to the range of temperatures that characterized the hydrothermal mineralizing fluids involved in ore formation (~80°–180°C; Richardson and others, 1988; Leach and others, 1991; Shelton and others, 1992). This similarity suggests that the fluids involved in diagenesis (i.e. dolomitization) may have been part of the same (paleo)hydrologic system responsible for MVT mineralization.

In shallower parts of the Illinois Basin, the maximum burial depths and temperatures of the St. Peter Sandstone have remained relatively unchanged relative to present-day conditions (<1,000 m; ~50°C; Cluff and Byrnes, 1991). Preliminary fluid inclusion data for planar and baroque dolospar in one of the more shallow wells on the north flank of the basin however, indicate that precipitation of some dolospar occurred at temperatures higher (T_h = 110°–115°C) than would be expected in these otherwise low-temperature rocks. Provided these values

are representative, the St. Peter Sandstone in the northern part of the basin was influenced by a heat source that was not solely burial related. In the absence of igneous activity, hydrothermal fluids related to ore formation in the UMV district could account for the elevated temperatures observed in the burial dolospar.

Composition of Dolomitizing Fluids

Oxygen isotope and fluid-inclusion data provide useful information about the composition of the pore fluids that precipitated the carbonate minerals during burial of the St. Peter Sandstone. The wide range in measured $\delta^{18}O$ ratios of individual carbonate cement types in the St. Peter Sandstone suggests they formed at different temperatures and/or from fluids of variable oxygen isotopic composition. Using the fractionation equations of Friedman and O'Neil (1977, for calcite) and Fritz and Smith (1970, for dolomite), possible temperature-$\delta^{18}O_{water}$ combinations for dolomite and calcite cements were constructed (Fig. 11). The minimum temperature of burial-carbonate cementation is presumed to be 65°C (see above); the maximum temperature of (late-diagenetic) carbonate precipitation is assumed to be ~140°C. As stated previously, this maximum temperature (~140°C) is an approximation based on paleothermal indicators and does not take into account heat transport by fluid flow. A temperature window of 65°–140°C overlaps fluid inclusion homogenization temperatures in dolospar (75°–115°C) in the

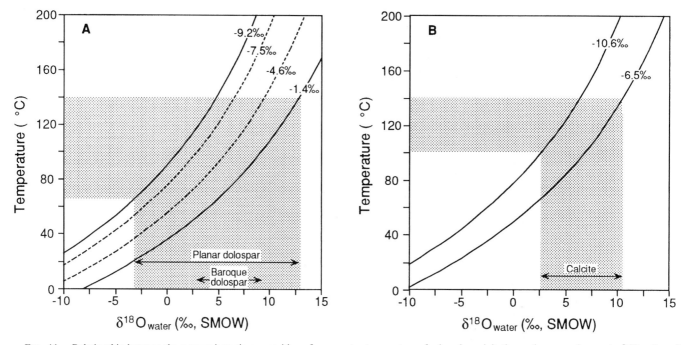

FIG. 11.—Relationship between the oxygen isotopic composition of pore water, temperature of mineral precipitation and measured range in δ^{18}O ratios of carbonate cements in the St. Peter Sandstone. (A) Diagram showing possible parent pore-water compositions in isotopic equilibrium with planar dolospar (continuous lines) and baroque dolospar (dashed lines) using the fractionation equation of Fritz and Smith (1970). For details about temperature assessment see text. (B) Same diagram for calcite based on the relationship of Friedman and O'Neil (1977). A 100°–140°C temperature window was assumed for late-stage calcite precipitation. See text for further details.

northern portion of the basin and quartz (90°–110°C) in the southern part of the basin. Figure 11 shows that calculated equilibrium δ^{18}O$_{water}$ compositions for planar dolospar, which cover the entire range of δ^{18}O ratios of all dolomite cements, vary from −3.3 to +12.7‰ SMOW. Clearly, most burial dolospar precipitated from fluids significantly enriched in [18]O relative to seawater. Assuming that planar dolospar samples with the lowest δ^{18}O ratios formed close to the upper temperature limit (i.e., ≤140°C), and, conversely, planar dolospar with the highest δ^{18}O ratios formed close to the lower temperature limit (i.e., ≤65°C), then the equilibrium δ^{18}O$_{water}$ compositions would be ~+4 to +5‰ SMOW (Fig. 11). Slightly higher temperatures for baroque dolospar (100°–140°C), based on it's relative position within the paragenetic sequence and it's confinement largely to the deeper portions of the basin, predict similar enriched δ^{18}O$_{water}$ compositions (+2.8 to +9.4‰ SMOW; Fig. 11). The estimated δ^{18}O$_{water}$ compositions for baroque dolospar (+2.8 to +9.4‰ SMOW) overlap the range of δ^{18}O values for the deeper formation waters in the Illinois Basin (~0 to +6‰ SMOW; Clayton et al., 1966; Stueber and Walter, 1991) and the main-stage, saline, ore-forming fluids in the Illinois/Kentucky fluorspar district (~0 to +8‰ SMOW; Richardson et al., 1988). Oxygen isotope data in the St. Peter Sandstone argue against a significant change in temperature between planar/baroque dolospar and calcite precipitation assuming a near-constant δ^{18}O$_{water}$ composition (~+4 to +5‰ SMOW) and a 2–3‰ fractionation between coexisting calcite and dolomite. If calcite formed at temperatures similar to those of quartz and baroque dolospar, then the parent pore waters also were isotopically fairly heavy (e.g., +2.7 to +10.5‰ SMOW at 100°–140°C; Fig. 11).

Geochemical and petrographic data provide additional evidence that saline fluids were involved in burial cementation of the St. Peter Sandstone. Salinity estimates based on preliminary fluid-inclusion data from planar and baroque dolospar (>20 wt% NaCl equivalent) are consistent with data from authigenic quartz (~15 wt% NaCl equivalent), suggesting these cements formed from brines concentrated beyond the stage of Ca-sulfate saturation but probably undersaturated with respect to halite (e.g., McCaffrey et al., 1987). A saline, sulfate-rich fluid source also is indicated by the occurrence of late-stage anhydrite in the sandstones. The salinity of the fluids that precipitated dolospar (and quartz) in Ordovician sandstones (~15–20 wt% NaCl equivalent) is comparable to the salinity of the brines responsible for MVT mineralization (>20 wt% NaCl equivalent; McLimans, 1977; Leach and Rowan, 1986; Richardson et al., 1988; Leach et al., 1991; Shelton et al., 1992; and others) and of modern formation waters in the basin (~20 wt% NaCl equivalent; Stueber and Walter, 1991). Fluid-inclusion data from baroque dolomite in the St. Peter Sandstone from the deep Michigan Basin suggest similar saline pore-water compositions (17–24 wt% NaCl eq., T$_h$ > 100°C; Drzewiecki et al., 1994; Barnes et al., 1992a). Saline pore fluids associated with a deep-burial environment also have been proposed for many other baroque dolomite occurrences (e.g., Radke and Mathis, 1980; Machel, 1987; Kaufman et al., 1990).

Sources and Pathways of Dolomitizing Fluids

Geologic and geochemical studies provide compelling evidence that high-temperature basinal brines expelled from tectonic terrane south of the Illinois Basin during Late Permian time formed the MVT deposits in the southern midcontinent region (Richardson and Pinkney, 1984; Leach and Rowan, 1986; Richardson et al., 1988; Rowan and Leach, 1989; Leach et al., 1991; Shelton et al., 1992; and others). A likely conduit for the passage of these fluids was the St. Peter Sandstone, which is an aquifer even today. The spatial distributions of (mean) $\delta^{18}O$ ratios of carbonate cements in the St. Peter Sandstone provide a record of the sources and migration pathways of dolomitizing fluids that passed through the basin during its burial history. Light $\delta^{18}O$ ratios in shallow-buried rocks on the south flank of the Michigan Basin and adjacent to localized ore districts along the southern margin of the Illinois Basin strongly suggest two flow vectors, one coming from the south and a smaller one originating from the north (Fig. 12). Fluid flow from the south likely included hydrothermal sources, intrabasinal sources (i.e., the Illinois Basin proper), sediments beneath extensively folded and faulted terrane along the southern margin of the basin, and meteoric water recharged from topographic highs. The flow vector from the north may have involved fluids derived in part from the deep Michigan Basin, as suggested by McLimans (1977). Alternatively, fluids from sources to the south were diverted along flow paths different from the regional conduit system that focused fluids through the basin.

Hydrologic models for the Illinois Basin have considered tectonic compression, sediment loading and uplift as mechanisms to transport large volumes of warm fluid through the Illinois Basin proper (Cathles and Smith, 1983; Bethke et al., 1988; Bethke and Marshak, 1990; Garven and others, 1993). Even in the absence of igneous activity (Zartman et al., 1967; McGinnis et al., 1976; Schwalb, 1982), warm, gravity-driven fluids (i.e., low-temperature hydrothermal brines) moving northward, updip from the southern tectonic margin of the Illinois Basin could account for much of the late-stage dolomite in the St. Peter Sandstone, particularly the isotopically light baroque dolospar in the Fairfield subbasin close to the Illinois/Kentucky Fluorspar

district. ^{18}O-depleted burial dolospar concentrated along the La Salle anticline in east-central Illinois suggests that this large structural feature was a major conduit for the movement of these hot fluids through the basin. It is conceivable that fluids gained access to aquifer sandstones by means of faults and fractures in the southern part of the basin and then moved northward precipitating significant amounts of dolomite along their flowpath. As fluids moved updip into the basin and mixed with formation waters, they gradually cooled which would account for the observed isotopically heavier $\delta^{18}O$ values northward along the anticline. The local focusing of hot fluids through the La Salle anticline also could explain the high temperature saline inclusions and ^{18}O-depleted dolospar ratios in shallow-buried sandstones in the northern part of the basin in the vicinity of the UMV district. The ultimate source of the hot fluids (see Fig. 12) might have been folded and faulted sediments associated with the Ouachita and Appalachian fold belt or the Reelfoot rift (Brecke, 1979; Farr, 1989, Garven et al., 1993). A component of these fluids also may have been expelled from the Arkoma and Black Warrior Basins underlying Mississippi-embayment sediments (Rowan and Leach, 1989). The fold and thrust belts and foreland basins were fluid-flow pathways during Late Pennsylvanian and Early Permian time because tectonic thrusting and compression in conjunction with the Ouachita orogeny significantly altered the basin hydrodynamics.

CONCLUSIONS

The St. Peter Sandstone in the Illinois Basin has undergone a complex diagenetic history involving formation of authigenic K-feldspar, multiple generations of carbonate cement, secondary quartz, anhydrite and illite. Mineral paragenesis in conjunction with isotopic age dates place constraints on the timing of carbonate emplacement and source(s) of fluids responsible for dolomitization. Results of the study show that different generations of dolomite formed from similar formation waters throughout the basin except early dolomicrospar, which formed by diagenetic alteration of a carbonate precursor mineral. In the southern part of the Illinois Basin, burial-history reconstruction indicates that precipitation of burial dolospar and associated mineral cements probably occurred in Late Pennsylvanian and Early Permian time when the St. Peter Sandstone was at or near maximum burial and major ore-forming events were taking place in the region. Several lines of evidence link burial-dolomite cements in the St. Peter Sandstone to the same hydrothermal fluid-flow event that caused MVT mineralization. These include (i) similar "cement stratigraphy", (ii) comparable stable isotopic compositions, and in particular, isotopically light $\delta^{18}O$ ratios in samples immediately north of the Illinois-Kentucky fluorspar district, (iii) high salinities based on fluid-inclusion studies, (iv) traces of fluorite and sphalerite in calcite cement and (v) largely coincident timing of late-stage dolomite precipitation in the St. Peter Sandstone and MVT mineralization and (vi) a comparable range of precipitation temperatures.

ACKNOWLEDGMENTS

This study was supported by the U. S. Geological Surveys' Evolution of Sedimentary Basins' Program. We thank T. B. Shaw for providing drill-core samples and V. Nuccio for generating a burial history model. The authors are indebted to M.

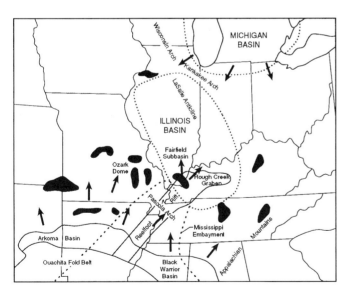

FIG. 12.—Paleoflow directions (shown as arrows) in the Illinois Basin area. Dominant flow was from the south; a smaller flow component or alternatively, a different flowpath was to the north. Areas shown in black are major ore districts.

B. Goldhaber and N. Fishman for reviewing the paper and providing helpful suggestions.

REFERENCES

BANNER, J. L., HANSON, G. N., AND MEYERS, W. J., 1988, Water-rock interaction history of regionally extensive dolomites of the Burlington-Keokuk Formation (Mississippian): isotopic evidence, *in* Shukla, V. and Baker, P. A., eds., Sedimentology and Geochemistry of Dolostones: Tulsa, Society for Sedimentary Geology Special Publication 43, p. 97–113.

BARBER, D. J., REEDER, R. J., AND SMITH, D. J., 1985, A TEM microstructural study of dolomite with curved faces (saddle dolomite): Contributions of Mineralogy and Petrology, v. 91, p. 82–92.

BARKER, C. E. AND PAWLEWICZ, M. J., 1994, Calculation of vitrinite reflectance from thermal histories and peak temperature: A comparison of methods, *in* Mukhopadhyay, P. K. and Dow, W. G., eds., Reevaluation of Vitrinite Reflectance: American Chemical Society Symposium Series 570, p. 216–229.

BARNES, D. A., LUNDGREN, C. E., AND LONGMAN, M. W., 1992a, Sedimentology and diagenesis of the St. Peter Sandstone, central Michigan Basin, United States: American Association of Petroleum Geologists Bulletin, v. 76, p. 1507–1532.

BARNES, D. A., GIRARD, J. P., AND ARONSON, J. L., 1992b, K-Ar dating of illite diagenesis in the middle Ordovician St. Peter sandstone, central Michigan basin, USA: Implications for thermal history, diagenesis, and petrophysics of clay minerals in sandstones, *in* Houseknecht, D. W. and Pittman, E. D., eds., Origin, Diagenesis, and Petrophysics of Clay Minerals in Sandstones: Tulsa, Society for Sedimentary Geology Special Publication 47, p. 35–48.

BARROWS, M. H. AND CLUFF, R. M., 1984, New Albany shale group (Devonian-Mississippian) source rocks and hydrocarbon generation, *in* Demaison, G. and Murris, R. J., eds., Petroleum Geochemistry and Basin Evaluation: Tulsa, American Association of Petroleum Geologists Memoir 35, p. 111–138.

BETHKE, C. M., 1985, A numerical model of compaction-driven groundwater flow and heat transfer and its application in the paleohydrology of intracratonic sedimentary basins: Journal of Geophysical Research, v. 90, p. 1817–1828.

BETHKE, C. M., 1986, Hydrologic constraints on the genesis of the upper Mississippi Valley mineral district from Illinois basin brines: Economic Geology, v. 81, p. 233–249.

BETHKE, C. M., HARRISON, W. J., UPSON, C., AND ALTANER, S. P., 1988, Supercomputer analysis of sedimentary basins: Science, v. 239, p. 261–267.

BETHKE, C. M. AND MARSHAK, S., 1990, Brine migrations across North America: The plate tectonics of groundwater: Annual Review of Earth Planetary Science, v. 18, p. 287–315.

BOND, D. C., ATHERTON, E., BRISTOL, H. M., BUSCHBACH, T. C., STEVENSON, D. L., BECKER, L. E., DAWSON, T. A., FERNALD, E. C., SCHWALB, H., WILSON, E. N., STATLER, A. T., STEARNS, R. G., AND BUEHNER, J. H., 1971a, Possible future petroleum potential of region 9—Illinois basin, Cincinnati arch, and northern Mississippi embayment, *in* Cram, L. H., ed., Future Petroleum Provinces of the U.S.: Their Geology and Potential: Tulsa, American Association of Petroleum Geologists Memoir 15, p. 1165–1218.

BOND, D. C., BUSCHBACH, T. C., BRISTOL, H. M., AND ATHERTON, E., 1971b, Background materials for symposium on future petroleum potential of NPC region 9 (Illinois basin, Cincinnati arch, and northern part of Mississippi Embayment): Illinois State Geological Survey, Illinois Petroleum 96, 63 p.

BRAILE, L. W., HINZ, W. J., KELLER, G. R., LIDIAK, E. G., AND SEXTON, J. L., 1986, Tectonic development of the New Madrid rift complex, Mississippi Embayment, North America: Tectonophysics, v. 131, p. 1–21.

BRANNON, J. C., PODOSEK, F. A., AND McLIMANS, R. K., 1993, Age and isotopic compositions of gangue versus ore minerals in the upper Mississippi Valley zinc-lead district, *in* Shelton, K. and Hagni, R., eds., Geology and Geochemistry of Mississippi-type ore deposits: Rolla, Proceedings Volume, University of Missouri-Rolla Press, p. 95–103.

BRECKE, E. A., 1979, A hydrothermal system in the Midcontinent region: Economic Geology, v. 74, p. 1327–1335.

BURKE, K. AND DEWEY, J. F., 1973, Plume-generated triple junctions: key indicators in applying plate tectonics to old rocks: Journal of Geology, v. 81, p. 406–433.

CATHLES, L. M. AND SMITH, A. T., 1983, Thermal constraints on the formation of Mississippi Valley-Type lead-zinc deposits and their implications for episodic basin dewatering and deposit genesis: Economic Geology, v. 78, p. 983–1002.

CHESLEY, J. T., HALLIDAY, A. N., KYSER, T. K., AND SPRY, P. G., 1994, Direct dating of Mississippi Valley-type mineralization: use of Sm-Nd in fluorite: Economic Geology, v. 89, p. 1192–1199.

CLAYTON, R. N., FRIEDMAN, I., GRAF, D. L., MAYEDA, T. K., MEENTS, W. F., AND SHRIMP, N. F., 1966, The origin of saline formation waters: I. Isotopic composition: Journal of Geophysical Research, v. 71, p. 3869–3882.

CLUFF, R. M. AND BYRNES, A. P., 1991, Lopatin analysis of maturation and petroleum generation in the Illinois Basin, *in* Leighton, M. W., Kolata, D. R., Oltz, D. F. and Eidel, J. J., eds., Interior Cratonic Basins, American Association of Petroleum Geologists Memoir 51, p. 425–454.

DAPPLES, E. C., 1955, General lithofacies relationships of the St. Peter Sandstone and Simpson Group: American Association of Petroleum Geologists Bulletin, v. 32, p. 1924–1947.

DRZEWIECKI, P. A., SIMO, J. A., BROWN, P. E., CASTROGIOVANNI, E., NADON, G. C., SHEPHERD, L. D., VALLEY, J. W., VANDREY, M. R., WINTER, B. L., AND BARNES, D. A., 1994, Diagenesis, diagenetic banding, and porosity evolution of the Middle Ordovician St. Peter Sandstone and Glenwood Formation in the Michigan Basin, *in* Ortoleva, P. J., ed., Basin Compartments and Seals: Tulsa, American Association of Petroleum Geologists Memoir 61, p. 179–199.

ELLIOTT, W. C. AND ARONSON, J. L., 1987, Alleghenian episode of K-bentonite illitization in the southern Appalachian Basin: Geology, v. 15, p. 735–739.

ELLIOTT, W. C., AND ARONSON, J. L., 1993, The timing and extent of illite formation in Ordovician K-bentonites at the Cincinnati Arch, the Nashville Dome and north-eastern Illinois Basin: Basin Research, v. 5, p. 125–135.

ERVIN, C. P. AND McGINNIS, L. D., 1975, Reelfoot rift: reactivated precursor to Mississippi Embayment: Geological Society of America Bulletin, v., 86, p. 1287–1295.

ESTEBAN, M. AND KLAPPA, C. F., 1983, Subaerial exposure environment, *in* Scholle, P. A., Bebout, D. G., and Moore, C. H., eds., Carbonate Depositional Environments: Tulsa, American Association of Petroleum Geologists Memoir 33, p. 1–95.

FARR, M. R., 1989, Compositional zoning characteristics of late dolomite cement in the Cambrian Bonneterre formation, Missouri: implication for brine migration pathways: Carbonates and Evaporites, v. 4, p. 177–194.

FRIEDMAN, I., AND O'NEIL, J. R., 1977, Compilation of stable isotope fractionation factors of geochemical interest, *in* Fleischer, M., ed., Data of Geochemistry, 6th edition: Washington, D.C., United States Geological Survey Professional Paper, v. 440-KK, 12 p.

FRITZ, P. AND SMITH, D. C. W., 1970, The isotopic composition of secondary dolomite: Geochimica et Cosmochimica Acta, v. 34, p. 1161–1173.

GARVEN, G. AND FREEZE, R. A., 1984a, Theoretical analysis of the role of groundwater flow in the genesis of stratabound ore deposits: 1. Mathematical and numerical model: American Journal of Science, v. 284, p. 1085–1124.

GARVEN, G. AND FREEZE, R. A., 1984b, Theoretical analysis of the role of groundwater flow in the genesis of stratabound ore deposits: 2. Quantitative results: American Journal of Science, v. 284, p. 1125–1174.

GARVEN, G., GE, S., PERSON, M. A., AND SVERJENSKY, D. A., 1993, Genesis of stratabound ore deposits in the midcontinent basins of North America. I. The role of regional groundwater flow: American Journal of Science, v. 293, p. 497–568.

GARVIN, P. L., LUDVIGSON, G. A., AND RIPLEY, E. M., 1987, Sulfur isotope reconnaissance of minor metal sulfide deposits fringing the Upper Mississippi Valley zinc-lead district: Economic Geology, v. 82, p. 1386–1394.

GREGG, J. M. AND SIBLEY, D. F., 1984, Epigenetic dolomitization and the origin of xenotopic dolomite texture: Journal of Sedimentary Petrology, v. 54, p. 908–931.

HALL, W. E., AND FRIEDMAN, I., 1969, Oxygen and carbon stable isotopic compositions of ore and host rock of selected Mississippi Valley deposits: Washington, D.C., United States Geological Survey Professional Paper, v. 650-C, p. C140–C148.

HARRIS, A. G., 1979, Conodont color alteration, an organo-mineral metamorphic index, and its applications to Appalachian Basin geology, *in* Scholle, P. A., and Schluger, P. R., eds., Aspects of Diagenesis: Tulsa, Society for Sedimentary Geology Special Publication 26, p. 3–16.

HAY, R. L., LEE, M., KOLATA, D. R., MATTHEWS, J. C., AND MORTON, J. P., 1988, Episodic potassic diagenesis of Ordovician tuffs in the Mississippi Valley area: Geology, v. 16, p. 743–747.

HEIDLAUF, D. T., HSUI, A. T., AND KLEIN, G. DE V., 1986, Tectonic subsidence analysis of the Illinois basin: Journal of Geology, v. 94, p. 779–794.

HOHOLICK, J. D., 1980, Porosity, grain fabric, water chemistry, cement, and depth of burial of the St. Peter Sandstone in the Illinois basin: Unpublised M.S. Thesis, University of Cincinnati, Cincinnati, 72 p.

HOHOLICK, J. D., METARKO, T., AND POTTER, P. E., 1984, Regional variations of porosity and cement: St. Peter and Mount Simon sandstones in Illinois basin: American Association of Petroleum Geologists Bulletin, v. 68, p. 733–764.

KAUFMAN, J., MEYERS, W. J., AND HANSON, G. N., 1990, Burial cementation in the Swan Hills Formation (Devonian), Rosevear field, Alberta, Canada: Journal of Sedimentology Petrology, v. 60, p. 918–939.

KOLATA, D. R., 1991, Tippecanoe I Sequence overview: Middle Ordovician Series through Late Devonian Series, in Leighton, M. W., Kolata, D. R., Oltz, D. F. and Eidel, J. J., eds., Interior Cratonic Basins: Tulsa, American Association of Petroleum Geologists Memoir 51, p. 87–99.

KOLATA, D. R., AND NELSON, W. J., 1991, Tectonic history of the Illinois basin, in Leighton, M. W., eds., Interior Cratonic Basins: Tulsa, American Association of Petroleum Geologists Memoir 51, p. 263–285.

KRUEGER, H. W. AND WOODARD, H. H., 1972, Potassium-argon dating of sanidine-rich beds in the St. Peter Sandstone, Wisconsin: Geological Society of American, Abstracts with Programs, v. 4, p. 568–569.

LEACH, D. L. AND ROWAN, E. L., 1986, Genetic link between Ouachita foldbelt tectonism and the Mississippi Valley-type lead zinc deposits of the Ozarks: Geology, v. 14, p. 931–935.

LEACH, D. L., PLUMLEE, G. S., HOFSTRA, A. H., LANDIS, G. P., ROWAN, E. L., AND VIETS, J. G., 1991, Origin of late dolomite cement by CO₂-saturated deep basin brines: evidence from the Ozark region, central United States: Geology, v. 19, p. 348–351.

LEE, W., 1994, Mechanisms of dolomitization in seawater and origin of middle Ordovician dolomites in the Illinois basin: Unpublished Ph.D. Thesis, University of Illinois, Urbana, 132 p.

LEE, M. K. AND ARONSON, J. L., 1991, Repetitive occurrence of potassic diagenesis in the region of the Upper Mississippi Valley (UMV) mineral district: Implications for a persistent paleo-hydrological setting favorable for diagenesis: Clay Mineral Society, 28th Annual Meeting, Programs with Abstracts, p. 98.

LO, H. B., 1993, Correction criteria for the suppression of vitrinite reflectance in hydrogen-rich kerogens: preliminary guidelines: Organic Geochemistry, v. 20, p. 653–657.

LOHMANN, K. C. AND WALKER, J. C. G., 1989, The δ¹⁸O record of Phanerozoic abiotic marine calcite cements: Geophysical Research Letters, v. 16, p. 319–322.

MACHEL, H. G., 1987, Saddle dolomite as a by-product of chemical compaction and thermochemical sulfate reduction: Geology, v. 15, p. 936–940.

MARSHALL, B. D., WOODARD, H. H., AND DEPAOLO, D. J., 1986, K-Ca-Ar systematics of authigenic sanidine from Waukau, Wisconsin, and the diffusivity of argon: Geology, v. 14, p. 936–938.

McCAFFREY, M. A., LAZAR, B., AND HOLLAND, H. D., 1987, The evaporation path of seawater and the coprecipitation of Br and K with halite: Journal of Sedimentary Petrology, v. 57, p. 928–937.

McGINNIS, L. D., HEIGOLD, C. P., ERVIN, C. P., AND HEIDI, M., 1976, The gravity field and tectonics of Illinois: Illinois State Geological Survey Circular, v. 494, 24 p.

McLIMANS, R. K., 1977, Geological, fluid inclusion, and stable isotope sudies of the Upper Mississippi Valley zinc-lead district, Southeast Wisconsin, Unpublished Ph.D. Thesis, Pennsylvania State University, city?, 175 p.

NELSON, W. J., 1991, Structural styles of the Illinois Basin, in Leighton, M. W., Kolata, D. R., Oltz, D. F. and Eidel, J. J., eds., Interior Cratonic Basins: Tulsa, American Association of Petroleum Geologists Memoir 51, p. 209–243.

NOWLAN, G. S. AND BARNES, C. R., 1987, Application of conodont color alteration indices to regional and economic geology, in Austin, R. L., ed.,

Conodonts: Investigative Techniques and Applications: Chichester, Ellis Horwood Limited, p. 188–202.

ODOM, I. E., WILLAND, T. N., AND LASSIN, R. J., 1979, Paragenesis of diagenetic minerals in the St. Peter Sandstone (Ordovician), Wisconsin and Illinois, in Scholle, P. A. and Schluger, P. R., eds., Aspects of Diagenesis: Tulsa, Society for Sedimentary Geology Special Publication 26, p. 425–443.

PAN, H., SYMONS, D. T. H., AND SANGSTER, D. F., 1990, Paleomagnetism of the Mississippi Valley-type ores and host rocks in the northern Arkansas and Tri-State districts: Canadian Journal of Earth Sciences, v. 27, p. 923–931.

RADKE, B. M. AND MATHIS, R. L., 1980, On the formation and occurrence of saddle dolomite: Journal of Sedimentary Petrology, v. 50, p. 1149–1168.

RETALLACK, G. J., 1992, What to call early plant formations on land: Palaios, v. 7, p. 508–520.

RICHARDSON, C. K. AND PINKNEY, D. M., 1984, The chemical and thermal evolution of fluids in the Cave-in-Rock fluorspar district, Illinois: Mineralogy, paragenesis, and fluid inclusions: Economic Geology, v. 79, p. 1833–1856.

RICHARDSON, C. K., RYE, R. O., AND WASSERMAN, M. D., 1988, The chemical and thermal evolution of the fluids in the Cave-in-Rock fluorspar district, Illinois: Stable isotope systematics: Economic Geology, v. 83, p. 765–783.

ROWAN, E. L. AND LEACH, D. L., 1989, Constraints from fluid inclusions on sulfide precipitation mechanisms and ore fluid migration in the Viburnum Trend lead district, Missouri: Economic Geology, v. 84, p. 1948–1965.

SCHWALB, H. R., 1982, Paleozoic geology of the New Madrid area: Washington, D.C., United States Nuclear Regulatory Commission, NUREG/CR-2909, 61 p.

SHAVER, R. H., 1985, Midwestern basin and arches region, in Lindberg, F.A., ed., Correlation of stratigraphic units of North America: Tulsa, American Association of Petroleum Geologists, COSUNA Chart Series, 1 sheet.

SHELTON, K. L., BAUER, R. M., AND GREGG, J. M., 1992, Fluid-inclusion studies of regionally extensive epigenetic dolomites, Bonneterre Dolomite (Cambrian), southeast Missouri: evidence of multiple fluids during dolomitization and lead-zinc mineralization: Geological Society of America Bulletin, v. 104, p. 675–683.

SIBLEY, D. F. AND GREGG, J. M., 1987, Classification of dolomite rock textures: Journal of Sedimentary Petrology, v. 57, p. 967–975.

SNEE, L. W. AND HAYES, T. S., 1992, ⁴⁰Ar/³⁹Ar geochronology of intrusive rocks and Mississippi-Valley-type mineralization and alteration from the Illinois-Kentucky fluorspar district, in Goldhaber, M. B. and Eidel, J. J., eds., Mineral Resources of the Illinois Basin in the Context of Basin Evolution: Washington, D.C., United States Geological Survey Open-File Report 92-1, Program and Abstracts, p. 59–60.

STACKELBERG, P. E., 1987, The role of intergranular pressure solution in the diagenesis of the St. Peter Sandstone along a traverse of the Illinois Basin: Unpublised M.S. Thesis, University of Columbia-Missouri, Columbia, 113 p.

STUEBER, A. M. AND WALTER, L. M., 1991, Origin and chemical evolution of formation waters from Silurian-Devonian strata in the Illinois Basin, USA: Geochimica et Cosmochimica Acta, v. 55, p. 309–325.

SWEENEY, J. J. AND BURNHAM, A. K., 1989, A chemical kinetic model of vitrinite maturation and reflection: Geochimica et Cosmochimica Acta, v. 53, p 2649–2657.

WALTERS, L. J., JR., CLAYPOOL, G. E., AND CHOQUETTE, P. W., 1972, Reaction rates and δ¹⁸O variation for the carbonate-phosphoric acid preparation method: Geochimica et Cosmochimica Acta, v. 36, p. 129–140.

WRIGHT, V. P., 1990, A micromorphological classification of fossil and recent calcic and petrocalcic microstructures, in Douglas, L. A., ed., Soil Micromorphology: A Basic and Applied Science: Developments in Soil Science 19, p. 401–407.

ZARTMAN, R. E., BROCK, M. R., HEYL, A. V., AND THOMAS, H. H., 1967, K-Ar and Rb-Sr ages of some alkalic intrusive rocks from central and eastern United States: American Journal of Science, v. 265, p. 858–870.

PRODUCTION-INDUCED DIAGENESIS DURING THERMAL HEAVY OIL RECOVERY: GRAIN SIZE AS A PREDICTOR OF RESERVOIR ALTERATION

MARY L. BARRETT

Department of Geology and Geography, Centenary College, Shreveport, Louisiana 71134

AND

RICHARD W. MATHIAS

GeoGraphix, Denver, Colorado 80202

ABSTRACT: Terrigenous clastic alteration during thermal heavy oil recovery processes has been documented in whole core and laboratory tests from four California fields. Sediments are subjected to either steam/hot water processes (up to 200°C) or in-situ combustion processes (up to 500°C) or both. Consideration of original and altered mineral compositions do not fully describe reservoir changes. Fabric and permeability changes can be understood in a predictive sense by comparison to original grain size and sorting.

Sand-size constituents are plagioclase feldspar and quartz, with lesser amounts of mica, volcanic rock fragments and volcanic glass. Original matrix is dominated by smectite, biogenic opal, zeolites and mica/illite. During hot water/steam alteration, primarily the fine-grained matrix is altered. Sediments with either finer grain sizes or increased original matrix have the highest magnitude of permeability decrease as compared to coarser, better sorted material. This pronounced permeability decrease is due to: (1) additional smectite and zeolite growth as the expense of original matrix components and (2) the dispersion, migration and pore-throat blockage by the matrix. The finer-grained sediments, characterized by smaller pore throats, are subjected to extensive pore blockage as unstable silt- and clay-size particles move (fines migration).

Sediment reaction during in-situ combustion follows the alteration pattern above. Fine-grained matrix continues to alter, with some smectite transforming to illite. Calcite and oil reaction rims form in the burn stage. The overall fabric is a lightly-consolidated sandstone. Only in the near-wellbore area of an combustion injector have partially-melted fabrics been recognized.

INTRODUCTION

Heavy oil production is becoming increasingly important as a component of domestic and world oil production. Abundant reserves require unique techniques to overcome many production problems. Different reservoirs are produced by various means, including primary production, cyclic steam stimulation, steam flooding, and in-situ combustion. Elevated reservoir temperatures and changing water compositions are capable of producing diagenetic changes similar in many aspects to geological burial conditions, yet at time-frames of days to a few years. The reservoir and its response to thermal enhanced recovery are of geological and engineering interest to understand flow rates and longer-term recovery rates during production. Concerns of both near-wellbore damage and detrimental reservoir alteration have resulted in concentrated efforts to characterize these sands in terms of reactivity potential during thermal operations. The objectives of this paper are to discuss observed thermal sediment alteration patterns in California heavy oil fields, plus relate the magnitude of alteration to a fundamental predictive characteristic, namely grain size.

California contains over 85% of the marketable heavy crude reserves in the continental United States. Heavy-oil production in Midway-Sunset, Lost Hills, South Belridge, and San Ardo fields is from poorly consolidated sands (Fig. 1). During thermal recovery operations, sediments may be exposed to two basic types of thermal alteration (Fig. 2). The first type is interaction with hot water/steam during either cyclic injection or steamdrive processes, where sediments are exposed to high temperatures (200°C) and pH values as high as 10 to 12 (Reed, 1980). The second type is in-situ combustion alteration, where sediments first go through similar hot water/steam reactions, but then are exposed to much greater temperatures (near 500°C) during the combustion process (Gates and Sklar, 1971).

The studied reservoirs range from Miocene to Pleistocene in age (Table 1). Midway-Sunset, Lost Hills and South Belridge fields are located in the San Joaquin Basin, while San Ardo field is in the Salinas Basin. Average present-day burial depths

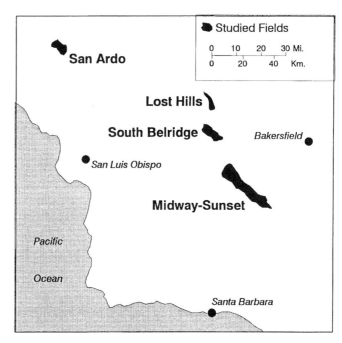

FIG. 1.—Location map of the studied heavy-oil fields, southern California.

of these sediments range from 165 to 850 m. Oil gravity varies from 9° to 14° API. Formation water compositions are generally brackish, ranging on the order of 5,000 to 11,000 mg/l dissolved constituents. The producing zones vary substantially in their original depositional settings and associated reservoir stratigraphy. The upper Miocene through Recent basin infill consists of an overall shallowing-upward pattern of submarine fan, shallow marine shelf, and modern alluvial environments. Despite their diverse depositional settings, these sands show an overall similarity in their patterns of sediment/water interaction and thermal alteration.

Siliciclastic Diagenesis and Fluid Flow: Concepts and Applications, SEPM Special Publication No. 55

Steamflooding

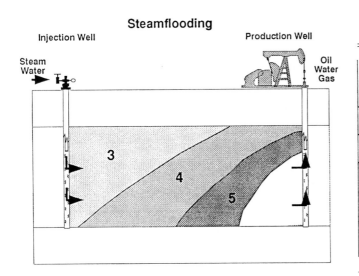

In-Situ Combustion

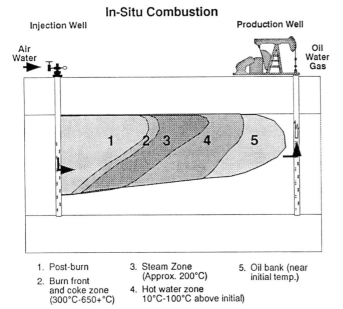

1. Post-burn

2. Burn front
   and coke zone
   (300°C-650+°C)

3. Steam Zone
   (Approx. 200°C)

4. Hot water zone
   10°C-100°C above initial)

5. Oil bank (near
   initial temp.)

FIG. 2.—Schematic of steamflood and in-situ combustion thermal processes.

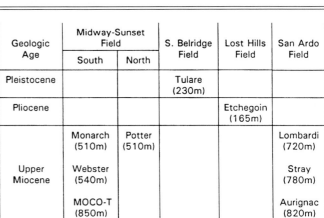

TABLE 1.—STUDIED HEAVY OIL RESERVOIRS AND THE AVERAGE MID-RANGE BURIAL DEPTHS

Geologic Age	Midway-Sunset Field		S. Belridge Field	Lost Hills Field	San Ardo Field
	South	North			
Pleistocene			Tulare (230m)		
Pliocene				Etchegoin (165m)	
Upper Miocene	Monarch (510m)	Potter (510m)			Lombardi (720m)
	Webster (540m)				Stray (780m)
	MOCO-T (850m)				Aurignac (820m)

The experimentally-derived relationships, although not perfect for buried sands and sandstones, have their closest geologic comparison in these shallow oil-cemented sands.

These relationships infer that coarser permeable sands experience a greater volume of reactive fluids during a given timeframe compared to finer-grained sediments. Grain size and sorting are also indicators of surface area and associated reactivity. Finer-grained sands (commonly with increased matrix) have a larger surface area for sediment/fluid reaction as compared to coarser sediments.

This also infers that for a given sorting parameter (or deviation from median grain size) and packing arrangement (reflected in burial depths), coarse sediments will have larger pore throats than fine-grain sediments. This directly affects the amount and size of fines (mobile matrix) that are able to migrate through pore throats. Pore throat blockage by migrating matrix material is a common detrimental process during production, significantly altering sediment permeability characteristics.

PROCEDURES OF STUDY

Pre- and post-thermal alteration of the heavy oil sediments was studied by core observation, laboratory coreflood tests, grain size analysis, thin section petrography, scanning electron microscopy and x-ray diffraction over a five-year period. Grain size (sieve and laser particle-size analyses) and water/steam coreflood analyses were performed by service company personnel. Additional near-wellbore material was described from samples collected from failed in-situ combustion injectors.

Approximately 650 x-ray diffraction mineralogical analyses were performed on both the original and altered reservoir sediments discussed here. X-ray diffraction mineralogy was determined on both a bulk and less-than-five-micron sample split. Samples were cleaned with solvents to remove heavy oil. Material was gently disaggregated, and two aliquot splits were made of each sample. One portion was used to make an oriented mount for analysis of clay minerals.

The bulk portion selected for analysis of whole rock mineralogy was ground to a standard size, and an internal standard (MgO) was added. By using the internal standard method, each

In any type of sediment alteration, whether geologic diagenesis or "production-induced" diagenesis ("artificial diagenesis" of Hutcheon, 1984) sand reactivity can be characterized in terms of grain size and sorting (controlling both surface area and permeability), composition, and conditions of alteration (temperature, pressure, fluid composition and flow patterns). As will be shown, the overall sand and matrix compositions of the different reservoirs are similar, generally feldspathic sands with a smectite-rich matrix. Secondly, the discussion of alteration is divided into hot water/steam reactions and in-situ combustion reactions to compare similar alteration conditions of the various sands. Lastly, sediment grain size allows comparison of sediment alteration patterns in these diverse stratigraphic reservoirs.

Texture (grain size and associated sorting) is a predictor of permeability in unconsolidated sands. Early workers (Krumbein and Monk, 1942) demonstrated that permeability is a function of grain size and sorting in unconsolidated artificial sand packs.

mineral was independently calculated and added to the total of all minerals found in the sample. When the result did not equal 100%, the "missing" amount was designated "unassigned" (UA). In these California sediments, this unassigned amount reflects the presence of amorphous biogenic silica and volcanic glass. The UA amount may also be associated with crystalline minerals not detected or inaccurately calculated by the analyst. The author's experience (R. W. Mathias) indicates the internal standard method is preferable to other XRD methods to detect and "semi-quantify" the amorphous material common in California sediments.

Special core tests were performed at overburden pressures on 2.5- to 3.8-cm (1.0- to 1.5-in.) plugs. Test temperatures ranged from 25° to 171°C. Cores undergoing specific water permeability tests were first removed of oil by low-rate miscible solvent injection ending with methanol. The compositions of the various testing waters ranged from distilled to the formation water composition. Formation water compositions are brackish, ranging from 5,628 mg/l (Potter Formation) to 11,139 mg/l (Monarch Formation). The term "brine" follows oil-field terminology for synthetic formation water compositions. Steam coreflood tests used native state cores which were oil re-saturated prior to being injected with 100% quality steam. The tests were conducted up to residual oil saturation. Typical fluid/steam amounts injected ranged from 100 to 130 pore volumes. Cores were then cleaned, dried and measured for air permeability (post-test) values under overburden conditions. Select post-test plugs were further studied by thin-section, scanning electron microscopy and x-ray diffraction data.

SEDIMENT CHARACTERIZATION

The textures of the producing reservoirs vary from very fine-grained sand up to pebble-size conglomerates. Permeable sand beds are often separated by siliceous mudstone barriers. Both vertical and lateral packaging of different sediment fabrics into sedimentary layering is a function of the depositional environment. However, the reactions and alteration of pore-size flow paths within layers can be understood in terms of grain size characteristics, allowing predictive relationships to be recognized which cross the boundaries of sedimentary facies.

The heavy-oil sands are poorly consolidated. Grains are sub-angular to sub-rounded with only point contacts between grains (Figs. 3A, B). Both consolidation and degree of compaction reflect their shallow burial nature. Sand-size components are dominated by quartz and plagioclase (Fig. 4), with lesser amounts of potassium feldspar, mica, and volcanic glass. Matrix amounts (clays, zeolites, and biogenic silica, Figs. 3C, D) vary from a few weight percent to over 30 weight percent. Matrix amounts correspond with grain sizes—highest matrix contents are in very fine-grain sands. Coarser sands with less matrix (generally 5 weight percent or less) contain smectitic clays and zeolite crystals as grain coatings.

Total clay mineral amounts in the sands range from 3 to 10 weight percent as determined by x-ray diffraction (Table 2). The clay mineral population is dominated by smectite. All sands are classified as "water sensitive" due to the presence of the swelling smectitic clays. Amorphous silica is present as biogenic opal matrix material and volcanic glass (sand and matrix). As discussed in the methods section, each x-ray diffraction analysis

has an internal standard to determine the amount of material unidentified by x-ray diffraction plus errors. These numbers, particularly over a few weight percent, are used to indicate the presence of an amorphous component. Average amounts of unassigned material range from 4 to 26 weight percent (Table 2).

Sediment permeability values vary directly with grain size. Figure 5 illustrates the pattern of higher permeability measurements in coarser-grained unaltered sediment; this relationship is derived by visual sample inspection using a grain size chart and a binocular microscope. Similar relationships exist in all studied units, in both original and thermally-altered sediments, and have been confirmed by more elaborate techniques of sieve and laser particle size diffraction analyses. Sorting differences within an average grain size are not large enough to break out controls from this variation.

HOT WATER AND STEAM ALTERATION

The patterns of mineralogical alteration are visible in both field core study and laboratory tests. Figure 6 summarizes these observations of dissolution and precipitation events in the heavy oil sands. Additional smectite and zeolite (heulandite) precipitation is common. Dissolution of opal, kaolinite, and calcite also occurs with the onset of hot water and steam movement through the sediments. However, these mineralogical changes, documented by many workers (for example, Day et al., 1967; Sedimentology Research Group, 1981; Hutcheon, 1984; Hutcheon and Abercrombie, 1990), do not fully describe how reservoir fabrics and flow properties are modified in sediments. Observations of grain size, petrographic characteristics of alteration, and associated permeability changes provide a predictive method to understand the magnitude of permeability decrease possible during exposure to hot water and steam.

Special Core Test Measurements

Grain size versus permeability relationships were applied to coreflood tests to determine the magnitude of reservoir property alteration. Figure 7, from the Potter Formation of north Midway-Sunset Field, illustrates the specific water permeability at 149°C to the measured post-test air permeability of the same plugs. The water ("brine") salinity was 5,628 mg/l. Recall that lower air permeability values (pre- and post-test) indicate generally finer-grained sediments. A linear relationship exists between the brine and air permeability values. In general, brine permeability numbers are lower than post-test air permeability numbers and cut across the one-to-one slope. Sediments with lower air permeability values (finer-grained sediment) have a much greater magnitude of hot brine permeability decrease as compared to higher air permeability sediments (the more coarse sand sizes).

Similar relationships occur in the Monarch zone of south Midway-Sunset Field (Figs. 8, 9). Plots illustrate a relationship of higher permeability values with larger grain sizes. Effective permeability (water permeability at residual oil saturation, Fig. 8) follows the expected pattern of lower values as compared to specific permeability (only water present in the pores, Fig. 9). Plots illustrate a relationship of higher permeability values with larger grain sizes. Monarch sediments with lower air permeability values, or finer-grained sediments, also have a greater order of magnitude permeability decrease to hot brine flow.

FIG. 3.—Typical unaltered sands. (A) Sand from the Potter zone. (B) Close-up of point contacts between grains and minor matrix in pores. (C) SEM photograph of smectite and zeolite grain coatings, Monarch zone. (D) SEM photograph of biogenic silica matrix, Etchegoin zone. Diatom fragments are most common.

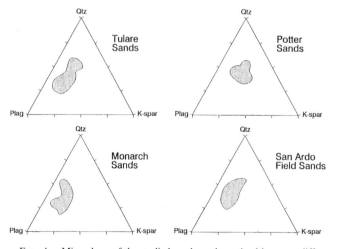

FIG. 4.—Mineralogy of the studied sands as determined by x-ray diffraction. Note the overall similarity of the sand compositions.

TABLE 2.—AVERAGE CLAY AND UNASSIGNED WEIGHT PERCENTAGES IN RESERVOIR SANDS AS DETERMINED BY X-RAY DIFFRACTION

Zone		Total Clay (Wt %)	% Smectite	UA (Amorphous) (Wt %)
I.	Midway-Sunset Field			
	A. Potter	5	66	7
	B. Monarch	6	48	19
	C. Webster	6	35	6
	D. MOCO-T	3	62	6
II.	S. Belridge Field			
	A. Tulare	5	66	15
III.	Lost Hills Field			
	A. Etchegoin	9	60	26
IV.	San Ardo Field			
	A. Lombardi	6	94	4
	B. Stray	10	92	19
	C. Aurignac	8	91	10

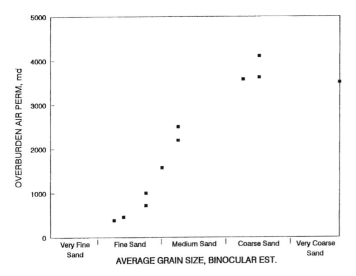

FIG. 5.—Visual grain size estimate versus measured overburden air permeability, Potter sand.

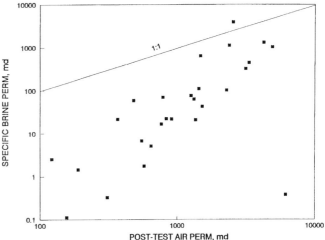

FIG. 7.—Hot (149°C) formation water permeability versus air permeability, Potter sand.

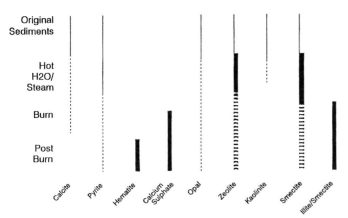

FIG. 6.—The proposed alteration patterns of original sediments exposed to the different thermal recovery processes. Thick line is precipitation; dashed line is dissolution.

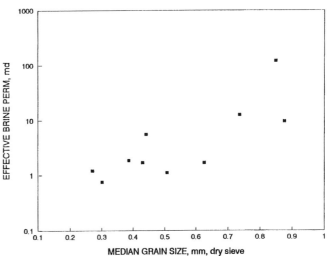

FIG. 8.—Effective formation water (149°C) permeability at residual oil saturation versus grain size, Monarch sand.

Permeability patterns during core steamflooding are illustrated in Figures 10–12 from the Lombardi unit of San Ardo Field. Effective steam permeability values again illustrate the relationship of increasing permeability with larger grain size (Fig. 10). A crude relationship between air permeability and measured grain size exists even after alteration (Fig. 11). The Lombardi sand measurements follow the pattern of lower-permeability finer-grained sands decreasing at a greater order of magnitude to steam (or hot water) injection as compared to coarser, more permeable sands (Fig. 12).

Permeability often changes over time at a constant flow rate, temperature, and water composition during coreflood testing. This is interpreted as migration of mobile matrix due to initial matrix instability and dispersion. The finer-grained sands often reach stable permeability readings sooner than coarse, permeable sands. Petrographic study indicates the mobile matrix ("fines") is mainly smectite, zeolite and biogenic silica (if still present). Permeability values discussed previously were recorded after permeability values stabilized.

Sediment Petrology and Mineralogy

The movement of hot water and steam through an unconsolidated heavy oil sand forms a clean, lightly-consolidated sand (Fig. 13A). Mineral alteration patterns reflect grain size control. The most pronounced mineral alterations are in the matrix with dissolution of opal, kaolinite and calcite (or dolomite), and associated precipitation of more smectite and zeolite (Fig. 6). From an equilibrium concept, other silicate minerals such as quartz and feldspars may be unstable under these thermochemical conditions (see Perkins et al., 1992, for detailed geochemical modeling of Monarch sediment alteration). However, in neither field core studies nor special core tests have the sand-size quartz and feldspars undergone visible dissolution related to thermal alteration. The only sand grains which clearly differ from pre- to post-coreflood testing are the volcanic rock fragments (with a glassy groundmass) and volcanic glass grains.

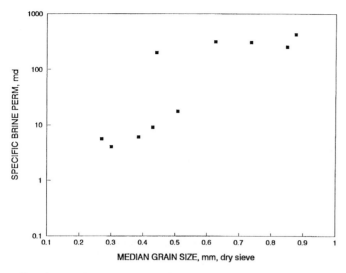

FIG. 9.—Hot formation water (149°C) permeability versus grain size, Monarch sand.

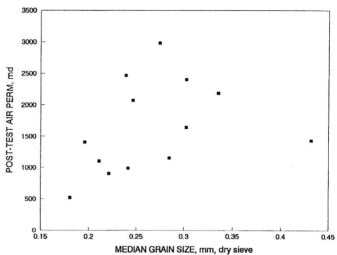

FIG. 11.—Air permeability (post-test steamflood) versus grain size, Lombardi sand.

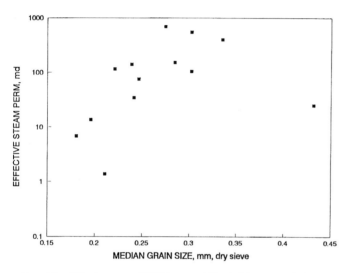

FIG. 10.—Effective steam (177°C) permeability (cold water equivalent) at residual oil saturation versus grain size, Lombardi sand.

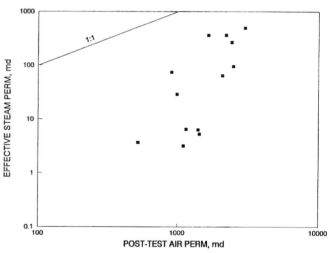

FIG. 12.—Effective steam (177°C) permeability (cold water equivalent) versus air permeability, Lombardi sand.

Finer-grained material is the most reactive during both laboratory time-frames (days) and field production time-frames (months to a few years). Longer time-frames of constant steam injection occur in the near-wellbore environment; here, sand and gravel dissolution has been documented (Day et al., 1967; Reed, 1980).

All sands are characterized as water-sensitive due to the natural occurrence of smectitic clays. Laboratory flow tests at various temperatures and water compositions (distilled to 11,000 mg/l) indicate that clay swelling, dispersion and migration of all the matrix are common occurrences. Hot water/steam injection and associated mineral reactions only contribute to matrix instability. In post-test (hot water and steam) plug studies by thin section and SEM, fines are commonly observed partially blocking pores (Fig. 13). The original grain coatings of smectite and zeolite are often stripped off grains, and the minerals occur as aggregates within pore spaces.

The presence of oil appears to reduce the amount of interaction between hot fluid/steam and the matrix. This pattern is suggested in Figure 14, where increased smectite growth in a cored Tulare sand appears related to decreased oil saturation from thermal recovery. Thus, the highest amount of fluid/sediment interaction and alteration is expected in sediments swept clean of oil.

IN-SITU COMBUSTION ALTERATION

In-situ combustion recovery methods involve air (and possibly water) injection, ignition of heavy oil in the reservoir, and the advancement of a combustion front (Fig. 2). In this manner, oil is displaced by the formation of steam and hot water ahead of the burning front. Figure 6 summarizes the observed mineral alteration associated with both steam (possibly 200°C) and later combustion (possibly 500°C) processes. All alteration stages

FIG. 13.—Sediment fabrics after steam/hot water interaction. (A) Clean sand from the Tulare. (B) Close-up of fines blocking a pore, Monarch test plug. (C) Migrated smectite and zeolite minerals, Monarch plug. (D) SEM photograph of migrated smectite and zeolite material, Monarch plug.

from dry (air injection only) in-situ combustion have been studied using whole and sidewall core from the Webster and MOCO-T zones, south Midway-Sunset Field. The Webster burn, its field-scale geometry and implications for later steamflood operating policy are described in Soustek et al. (1993).

Reservoir Alteration

The sediment appearance exhibits zonation from the in-situ burn stages. The first evidence of alteration is seen as reduced oil saturation, recording passage of hot water and steam. Next, the combustion zone is identified by the presence of black coke in sediment pores. The final alteration product after total burn is seen macroscopically by a clean, reddish sediment. These mappable zonations are used to schematically draw the shape of an advancing combustion front (Fig. 15). Similar macroscopic zones are described by Tilley and Gunter (1988) from a combustion pilot in Alberta.

The advancement of major burn fronts is associated with thick, permeable sand bodies. The slower-moving sections of

the burn are reflected in the interbedding of slightly- to non-altered sediments with totally swept sections. Less-altered sections are within finer-grained thin- to medium-bedded sands interbedded with siliceous mudstones. This pattern also holds true at the field scale, where mapping of the burn using modified electric log signatures reveals this permeability control on burn advancement (Soustek et al., 1993).

The mineralogical and fabric alteration from hot water and steam processes records the earliest evidence of a moving burn front. These alterations are similar to those observed in steam flood recovery. The final product of a completed burn in heavy oil sand is a clean consolidated sandstone. The overall permeable fabric of the post-burn sand does not differ much from clean sands related to steam/hot water injection only. Petrographic study of all combustion zones indicate that most sand-size components are not visibly altered with the exception of carbonate grains (Figs. 16A, B). The carbonate grains show various stages of alteration, from thin reaction rims of gypsum surrounding the grains to total grain dissolution with remnants

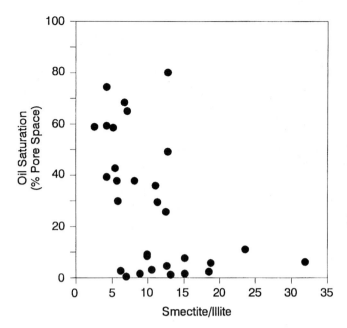

FIG. 14.—Smectite/illite ratio as function of remaining oil, steamed Tulare cored reservoir (data from J. D. Cocker, pers. commun., 1987). Note the change in smectite amounts with less than 20% oil saturation.

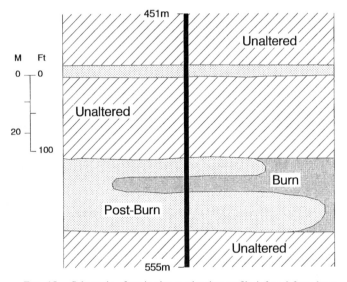

FIG. 15.—Schematic of an in-situ combustion profile inferred from burn patterns in a Webster core.

of a thin reaction rim. The gypsum rim appears related to a calcite and oil reaction during the burn (Fig. 16B).

The finer-grain sediment fraction undergoes the greatest amount of mineralogical change during in-situ combustion. This mineralogical alteration occurs in the sand matrix component or the interbedded mudstones (Table 3). Comparison of x-ray diffraction data from unaltered and burned sediment indicates opal, zeolites, clays and possibly pyrite have decreased after in-situ combustion processes (Fig. 6). Diffraction patterns identify mixed-layer illite/smectite confined to burned and post-burned sediments (Fig. 17), suggesting some smectite loss due

to illite/smectite conversion. However, most smectite remains and easily re-hydrates upon exposure to water. Opal and zeolites exhibit clear dissolution fabrics under the SEM (Figs. 16C, D). Possible opal-CT recrystallization has also been recognized in burned siliceous mudstone (Fig. 18). Recrystallization is indicated by the slight shift of d-spacing from 4.09 to 4.08 angstroms in the opal-CT diffraction peak. Hematite and possible amorphous iron oxide are present in post-burn oxidized sediments and result in a slightly reddish tint. However, much of the original sand fabric is preserved.

Failed Air Injector Fabrics

The above fabrics from inter-well combustion areas contrast significantly from bailer run samples collected from a failed air injector completed within the MOCO-T zone of south Midway-Sunset field. Records indicate water was injected after 3 years of air injection and was thought to be the possible cause of later air injection problems. Loose material was detected in the borehole, and a bailer run collected pebble-size rock material and scale. Collected material was examined by thin-section petrography and XRD.

Altered fabrics, first described by J. D. Cocker (pers. commun., 1987), clearly reflect high-temperature alteration of sediments. Fabrics are composed of various textures indicating all stages of sediment melting (Fig. 19). The least altered fabrics are characterized by glassy matrix and preserved sand-size grains possessing reaction rims. Highest alteration consists of a glassy groundmass with vesicles, some remnant sand grains and often new precipitates ("scale"). Mineralogical study identified anhydrite, wollastonite, iron oxide, amorphous glass and cristobalite as new products.

SUMMARY AND CONCLUSIONS

The heavy oil sands of Midway-Sunset, South Belridge, Lost Hills and San Ardo fields are characterized as feldspathic, poorly-consolidated sands. Although depositional facies and reservoir stratigraphy vary in the different producing zones, their patterns of sediment alteration during thermal processes are very similar. Predictions can be made of how these different sediments alter by characterizing both sediment composition and textures.

Hot Water and Steam Alteration

Sands are already water-sensitive due to the presence of smectitic clays. Common mineral alterations during hot water/steam interaction include further smectite and zeolite growth, plus the dissolution of silica, carbonate and other clays. Reaction rates of specific minerals are also grain-size dependent. The finest-grained matrix material is the most altered during the relatively short time-frame of laboratory testing or field production. The volcanic rock and glass fragments are the most reactive sand-size grains. However, under conditions of long-term flow and exposure to steam condensate (such as the near wellbore area of a steam injector), sand-size quartz and feldspar grains are expected to break down.

The degree of fabric alteration is also reflected by changes in permeability values. Permeability reduction related to hot water or steam interaction is related to the sediment texture.

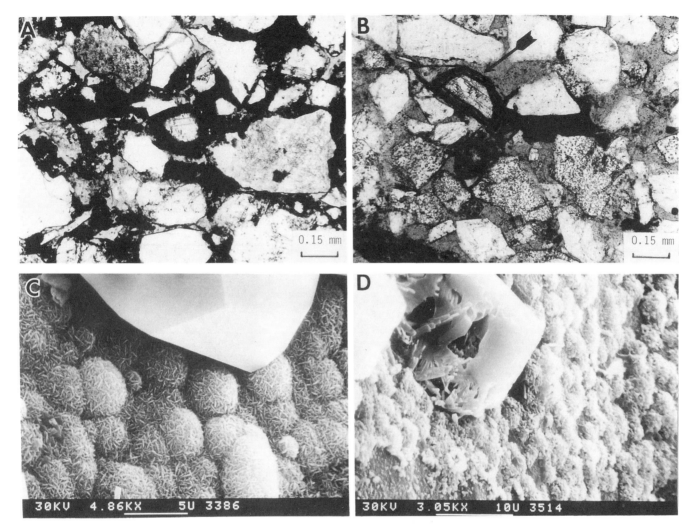

FIG. 16.—Fabrics of in-situ combustion sands, Webster zone. (A) Black coke within sand pores. (B) Post-burn clean sand; arrow points to reaction rim around a calcite grain. (C) Opal-CT and zeolite, unaltered layer. (D) Partially dissolved opal-CT and zeolite, burned layer.

The finer-grained matrix-rich sands show the greatest permeability drop during special core testing while coarser sands exhibit a lower permeability drop under the same conditions. Proposed reasons for this grain size patterns are: (1) finer-grained sediments have more surface area and higher matrix for reaction with hot fluids, thus more smectite and zeolites are precipitated within pore space from the original matrix, and (2) finer-grained lower-permeability sands have smaller pore throats which may become blocked during clay swelling, dispersion and fines migration of the previously stable matrix. Both processes acting together result in decreased permeability.

The more coarse and permeable sediments often experience an extended period of continual fines migration during flow tests (reflected in changing permeability readings). In contrast, finer-grained lower-permeability sediments appear to stabilize earlier. Limited fines (unstable matrix) migration in finer-grained sediments is probably due to smaller pore and pore-throat sizes, where clay swelling and fines dispersion is quickly followed by pore blockage. This conclusion is supported by studies of entrainment and fines deposition in porous media

(Gruesbeck and Collins, 1982). Pore size and the particle size of migrating matrix determine which pores are flow pathways. For a fixed flow rate, unstable matrix movement and deposition varied with the median grain size (thus pore size), and blockage occurred first in smaller pores (or smaller grain sizes) given similar particle sizes of the mobile matrix.

In-Situ Combustion Alteration

The earliest sediment alteration in in-situ combustion reactions is similar to hot water/ steam reactions. In both reactions, the advancement of the burn front and sediment alteration is controlled by grain size and permeability. The burn front laterally advances the fastest in more permeable sediments. Within these sands, it is the matrix which continues to react after hot water and steam interaction. Opaline silica continues to dissolve, but there is also evidence for opal-CT recrystallization at higher temperatures. Zeolites break down, and smectite begins to show evidence of recrystallization into a smectite/illite. However, much smectite remains and is fully expandable once it is

TABLE 3.—XRD MINERALOGY OF SEDIMENTS FROM THE WEBSTER ZONE. "ORIGINAL" REFERS TO NO THERMAL EXPOSURE, WHILE "BURNED" REFERS TO SEDIMENTS IN THE COKE OR POST-BURN STAGE OF IN-SITU COMBUSTION

I. Sands

	Qtz	Feld	Gyp	Pyr	Hem	Opal	Cal	Kaol	Smec	Ill/Smec	Ill/Mica
Original	41-50	39-50	ND	ND-1	ND	ND-19	ND-2	tr	2-6	ND	1
Burned	40-48	36-52	ND-3	ND-1	ND-1	ND-11	ND-1	ND-tr	3-7	ND-4	1
Significant	N	N	Y	N	Y	Y	Y	Y	Y	Y	N

II. Mudstones

	Qtz	Feld	Gyp	Pyr	Hem	Opal	Cal	Kaol	Smec	Ill/Smec	Ill/Mica
Original	10-18	10-25	ND	ND-3	ND	35-65	ND	tr	10-20	ND	ND-1
Burned	8-20	9-31	ND-12	ND-4	ND-10	20-70	ND	ND	9-20	ND-9	1-2
Significant	N	N	Y	?	Y	?	N	Y	?	Y	?

NOTES --

Numbers in weight%
ND = not detected
"significant" (yes or no) means number variability indicates rock reactions rather than original compositional variability. From combined XRD, SEM, thin section.

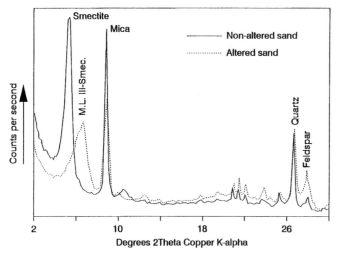

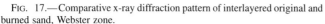

FIG. 17.—Comparative x-ray diffraction pattern of interlayered original and burned sand, Webster zone.

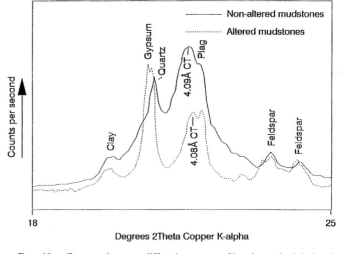

FIG. 18.—Comparative x-ray diffraction pattern of interlayered original and burned siliceous mudstones, Webster zone. Note the shift in the opal peak plus the presence of gypsum in the burned mudstone.

contacted again with water. The only sand-size grain that reacts is calcite, commonly by dissolution and reaction rim development. These rims record a calcite and oil (coke) interaction and the precipitation of gypsum. During post-burn air injection, hematite and iron oxide precipitate to give the sediment a reddish appearance. The end-product of a combustion front moving through the heavy oil sediments is a clean, reddish, lightly-consolidated sandstone. Similar sediment fabrics have been described in the Canadian in-situ combustion projects (Hutcheon, 1984).

This fabric differs greatly from highly-altered samples recovered from a failed air injector. Here, the sediments have been clearly exposed to longer-term, higher temperatures. Fabrics show evidence of melting and greatly-reduced permeability.

Minerals such as wollastonite and cristobalite confirm the extreme conditions of the near-wellbore alteration. Probable temperatures of 800°C are proposed (J. D. Cocker, pers. commun., 1987) to cause these fabric and mineral assemblages.

Well records indicate the melted fabrics were possibly related to a specific time of water injection. The presence of water may have greatly increased the rate of alteration at high temperatures. If water injection was responsible for creating these highly-altered fabrics, this would suggest that wet combustion processes (or alternating air and water injection) have a greater probability of similar near-wellbore glassy fabrics than air injection alone.

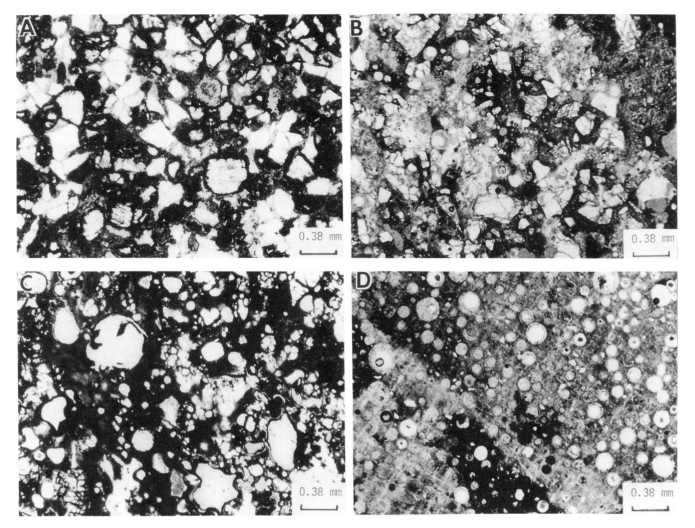

FIG. 19.—Highly altered MOCO-T sand, failed air injector. (A) Partially melted sand. The white areas are remnant grains, the black area is glass. (B) Increased molten fabric. (C) Fabric of glass (black) and rounded vesicles (light). (D) Precipitated "scale"; XRD identified anhydrite, wollastonite, and iron oxides.

The development of highly-altered and different fabrics is also a function of time and temperature. Different layers igniting and burning at different times and rates might have allowed the near-wellbore area to be exposed to high temperatures for longer time periods (a few months or more); this contrasts with fabrics from a moving combustion front where the burn advances through sediment at a rate of a few centimeters per day. Additionally, the ignition of air injectors may record very high short-term temperatures (up to 800°C), where a stable moving burn front does not reach such temperatures (Gates and Sklar, 1971).

IMPLICATIONS FOR RESERVOIR ALTERATION PREDICTION

The ability to understand permeability alteration and potential formation damage in terms of textures is a predictive tool for Californian poorly-consolidated sands and other similar sands. Both original and alteration mineral products used alone are insufficient to understand changing reservoir properties during thermal alteration processes. The degree of sediment alter-

ation in hot water or steam permeability tests is a function of grain size. The fine-grained sediment is the most visibly altered during all stages of thermal alteration. Thus, relating textural characteristics to quantitative measurements supplies a fundamental piece of information to both interpret measured values and build a predictive data base for a specific reservoir over time.

ACKNOWLEDGMENTS

This information was first presented at the Society of Petroleum Engineers International Symposium on Formation Damage Control, held in Lafayette, Louisiana, February 26–27, 1992. Gratitude is extended to SPE for release of journal-publication rights for this work. California heavy oil study was pursued during employment at Mobil Exploration and Producing U.S. Inc., Denver, Colorado. We extend our appreciation to Mobil Oil for permission to publish this work, plus the interaction of many Mobil geologists and engineers who made these results possible.

REFERENCES

DAY, J. J., MCGLOTHLIN, B. B., AND HUITT, J. L., 1967, Laboratory study of rock softening and means of prevention during steam or hot water injection: Journal of Petroleum Technology, v. 19, p. 703–711.

GATES, C. F. AND SKLAR, I., 1971, Combustion as a primary recovery process Midway-Sunset Field: Journal of Petroleum Technology, v. 23, p. 981–986.

GRUESBECK, C. AND COLLINS, R. E., 1982, Entrainment and deposition of fine particles in porous media: Society of Petroleum Engineers Journal, December, p. 847–856.

HUTCHEON, I., 1984, A review of artificial diagenesis during thermally enhanced recovery, *in* McDonald, D. A. and Surdam, R. C., eds., Clastic Diagenesis: Tulsa, American Association of Petroleum Geologists Memoir 37, p. 413–429.

HUTCHEON, I. AND ABERCROMBIE, H. J., 1990, Fluid-rock interactions in thermal recovery of bitumen, Tucker Lake pilot, Cold Lake, Alberta, *in* Meshri, I. D. and Ortoleva, P. J., eds., Prediction of Reservoir Quality Through Chemical Modeling: Tulsa, American Association of Petroleum Geologists Memoir 49, p. 161–170.

KRUMBEIN, W. C. AND MONK, G. D., 1942, Permeability as a function of the size parameters of unconsolidated sand: American Institute of Metallurgical Engineers Transactions, v. 151, p. 153–162.

PERKINS, E. H., GUNTER, W. D., BARRETT, M. L., AND TURNER, J., 1992, Rock-steam reactions at a south Midway-Sunset heavy oil reservoir: Symposium on Exploration, Characterization and Utilization of California Heavy Fossil Fuel Resources, American Chemical Society, San Francisco Meeting, April 5–10, p. 874–885.

REED, M. G., 1980, Gravel pack and formation sandstone dissolution during steam injection: Journal of Petroleum Technology, v. 32, p. 941–949.

SEDIMENTOLOGY RESEARCH GROUP, 1981, The effects of in situ steam injection on Cold Lake oil sands: Bulletin of Canadian Petroleum Geology, v. 29, p. 447–478.

SOUSTEK, P. G., EAGAN, J. M., NOZAKI, M. A., AND BARRETT, M. L., 1993, After the fire is out: a post in-situ combustion audit and steamflood operating strategy for a heavy oil reservoir, San Joaquin valley, CA, *in* Linville, B., ed., Reservoir Characterization III: Tulsa, PennWell Books, p. 647–673.

TILLEY, B. J. AND GUNTER, W. D., 1988, Mineralogy and water chemistry of the burnt zone from a wet combustion pilot in Alberta: Bulletin of Canadian Petroleum Geology, v. 36, p. 25–36.

INDEX

INDEX